AF411359

Lake and River Restoration

Lake and River Restoration

Method, Evaluation and Management

Special Issue Editors

Gang Pan
Lirong Song
Qiuwen Chen
Tao Lyu

MDPI • Basel • Beijing • Wuhan • Barcelona • Belgrade • Manchester • Tokyo • Cluj • Tianjin

Special Issue Editors
Gang Pan
Nottingham Trent University
UK

Qiuwen Chen
Hydraulic Research Institute
China

Lirong Song
Chinese Academy of Science
China

Tao Lyu
Cranfield University
UK

Editorial Office
MDPI
St. Alban-Anlage 66
4052 Basel, Switzerland

This is a reprint of articles from the Special Issue published online in the open access journal *Water* (ISSN 2073-4441) (available at: https://www.mdpi.com/journal/water/special_issues/lake_river_restoration).

For citation purposes, cite each article independently as indicated on the article page online and as indicated below:

LastName, A.A.; LastName, B.B.; LastName, C.C. Article Title. *Journal Name* **Year**, *Article Number*, Page Range.

ISBN 978-3-03936-042-0 (Pbk)
ISBN 978-3-03936-043-7 (PDF)

Cover image courtesy of Gang Pan.

Contents

About the Special Issue Editors

Gang Pan is Full Professor and Ex-Associate Dean of Research for the School of Animal, Rural and Environmental Sciences. He is now the Director of the Centre of Integrated Water–Energy–Food (iWEF) studies at Nottingham Trent University. Prof. Pan is an environmental, ecological, material, and physical chemist. He pioneers in geoengineering technologies for natural water restoration and eutrophication control, sediment remediation, environmental nanotechnology, surface adsorption, algal biotechnology, and synchrotron techniques. He was awarded National First Prize of Environmental Science and Technology of China (2009); "Scientist of the Year" (2010, Chinese national annual award); 2014 Excellence in Review Awards (ES&T), and the First Prize of "Dayu Award" in 2017 (the highest Chinese national award in hydrology and ecology). He has published over 300 peer reviewed papers (h-index of 48, Google Scholar) and filed over 50 patents. His current research focuses on algal biotechnology and nanobubble technology.

Lirong Song is Full Professor at Institute of Hydrobiology, Chinese Academy of Sciences, China. His research areas include algae biology and algae physiology, cyanobacteria bloom ecology, and the application of algae resources. He has published nearly 300 peer reviewed papers. He is a member of Editorial Boards of Journal of Environmental Sciences, Algae, Ecological Informatics, Aquatic Biology, and Lake Science. He is currently Director of the Freshwater Algae Bank of the Chinese Academy of Sciences and Deputy Director of the Key Laboratory of Algae Biology of the Chinese Academy of Sciences. He has previously served as Chairman of the Algae Branch of the Chinese Society of Marine Limnology (2011–2015). He has served as Chief Scientist of several national lake restoration projects (e.g., the 973 Program and 863 Program) in China. He is an invited keynote speaker of several domestic and international academic conferences. In October 2016, he successfully hosted the 10th International Conference on Toxic Cyanobacteria in Wuhan.

Qiuwen Chen is Full Professor at Nanjing Hydraulic Research Institute, China. He was awarded his Ph.D. from Delft University of Technology in the Netherlands. He is currently Director, Researcher, and Doctoral Tutor at the Research Center for Ecological Environment. His research focuses on ecological hydraulics and environmental water informatics. He is winner of the National Outstanding Youth Science Fund, Young and Middle-Aged Technological Innovation Leading Talents of the Ministry of Science and Technology, Hundred Talents Program of the Chinese Academy of Sciences, and Double Innovation Talent of Jiangsu Province. He is the Associate Editor of Ecological Informatics and Journal of Ecohydraulics, and member of the Editorial Boards of Journal of Hydroinformatics, Journal of Water Resources, Journal of Ecology, Journal of Water Resources, and Water Transport Engineering. He has published more than 190 academic journal papers, 3 monographs, and 17 authorized patents. He has won 3 first prizes, 3 second prizes, and 1 international academic achievement prize at the provincial and ministerial levels.

Tao Lyu is Lecturer (Assistant Professor) at Cranfield Water Science Institute, Cranfield University. Prior to joining Cranfield University, Dr. Lyu was a Research Fellow and Module Leader at Nottingham Trent University. His Ph.D. degree in Environmental Science and Engineering was awarded from Aarhus University, Denmark. He has been invited to chair a Conference Session at WETPOL 2019, Aarhus, Denmark. His research spans different research areas of Engineering (wastewater treatment and natural water restoration) and Energy (Bioenergy). He has published 50 peer reviewed journal papers, 27 conference papers, 1 book chapter, and authorized 7 patens (665 citations, h-index of 15 in Google Scholar). His current research focuses on algal biotechnology and nanobubble technology and constructed wetland technology.

Preface to "Lake and River Restoration"

Harmful algal blooms (HAB) are one of the most notorious consequences of eutrophication in natural waters, e.g., lakes and rivers, and pose serious threats to water quality, human health, economic development, ecological balance, landscape aesthetics, and social stability. Both external loading of pollutants from anthrophonic discharge and internal loading of pollutants from sediments are expected to further increase HAB occurrence and provide continuous pressure on river and lake ecosystems over the coming decades. Thus, it has become urgent to draw the attention of researchers around the world toward making great efforts toward lake and river restoration to eliminate HAB threats. Lake and river restoration heavily depend on integrated basin management and technical developments. Integrated water restoration management aims to promote coordinated welfare in an equitable manner without compromising the sustainability of vital ecosystems. Nutrient recovery (e.g., of phosphorous) and residue biomass (e.g., algae biomass) are expected to be valuable resources to promote agricultural sustainability and aquatic ecology. Long-term monitoring of water quality and ecological responses as well as whole water experiments are necessary for a comprehensive evaluation of innovative restoration methods. In this regard, the Special Issue "Lake and River Restoration: Methods, Evaluation and Management" brings together recent research findings from scientists in this field. These papers emphasize the research foci within the three stages of restoration, i.e., background investigation before restoration, various strategies selected for restoration, and long-term monitoring after the treatment. We believe this Special Issue will not only indicate coherent directions of research but will also support policy makers toward more sustainable and effective water restoration and assessment.

Gang Pan, Lirong Song, Qiuwen Chen, Tao Lyu
Special Issue Editors

Editorial

Lake and River Restoration: Method, Evaluation and Management

Tao Lyu [1,2,3], Lirong Song [4], Qiuwen Chen [5] and Gang Pan [1,2,*

[1] School of Animal, Rural, and Environmental Sciences, Nottingham Trent University, Brackenhurst Campus, Nottinghamshire NG25 0QF, UK; tao.lyu@ntu.ac.uk

[2] Centre of Integrated Water-Energy-Food studies (iWEF), Nottingham Trent University, Nottinghamshire NG25 0QF, UK

[3] Cranfield Water Science Institute, Cranfield University, College Road, Cranfield, Bedfordshire MK43 0AL, UK

[4] State Key Laboratory of Freshwater Ecology and Biotechnology, Institute of Hydrobiology, Chinese Academy of Sciences, Wuhan 430072, China; lrsong@ihb.ac.cn

[5] Center for Eco-Environmental Research, Nanjing Hydraulic Research Institute, Nanjing 210029, China; qwchen@nhri.cn

* Correspondence: gang.pan@ntu.ac.uk

Received: 17 March 2020; Accepted: 27 March 2020; Published: 30 March 2020

Abstract: Eutrophication has become one of the major environmental issues of global concern due to the adverse effects on water quality, public health and ecosystem sustainability. Fundamental research on the restoration of eutrophic freshwaters, i.e., lakes and rivers, is crucial to support further evidence-based practical implementations. This Special Issue successfully brings together recent research findings from scientists in this field and assembles contributions on lake and river restoration. The 12 published papers can be classified into, and contribute to, three major aspects of this topic. Firstly, a background investigation into the migration of nutrients, and the characteristics of submerged biota, will guide and assist the understanding of the mechanisms of future restoration. Secondly, various restoration strategies, including control of both external and internal nutrients loading, are studied and evaluated. Thirdly, an evaluation of the field sites after restoration treatment is reported in order to support the selection of appropriate restoration approaches. This paper focuses on the current environmental issues related to lake and river restoration and has conducted a comprehensive bibliometric analysis in order to emphasise the fast-growing attention being paid to the research topic. The research questions and main conclusions from all papers are summarised to focus the attention toward how the presented studies aid gains in scientific knowledge, engineering experience and support for policymakers.

Keywords: eutrophication control; external loads; harmful algal blooms (HABs); phosphorous recovery; sediment load control

---

1. Introduction

Harmful algal blooms (HABs) are one of the most notorious consequences of eutrophication of natural waters, e.g., lakes and rivers, and pose serious threats to water quality, human health, economic development, ecological balance, landscape aesthetics, and social stability [1]. Owing to rapid population growth and economic development, various human activities in industry, agricultural and transport sectors have deteriorated and globally intensified the freshwater eutrophication (Figure 1) [2–5]. In addition to the external loading of pollutants from anthropogenic discharge, the internal loading of pollutants from sediments is expected to further increase the occurrence of HABs and deliver continuous pressure on river and lake ecosystems over the coming decades [6].

Thus, it is urgent to draw the attention of researchers around the world in order to make great efforts towards lake and river restoration in order to eliminate the threat of eutrophication.

Figure 1. Cases of freshwater eutrophication in (**a**) China [2], (**b**) USA [3], (**c**) UK [4], and (**d**) South Africa [5]. Note: the photos were derived with modification from the original sources.

Both lake and river restoration depend heavily on integrated basin management and technical developments [7]. Integrated water restoration management aims to promote the coordinated development and management of water, land, and related resources in order to maximize economic and social welfare in an equitable manner without compromising the sustainability of vital ecosystems. Nutrient recovery, e.g., phosphorous and residue biomass, e.g., algae biomass, are expected to be valuable resources to promote agricultural sustainability and aquatic ecology [8]. Long-term monitoring of water quality and ecological responses, as well as whole water experiments, are necessary for a comprehensive evaluation of innovative restoration methods [9].

This Special Issue of *Water* aims to compile the latest advances in lake and river restoration technology, in terms of advanced materials, applications, evaluation, and management. The research areas of the 12 papers cover site investigation prior to restoration, evaluations of various technologies for restoration, and long-term monitoring strategies post-treatment. These papers have successfully introduced studies to tackle the knowledge gaps between basic research and engineering implementation, which could significantly contribute to eutrophication control, natural water sustainability, and ecological restoration.

2. Research Development and Current Status

A comprehensive bibliometric analysis was conducted in order to understand the history and current trends of research into lake and river restoration. The Science Citation Index Expanded (SCI-EXPANDED) database, from the Web of Science, was used to search the following words and phrases: "lake restoration", "river restoration", "harmful algal blooms", and "eutrophication", to compile a bibliography of all papers before 2020 related to the research topic for this Special Issue.

A total of 35,305 publications resulted from the screening, and only research articles published in the English language were selected for further analysis.

The adjective eutrophe (literally "well fed") was first used by the German botanist Weber in 1907, to describe the initially high nutrient conditions that occur in some types of ecosystems at the start of secondary succession [10]. Following the development of industrial and agricultural activities, excess nutrients entered into bodies of water and caused enhanced biomass and/or growth rates of algae and changes in water quality. Thus, the term "eutrophication" came into common usage and has been a particular concern when public awareness of the problem was heightened by widespread HABs. The earliest two research articles from the database were both published in 1975 with the titles "Eutrophication of Microponds" and "Eutrophication", demonstrating the historic concern of this environmental issue.

There is little reported research on river and lake restoration before 1990 (Figure 2a), however, the article tally increased markedly from 1991 and has exhibited a more marked increase in the present. This may be because the pollution of natural waters and their restoration has become a global issue, now attracting widespread interest worldwide. It should be noted that article abstract information was also not available in the SCI-EXPANDED database prior to 1991, which may contribute to the fact that only 1.2% (440 papers) of articles were collated through the literature search. Therefore, the cumulative numbers of articles from 1991 were further examined by model simulation (Figure 2b). From 1991 to 2019, the number of articles increased from 187 in 1991 to 3,251 in 2019. Two exponential models were established to describe the relationships between the annual cumulative number of articles, and the years in which they were published for the two periods (1991–2000, and 2001–2019), respectively. The rate of the increase in the number of articles published (0.2003) during 2001–2019 is slightly lower than that (0.1145) during 2001–2019, which may be attributed to more engineering projects and technology implementations being submitted in recent years, written with the aim of solving real environmental and public health problems, rather than the publishing of, exclusively, research papers. It is inappropriate to distinguish the publication numbers related to lakes or river systems in the current bibliometric analysis, because many laboratory-scale studies, e.g., HAB flocculation material development, can be applied in both systems. Nevertheless, it can be concluded from the literature that the research development on eutrophication restoration in both lake and river systems started from a mechanistic basis and evolved to an evidence-based implementation approach, e.g., integrated mitigation technology and management. Moreover, the exponential increase in published articles addressing lake and river restoration research indicates the current growing attention being paid to environmental protection, which also promises the publishing of more papers, and the recruitment of more researchers, in the near future.

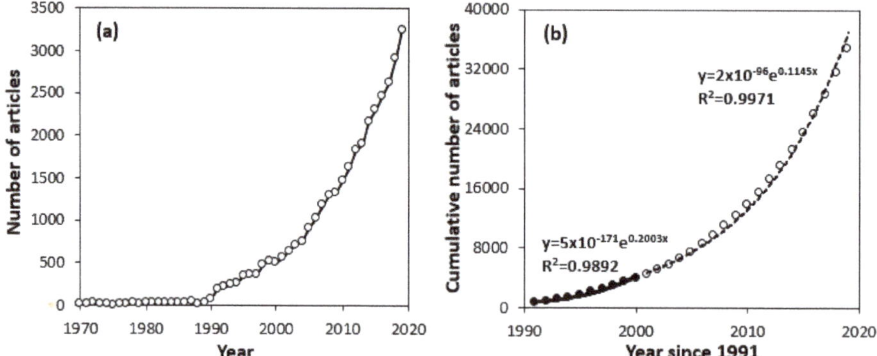

Figure 2. Publication patterns and modelled characteristics of papers addressing river and lake restoration ((**a**) the number of articles from 1970 to 2019; (**b**) the relationship between the cumulative number of articles and year published, since 1991).

3. Overview of the Special Issue

The top five most frequently used keywords in all of the articles selected from the database are "Eutrophication", "Phosphorus", "Algae" "Water quality" and "Nitrogen", which, coincidentally, show high correlation with the frequent keywords used in the 12 papers from this Special Issue (Figure 3a). Phosphorus and nitrogen have long been recognised as crucial nutrients for harmful algal growth in eutrophic waters [11]. In addition to the input of external pollutants from wastewater or runoff waters, the embedded sediment-bonded nutrients can be released back to the water column, especially under hypoxic conditions at sediment–water interfaces [12]. Thus, integrated restoration strategies always consider the control of both external and internal nutrient loadings. In order to select the appropriate restoration technology and estimate the potential risk to the surrounding biota, it is crucial to conduct a background site investigation and understand the potential effects of pollutant migration. Without a long-term monitoring plan and evaluation of the final restoration, it is impossible to assess the success or failure of the treatment. The studies from the current 12 papers in this Special Issue perfectly cover these three aspects (Figure 3b): background investigation, restoration strategies and post-treatment evaluation.

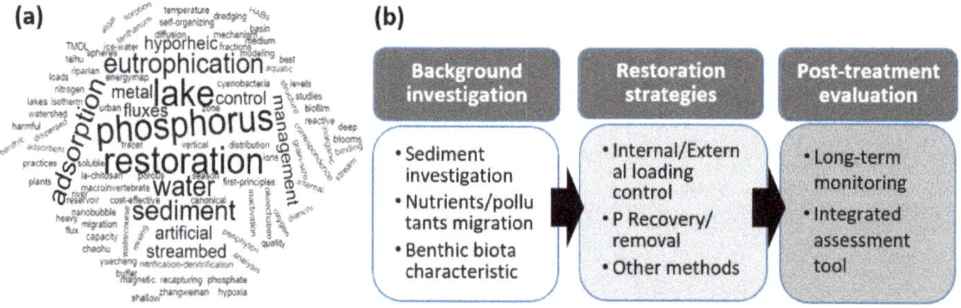

Figure 3. Distribution of keywords used in the 12 papers in the Special Issue (**a**), and the summarised procedure for research into (and applications in) lake and river restoration (**b**).

3.1. Background Investigation

Dang et al. [13] investigated the phosphorus (P) fractions and the relationships between environmental factors of a field reservoir, a main local water supply source of the municipal, industry and agricultural sectors. Their research aims to enhance the knowledge of P cycling in a high P concentration basin, so that water quality and trophic status can be better controlled in the future. The results demonstrated that, although the implementation of a large number of water conservancy projects in the upper reaches of the river resulted in a decrease in inflow runoff, pollution from terrestrial plants or materials played a key role in the dynamics of the sediment P fraction, highlighting that they should be emphasised in the environmental management of river basins.

Understanding the drivers of macroinvertebrate community structure is fundamental for adequately controlling pollutants and managing aquatic ecosystems under global change. Liu et al. [14] analysed the characteristics of the environment and macroinvertebrate community in a river basin over time. They have established an approach by which to reveal community–environment relationships and identified priority pollutants that need to be controlled from the perspective of water ecological health protection. The results could be used to control river pollutants, improve river health management and guide ecological restoration.

Lake freezing in natural conditions is a macroscopic and complex process, which may influence the translocation of pollutants between solid ice and liquid water in an aquatic system. In order to explore the mechanism for the migration of heavy metal ions (HMIs) from ice to water in a lake, Sun et al. [15] carried out a laboratory freezing experiment and simulated the distribution and migration of HMIs (Fe, Cu, Mn, Zn, Pb, Cd, and Hg) under different conditions. The results demonstrated that HMIs could migrate from ice to water as the lake froze, due to the binding energy in water being smaller than that in ice. The study is important for the better understanding of the mechanisms of lake ecosystems in cold regions, especially as a consequence of climate change.

3.2. Restoration Strategies

3.2.1. External Loading Control

Intercepting nutrients entering lakes from point or non-point sources is important for eutrophication control. Cao et al. [16] examined the efficiency of nutrient removal, including the quantity of nutrients and resultant water quality in the littoral zone of different types of riparian buffers in the watershed around a eutrophic lake, and estimated the optimal width for different types of riparian buffers for effective nutrient removal. Different riparian buffers contribute numerous efforts to the removal of nutrients due to vegetation and soil types. Overall, construction of wetland or grass/forest and grass riparian buffer strips is strongly recommended for effective total phosphorous (TP) and total nitrogen (TN) removal, respectively.

3.2.2. Internal Loading Control

Dredging is one of the few options available for the improvement of the ecological balance of lakes by removing contaminated sediments towards reduction in the internal loading of nutrients. However, determination of the season during which any dredging activity could provide the maximum effect, but minimise any adverse effects, is not clear. Zhong et al. [17] conducted experiments in all four seasons in order to evaluate the effects of dredging on the internal loading, and release from sediment of N and P. The results indicated that dredging could be a useful approach for decreasing internal loading and that seasons with low temperature (non-growing seasons) are suitable for performing these operations.

Considering the relatively high cost of dredging operations, in situ treatment through flocculating HABs and locking them onto the surface of sediment has been demonstrated to be sufficient in order to mitigate the loading of internal nutrients [18]. Following on from this concept, Pan et al. [19] summarised the main features of Modified Local Soil (MLS) technology by investigating the effect of this treatment in five pilot-scale whole-pond field experiments. Results showed that, combined with the integrated management of external loads, MLS can be used as an in-lake restoration technology through multiple functions of water quality improvement, sediment remediation, and ecological restoration.

These findings can serve as useful guidelines for researchers, engineers, and local governments for controlling eutrophication and accelerating lake restoration in an eco-friendly and sustainable manner.

3.2.3. Precapture from Water

The removal of P from the water column is crucial and effective for the control of eutrophication, and adsorption is one of the most effective treatment processes. Cheng et al. [20] synthesised a P adsorbent of lanthanum–chitosan magnetic spheres, and evaluated its capability under different conditions. A superior P adsorption ability was demonstrated and, furthermore, the adsorption kinetics, isotherms, and thermodynamic analyses of P by as-prepared adsorbents were studied to reveal the underpinning mechanisms. Owing to their unique hierarchical porous structures, high adsorption capacity and low cost, lanthanum–chitosan magnetic spheres are potentially applicable for eutrophic water treatment.

To study the application of an adsorbent in natural waters in order to achieve very low levels of P (10 μg L^{-1}) for potential eutrophication control, Pan et al. [21] evaluated a combined external and internal P control approach in a simulated pilot-scale river–lake system. Under such a strategy, a granulated lanthanum/aluminium hydroxide composite (LAH) adsorbent was demonstrated to be an effective material for maintaining the P concentration below 10 μg L^{-1} under low levels of additional P input. The results demonstrated that the synergy of external and internal phosphorus recapture by the LAH adsorbent material could be an effective and cost-effective strategy for the potential management of eutrophication under additional, mesotrophic, P inputs.

3.2.4. Other Methods

Stream restoration, designed specifically to enhance hyporheic processes, has seldom been contemplated. Bakke et al. [22] conducted such a project through the engineering of a streambed using a gravel mixture formulated to mimic natural streambed composition, filling an over-excavated channel to a minimum depth of 90 cm. Specially designed plunge pool structures, built with subsurface gravel extending down to 2.4 m, promoted greatly enhanced hyporheic circulation, path length, and residence time. Results from post-restoration monitoring demonstrated that this approach to enhanced hyporheic design successfully restored hyporheic processes and yielded significant water quality improvements compared to control and pre-project conditions.

In order to achieve the restoration of submerged vegetation and HABs mitigation in shallow eutrophic lakes, Wu et al. [23] introduced artificial aquatic plants (AAPs) into enclosures in the eutrophic lake and investigated their effect on the reduction in cyanobacterial blooms and promotion of the growth of submerged macrophytes. On the 60th day after the AAPs were installed, turbidity, total nitrogen (TN), total phosphorous (TP), and the cell density of phytoplankton (especially cyanobacteria) of the treated enclosures were significantly reduced compared with the control enclosures. The results supported that the application of AAPs with incubated periphyton could be an effective and environmentally friendly solution for the reduction in nutrients and for the control of phytoplankton, thereby promoting the restoration of submerged macrophytes in shallow eutrophic waters.

3.3. Post-Treatment Evaluation

Long-term monitoring following restoration engineering is essential in order to evaluate the feasibility and sustainability of a certain strategy. Długie Lake (Olsztyńskie Lakeland, Poland) was restored using both artificial mixing and phosphorus inactivation methods by aluminum compounds. Fifteen years after the termination of the restoration procedure, the investigation conducted by Augustyniak et al. [24] demonstrated that the alum-modified "active" sediment layer still possessed substantial P adsorption abilities, which could limit the internal loading of P. Moreover, research into the adsorptive properties of sediment can be used as a tool for the evaluation of lake restoration effects.

Khare et al. [25] developed a phased scenario analysis approach for the identification of a cost-effective restoration alternative using four TP control strategies—Best Management Practices

(BMPs), Dispersed Water Management (DWM), Wetland Restoration, and Stormwater Treatment Areas (STAs)—to achieve a flow-weighted mean TP concentration of 40 µg/L at lake inflow points. A Watershed Assessment Model was utilised to simulate flow and phosphorus dynamics and the results from a 10-year data collection, which supported that STAs are necessary components to achieve restoration of the Everglades' ecosystem and sustainability in south Florida.

4. Conclusions

This Special Issue concerns the environmental problems of freshwater eutrophication and focuses on solutions for lake and river restoration. Based on the bibliographic investigation, clearly, increasing attention has been attracted by scientists with the aim of solving this environmental problem, reflected by the exponential rise in article numbers. The 12 papers emphasise the research foci within three stages of the restoration, i.e., background investigation before restoration, various strategies selected for restoration, and long-term monitoring after the treatment. We believe this Special Issue will not only indicate coherent directions of research but will also support policy makers toward more sustainable and effective water restoration and assessment.

Author Contributions: Conceptualization, G.P., T.L.; writing—original draft preparation, T.L.; writing—review & editing, G.P., L.S., Q.C. All authors have read and agreed to the published version of the manuscript.

Funding: This research received no external funding.

Acknowledgments: We highly appreciate the journal editors, authors of the 12 papers in this Special Issue, and referees who contributed to paper revision and improvement.

Conflicts of Interest: The authors declare no conflict of interest.

References

1. Conley, D.J.; Paerl, H.W.; Howarth, R.W.; Boesch, D.F.; Seitzinger, S.P.; Havens, K.E.; Lancelot, C.; Likens, G.E. Controlling eutrophication: Nitrogen and phosphorus. *Science* **2009**, *323*, 1014–1015. [CrossRef] [PubMed]
2. Paerl, H.; Gardner, W.; McCarthy, M.; Peierls, B.; Wihelm, S. Algal blooms: Noteworthy nitrogen. *Science* **2014**, *346*, 175. [CrossRef] [PubMed]
3. Erickson, J.; Phillips, B. "Large summer harmful algal bloom predicted for western Lake Erie". Michigan News, University of Michigan. 11 July 2019. Available online: https://news.umich.edu/large-summer-harmful-algal-bloom-predicted-for-western-lake-erie/ (accessed on 26 March 2020).
4. Carvalho, L. "Bloomin' Algae! A new app to help reduce public health risks from harmful algal blooms". UK Centre for Ecology & Hydrology, 3 July 2017. Available online: https://www.ceh.ac.uk/news-and-media/news/bloomin-algae-new-app-help-reduce-public-health-risks-harmful-algal-blooms (accessed on 26 March 2020).
5. Harding, W. Living with eutrophication in South Africa: A review of realities and challenges. *R. Soc. S. Afr.* **2015**, *70*, 155–171. [CrossRef]
6. Zhang, H.; Shang, Y.; Lyu, T.; Chen, J.; Pan, G. Switching harmful algal blooms to submerged macrophytes in shallow waters using geo-engineering methods: evidence from a ^{15}N tracing study. *Environ. Sci. Technol.* **2018**, *52*, 11778–11785. [CrossRef]
7. LoSchiavo, A.; Best, R.; Burns, R.; Gray, S.; Harwell, M.; Hines, E.; McLean, A.; Clair, T.; Traxler, S.; Vearil, J. Lessons Learned from the First Decade of Adaptive Management in Comprehensive Everglades Restoration. *Ecol. Soc.* **2013**, *18*, 70. [CrossRef]
8. Pan, G.; Lyu, T.; Mortimer, R. Comments: Closing phosphorus cycle from natural waters: Re-capturing phosphorus through an integrated water-energy-food strategy. *J. Environ. Sci.* **2018**, *65*, 375–376. [CrossRef]
9. Søndergaard, M.; Jeppesen, E.; Lauridsen, T.; Skov, C.; Van Nes, E.; Roijackers, R.; Lammens, E.; Portielje, R. Lake restoration: successes, failures and long-term effects. *J. Appl. Ecol.* **2007**, *44*, 1095–1105. [CrossRef]
10. Harper, D. *What is Eutrophication? Eutrophication of Freshwaters*; Springer: Dordrecht, Germany, 1992; Chapter 1; p. 2.
11. Lewis, W.; Wurtsbaugh, W.; Paerl, H. Rationale for Control of Anthropogenic Nitrogen and Phosphorus to Reduce Eutrophication of Inland Waters. *Environ. Sci. Technol.* **2011**, *45*, 10300–10305. [CrossRef]

12. Zhang, H.; Lyu, T.; Bi, L.; Tempero, G.; Hamilton, D.; Pan, G. Combating hypoxia/anoxia at sediment-water interfaces: A preliminary study of oxygen nanobubble modified clay materials. *Sci. Total Environ.* **2018**, *637*, 550–560. [CrossRef]

13. Dang, C.; Lu, M.; Mu, Z.; Li, Y.; Chen, C.; Zhao, F.; Yan, L.; Cheng, Y. Phosphorus Fractions in the Sediments of Yuecheng Reservoir, China. *Water* **2019**, *11*, 2646. [CrossRef]

14. Liu, X.; Zhang, J.; Shi, W.; Wang, M.; Chen, K.; Wang, L. Priority Pollutants in Water and Sediments of a River for Control Basing on Benthic Macroinvertebrate Community Structure. *Water* **2019**, *11*, 1267. [CrossRef]

15. Sun, C.; Li, C.; Liu, J.; Shi, X.; Zhao, S.; Wu, Y.; Tian, W. First-Principles Study on the Migration of Heavy Metal Ions in Ice-Water Medium from Ulansuhai Lake. *Water* **2018**, *10*, 1149. [CrossRef]

16. Cao, X.; Song, C.; Xiao, J.; Zhou, Y. The Optimal Width and Mechanism of Riparian Buffers for Storm Water Nutrient Removal in the Chinese Eutrophic Lake Chaohu Watershed. *Water* **2018**, *10*, 1489. [CrossRef]

17. Zhong, J.-C.; Yu, J.-H.; Zheng, X.-L.; Wen, S.-L.; Liu, D.-H.; Fan, C.-X. Effects of Dredging Season on Sediment Properties and Nutrient Fluxes across the Sediment–Water Interface in Meiliang Bay of Lake Taihu, China. *Water* **2018**, *10*, 1606. [CrossRef]

18. Jin, X.; Bi, L.; Lyu, T.; Chen, J.; Zhang, H.; Pan, G. Amphoteric starch-based bicomponent modified soil for mitigation of harmful algal blooms (HABs) with broad salinity tolerance: Flocculation, algal regrowth, and ecological safety. *Water Res.* **2019**, *165*, 115005. [CrossRef]

19. Pan, G.; Miao, X.; Bi, L.; Zhang, H.; Wang, L.; Wang, L.; Wang, Z.; Chen, J.; Ali, J.; Pan, M.; et al. Modified Local Soil (MLS) Technology for Harmful Algal Bloom Control, Sediment Remediation, and Ecological Restoration. *Water* **2019**, *11*, 1123. [CrossRef]

20. Cheng, R.; Shen, L.-J.; Zhang, Y.-Y.; Dai, D.-Y.; Zheng, X.; Liao, L.-W.; Wang, L.; Shi, L. Enhanced Phosphate Removal from Water by Honeycomb-Like Microporous Lanthanum-Chitosan Magnetic Spheres. *Water* **2018**, *10*, 1659. [CrossRef]

21. Pan, M.; Lyu, T.; Zhang, M.; Zhang, H.; Bi, L.; Wang, L.; Chen, J.; Yao, C.; Ali, J.; Best, S.; et al. Synergistic Recapturing of External and Internal Phosphorus for In Situ Eutrophication Mitigation. *Water* **2020**, *12*, 2. [CrossRef]

22. Bakke, P.D.; Hrachovec, M.; Lynch, K.D. Hyporheic Process Restoration: Design and Performance of an Engineered Streambed. *Water* **2020**, *12*, 425. [CrossRef]

23. Wu, Y.; Huang, L.; Wang, Y.; Li, L.; Li, G.; Xiao, B.; Song, L. Reducing the Phytoplankton Biomass to Promote the Growth of Submerged Macrophytes by Introducing Artificial Aquatic Plants in Shallow Eutrophic Waters. *Water* **2019**, *11*, 1370. [CrossRef]

24. Augustyniak, R.; Grochowska, J.; Łopata, M.; Parszuto, K.; Tandyrak, R.; Tunowski, J. Sorption Properties of the Bottom Sediment of a Lake Restored by Phosphorus Inactivation Method 15 Years after the Termination of Lake Restoration Procedures. *Water* **2019**, *11*, 2175. [CrossRef]

25. Khare, Y.; Naja, G.M.; Stainback, G.A.; Martinez, C.J.; Paudel, R.; Van Lent, T. A Phased Assessment of Restoration Alternatives to Achieve Phosphorus Water Quality Targets for Lake Okeechobee, Florida, USA. *Water* **2019**, *11*, 327. [CrossRef]

Article

Phosphorus Fractions in the Sediments of Yuecheng Reservoir, China

Chenghua Dang [1], Ming Lu [1,2,*], Zheng Mu [1], Yu Li [1], Chenchen Chen [1], Fengxia Zhao [1], Lei Yan [1] and Yao Cheng [1,*]

[1] School of Water Conservancy and Hydroelectric Power, Hebei University of Engineering, Handan 056002, China; dangchenghua@sohu.com (C.D.); muzheng1981@hebeu.edu.cn (Z.M.); liyu199607@163.com (Y.L.); ccc2019szy@163.com (C.C.); zhaofx1123@163.com (F.Z.); yanl@whu.edu.cn (L.Y.)

[2] Department of Earth Sciences, University of the Western Cape, Cape Town 7535, South Africa

* Correspondence: luming_hd@163.com (M.L.); chengyao1108@163.com (Y.C.)

Received: 16 September 2019; Accepted: 12 December 2019; Published: 15 December 2019

Abstract: As a result of the inexorable development of the economy and the ever-increasing population, the demand for water in the urban and rural sectors has increased, and this in turn has caused the water quality and eutrophication of the reservoir to become a legitimate concern in the water environment management of river basins. Phosphorus (P) is one of the limiting nutrients in aquatic ecosystems; P in the sediment is a primary factor for eutrophication. Yuecheng Reservoir is located in one of the most productive and intensively cultivated agricultural regions in North China. Detailed knowledge of the sediment is lacking at this regional reservoir. The first study to look into the different P fractions and its diffusion fluxes at the water sediment interface of the Yuecheng Reservoir makes it possible to learn about the internal P loading. According to the results, the concentrations of total phosphorus (TP) ranged from 1576.3 to 2172.6 mg kg and the P fraction concentration sequence is as follows: P associated with calcium (Ca–Pi) > organic P (Po) > P bound to aluminum (Al), ferrum (Fe) and manganese (Mn) oxides and hydroxides (Fe/Al–Pi). The results demonstrated that, although the construction of a large number of water conservancy projects in the upper reaches of the river resulted in the decrease of inflow runoff, the pollutions from terrestrial plants or materials played a key role in the sediment phosphorus fraction, and they should be emphasized on the water environment management of river basin.

Keywords: phosphorus fractions; diffusion fluxes; sediment; Zhangweinan River Basin; Yuecheng Reservoir

---

1. Introduction

Phosphorus (P) is one of the limiting nutrients in aquatic ecosystems [1,2]. Water trophic status and phytoplankton growth are closely related to P in the water column and surface sediment. P in the sediment is a primary factor for eutrophication [3], which is one of the most serious problems in reservoirs, lakes and rivers. The sediment P adsorption and cycling in different regions of the river basin is important for river environment management [4,5]. There are important influence factors, such as the characteristics of sediments, environmental factors and the concentration of P in the overlying water. All of these factors play a pivotal role in the transfer direction of P at the sediment–water interface [6,7]. As the external loading of P is increased, the sediments will absorb it, while if the external loading is decreased, the sediments release the absorbed P into the water in the long-term [8]. Accordingly, the most common strategy for the restoration of eutrophic reservoirs and lakes is focused on reducing external P loading. This method, however, achieves limited effects because bottom sediments can also release phosphorus to the overlying water, especially when the

P input is reduced [9,10]. Therefore, systematic studies of internal P loading are indispensable to understand the process of P circulation and support effective policies to manage eutrophication [11,12].

The Yuecheng Reservoir is located in the south of Handan, Hebei province, which is the main local water supply of the municipal, industry and agricultural sectors [13]. Today, the precipitation is decreasing and the number of reservoirs and diversion channels is increasing in the upper reaches. As a result, the inflow of the Yuecheng Reservoir is becoming much less than a decade ago. Moreover, the inexorable development of the economy and the ever-increasing population have increased the demand for water in the urban and rural sectors, and these in turn cause the water quality and eutrophication of the reservoir to become a legitimate concern in the water environment management of the river basin.

Consequently, in this study, the P fractions and the relationships with environmental factors of the reservoirs have been investigated for the first time to enhance the knowledge of P cycling in this high P concentration basin, so that the water quality and eutrophication can be controlled better in the future.

2. Materials and Methods

2.1. Study Area and Sampling Sites

As the main reservoir of flow regulation for the water supply in Zhangweinan Basin [13], Yuecheng Reservoir has a capacity of 1.3 billion m^3 and a drainage area of 18,100 km^2, which provides the water supply for Anyang city and Handan city and irrigation for the two large agricultural areas of Minyou and Zhangnan. The average depth of the reservoir is 20 m and the maximum depth is 37 m. Algae may appear occasionally in some areas of the reservoir between April and September, with a maximum concentration of 40 $\mu g \cdot L^{-1}$.

The longitude and latitude of the Zhangweinan River Basin range from 112 to 118° E and from 35 to 39° N, respectively, and it is located in North China. It consists of five main rivers, including the Zhang River, Wei River, Wei Canal, Zhangweixin River and Nan River [14]. This basin is comprised of roughly 25,466 km^2 mountainous area (i.e., Taihang Mountains and Taiyue Mountain) and about 12,234 km^2 plain area. Thus, the total area of this basin is approximately 37,700 km^2.The basin is characterized by semiarid, semi-humid climatic conditions and its average annual precipitation is 608.4 mm, with a mean annual temperature of 14 °C. The distribution of temporal precipitation varies greatly among the four distinct seasons. For example, in July and August, more than 50% of precipitation falls, while in the spring, autumn, and winter seasons, the rainfall occupies 8%–16%, 13%–23%, and 2% of the total for one year in order. As one of the most productive and intensively cultivated agricultural regions in North China, the basin has a population of about 30 million and has 293×10^3 ha cultivated area. The main cropland features various kinds of cultivated crops, including wheat, maize, cotton, rice, bean, oilseed, and vegetables, and about 75% of the cropland is irrigated, which consumes 70%–80% of the total water resources.

Duplicated samples of superficial sediments (about 0–5 cm deep) and the corresponding overlying water (about 0–4 cm above the interface) were collected using a gravity corner (diameter 6.5 cm and length 60 cm) [15] at five points (YR1–YR5) in Yuecheng Reservoir (Figure 1) in July 2017. Stratified overlying water was injected into PVC bottles at an interval of 2 cm. The sediments were sampled into sections of 2 cm length and packed in polyethylene centrifuge tubes and sealed to avoid sediment oxidation. All samples were collected in triplicate, taken in air-sealed plastic bags and kept at 4 °C until analysis (within 24 h).

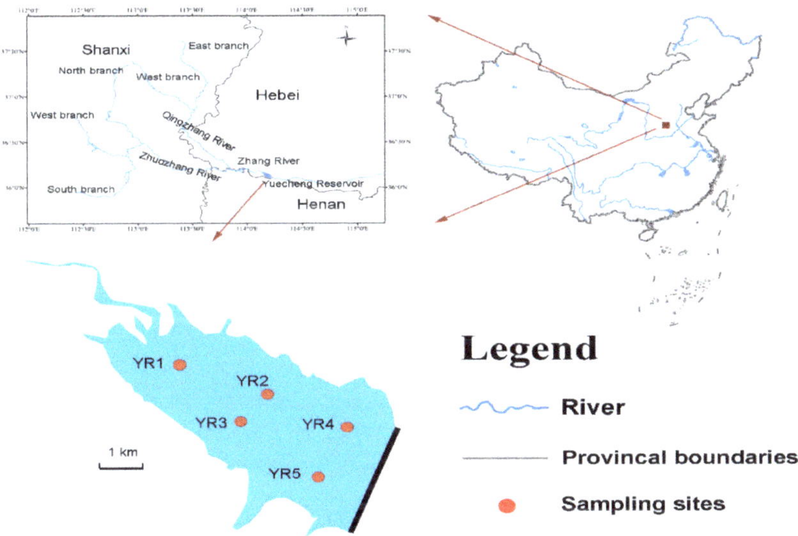

Figure 1. Location of the Yuecheng Reservoir and sampling sites.

2.2. Water Chemistry

The parameters of water, such as the temperature (T), dissolved oxygen (DO), electrical conductivity (EC), oxidation reduction potential (ORP), and pH values, were measured in situ using a YSI (Yellow Springs Instruments Inc., Yellow Springs, OH, USA) EXO2 multisensor sonde. The overlying water and pore water (extracted from sediment by centrifuging at 4000 rpm for 30 min) samples were filtered through 0.45 μm GF/C filter membranes. The concentration of orthophosphate (PO_4^{3-}–P), the most bioavailable P form, was determined by the molybdenum blue method [16].

2.3. Sediment. Sample Analysis

One sample of sediments was used to calculate the water content by weighing the weight loss after drying the sediments at 105 °C. Meanwhile, the water volume of the sediment was approximately regarded as the pore volume, and the porosity of sediment was the ratio between pore volume and total sedimentary volume. The dried samples were homogenized and separated by a laser diffraction particle size analyzer (LS 13 320 MW, Beckman Coulter Company, Erie, PA, USA) into three grain size fractions: the sand fraction (62.5–500 μm), the silt fraction (3.9–62.5 μm), and the clay fraction (0.5–3.9 μm).

Other sediment samples were freeze-dried, homogenized and sieved through a 100-mesh sieve. Total organic carbon (TOC) and total nitrogen (TN) were detected by an elemental analyzer (Vario EL III, Elementar Company, Langenselbold, Germany) after pretreatment in 1 mol L^{-1} hydrochloric acid (HCl) to remove inorganic carbon. P fractions were classified into TP, inorganic P (Pi), organic P (Po), P associated with calcium (Ca) (Ca–Pi) and P bound to aluminum (Al) and ferrum (Fe) oxides and hydroxides (Fe/Al–Pi), and determined by the SMT (Standards Measurements and Testing) protocol [17]. For all samples, triplicates were analyzed and the average or "mean value ± standard deviation" of data were reported.

2.4. Flux Estimation and Data Analysis

Fick's first law, based on the principle that the concentration gradient initiates the exchange, was used to estimate the theoretical release flux of PO_4^{3-}–P at the interface of water and sediment [18].

$$J = \varphi_0 D_s \left. \frac{\partial C}{\partial x} \right|_{x=0} \tag{1}$$

where J is the diffusion flux (mg m^{-2} d^{-1}); φ_0 is the porosity of surface sediment; and $\left. \frac{\partial C}{\partial x} \right|_{x=0}$ is the concentration gradient at the water–sediment interface. It takes the slope value of the best fitting line of the distribution profile of PO_4^{3-}–P concentrations in the overlying water and sediment. D_s is the effective diffusion coefficient, which is calculated from D_0 by the following equations. D_0 (cm^2 s^{-1}) is the theoretical diffusion coefficient of infinite dilution solution [19]. T in Equation (4) is the temperature of the overlying water (°C).

$$D_s = \varphi_0 D_0 \qquad \varphi_0 < 0.7 \tag{2}$$

$$D_s = \varphi_0^2 D_0 \qquad \varphi_0 > 0.7 \tag{3}$$

$$D_0\left(\times 10^{-6}\right) = 7.37 + 0.16 \times (T - 25\,°C) \tag{4}$$

The mean values and standard deviations were calculated by Microsoft Excel 2013, and correlation statistical analyses were tested with SPSS 22 (Statistical Program for Social Sciences 22).

3. Results and Discussion

3.1. Physicochemical Properties of Overlying Water and Interstitial Water

The physicochemical properties of overlying water and PO_4^{3-}–P concentrations of interstitial water are listed in Table 1. The parameters gave pH values that ranged from 7.92 to 8.44. EC, a measure of the ionic strength, ranged from 462.3 to 479.6 µS cm^{-1}. The ORP values were found to be 123.3 ± 21.7 mv, and the DO concentrations were 6.3 ± 0.3 mg L^{-1}. The PO_4^{3-}–P concentrations (mg L^{-1}) of the overlying water and the interstitial water ranged from 0.14 ± 0.03 to 0.55 ± 0.04 and from 0.34 ± 0.02 to 0.67 ± 0.04, respectively.

Table 1. Characteristics of overlying water in the Yuecheng reservoir.

Sample	PO_4^{3-}–P (mg L^{-1}) [a]	PO_4^{3-}–P (mg L^{-1}) [b]	T (°C)	DO (mg L^{-1})	Ph	EC (µS cm^{-1})	ORP (mv)
YR1	0.35 ± 0.02	0.44 ± 0.01	18.7	6.1	8.42	479.6	117.9
YR2	0.14 ± 0.03	0.34 ± 0.02	18.5	6.5	7.92	462.3	148.6
YR3	0.24 ± 0.03	0.57 ± 0.02	18.4	5.2	8.07	468.9	124.7
YR4	0.55 ± 0.04	0.67 ± 0.04	18.8	6.7	8.26	464.7	137.2
YR5	0.37 ± 0.01	0.54 ± 0.02	18.7	5.8	8.22	470.4	132.8

[a] Represents the mean values of PO_4^{3-}–P concentrations of the overlying water (0–4 cm). [b] Represents the mean values of PO_4^{3-}–P concentration of interstitial water of the surface sediment (0–4 cm). DO: dissolved oxygen; EC: electrical conductivity; ORP: oxidation reduction potential.

3.2. Characteristics of the Surface Sediment

The physical and chemical properties of sediments are shown in Table 2. The water content of sediment reflects its resuspension ability: the higher the value is, the easier it is to suspend. According to Table 2, the surface sediment contained a large percentage of water (63.4 ± 1.9%) which facilitates the diffusion of P through its suspension under external disturbance. The average value of the porosity of the surface sediment was 82.2 ± 1.3 %. Particle size analysis indicated that most surface sediments were high in terms of the proportion of clay and silt fractions; on average, at 97.1 %. Higher porosities and smaller sediment particles are more reactive due to the increased surface area [20].

Table 2. Physical and chemical characteristics of the sediments. TP: total phosphorus; TN: total nitrogen; TOC: total organic carbon.

Sample	Moisture (%)	Porosity (%)	Clay (%)	Silt (%)	Sand (%)	TP (mg·kg^{-1})	TN (g·kg^{-1})	TOC (g·kg^{-1})
YR1	61.1	80.3	29.5	66.7	3.8	2172.6	8.2	109.6
YR2	62.2	81.5	43.5	53.2	3.3	1576.3	7.0	75.4
YR3	64.7	83.1	40.3	57.4	2.3	1972.8	8.1	108.6
YR4	65.7	83.3	42.3	55.2	2.5	1627.3	7.7	95.6
YR5	63.1	82.7	44.3	52.8	2.9	1844.7	7.4	97.5

As presented in Table 2, TN varied from 7.0 to 8.2 g·kg^{-1} and TOC varied from 75.4 to 109.6 g·kg^{-1}. However, TP ranged from 1576.3 to 2172.6 mg·kg^{-1}. The C/N (TOC/TN) ratios ranged from 12.6 to 15.7 with an average of 14.8. The C/N ratio varied from 2.6 to 4.3 in bacteria, and 7.7 to 10.1 in aquatic plants, whereas this ratio of terrestrial plants or materials was higher than 20 [21]. Almost all C/N ratios in the sampling sediments of Yuecheng Reservoir were nearly 15, which illustrated that both endogenetic sources and terrestrial plants or materials had a pivotal role in sedimentary P.

3.3. Phosphorus Fraction Composition

As presented in Figure 2, the fractions of different P values varied greatly. For five sampling sites, the rank order of P fractions concentrations was Ca–Pi > Po > Fe/Al–Pi.

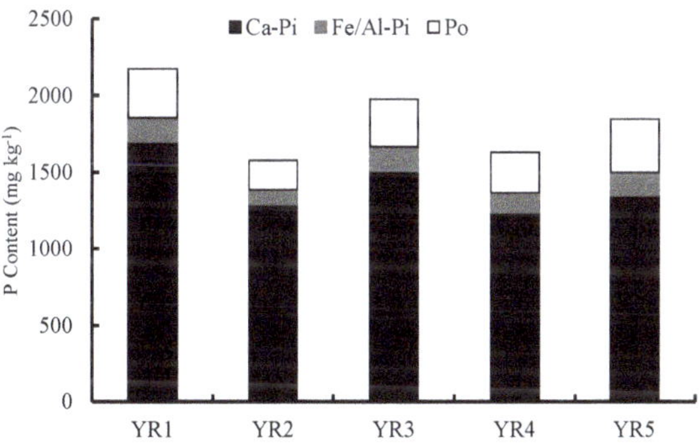

Figure 2. Concentrations of different P forms in surface sediments in Yuecheng Reservoir.

Po showed the P fraction bound to organic matters and its stability were relative to the different Po structures [12]. Po contents in the surface sediments varied from 188.6 to 358.8 mg kg^{-1}, and this contributed 12.1%–19.2% to the TP (Figure 2), which mainly resulted from the difference of TP contents. Fe/Al–Pi, which was considered as algae-available, could be easily affected by pH and ORP, and consequently was released into pore water under reductive conditions [22]. About 6.1%–8.9% of the sedimentary TP was Fe/Al–Pi, which ranged from 96.4 to 178.8 mg kg^{-1}. Ca–Pi was assumed to mainly consist of apatite P (natural and detritus) [22]. The Ca–Pi contents ranged from 1192.5 to 1703.6 mg kg^{-1}, and the relative contribution ranged between 72.4%–81.9% of TP. This inorganic fraction was a subject of debate because it had long been considered to have little mobilization. However, much research has shown that this fraction could be mobilized with decreased pH and specific biochemical effects [12,22].

3.4. Estimated Diffusion Fluxes of PO_4^{3-}–P

Figure 3 showed the typical distribution profiles of PO_4^{3-}–P concentrations in the overlying water and sediments of five sampling sites. The PO_4^{3-}–P in sediment is higher than overlying water, which indicates that the sediment was releasing P in all sections of the reservoir. The variations of PO_4^{3-}–P diffusion fluxes across the water–sediment interface in the sampling sites are presented in Figure 4. The fluxes ranged from 0.10 to 0.31 mg m^{-2} d^{-1}. Ca–Pi is more stable in the sediment and less easily released into overlying water compared with Po and Fe/Al–Pi. Although the TP was high at YR3, the flux was not the highest.

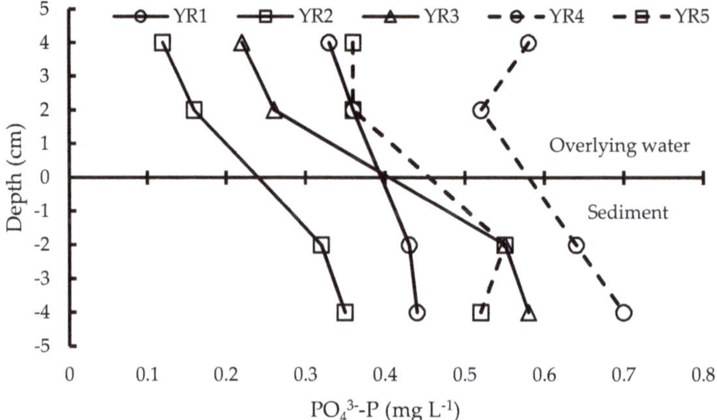

Figure 3. PO_4^{3-}–P concentrations of pore water and the corresponding overlying water.

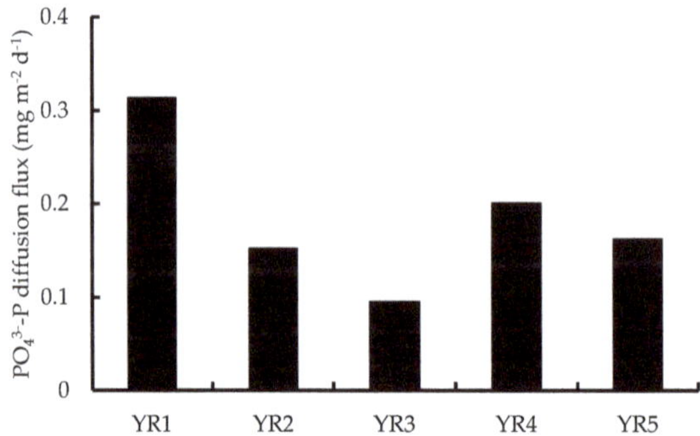

Figure 4. Variations of PO_4^{3-}–P diffusion fluxes at the water–sediment interface.

The threshold ratio of Fe to TP in the sediment is 15 for P retention under aerobic conditions. If this ratio holds true across the reservoir, the internal P loading could be controlled by keeping the surficial sediments oxidized. However, the ratios of Fe to TP of the sediment (6.1–8.9) were below that threshold, which indicated that the sediments could not adsorb more PO_4^{3-}–P of the overlying water. The microbial degradation of Ca–Pi probably contributes to the PO_4^{3-}–P diffusion due to their acclimation of living environment under this high Ca–Pi concentration background. Perez [23] detected that a total of 130 heterotrophic bacteria showed different degrees of mineral tri-calcium phosphate $(Ca_3(PO_4)_2)$-solubilizing activities.

Compared with other studies (Table 3), the TP concentration in the sediments was much smaller than that of almost all typical eutrophic lakes. Compared to Dianchi and Taihu, a greater proportion (more than 70%, Figure 2) of TP in the sediments was made up of Ca–Pi, which was considered to have little mobilization under physical–chemical conditions compared to Fe/Al–Pi [12,22]. Moreover, it is speculated that the behavior of the microbial degradation of Ca–Pi is driven by the microorganism to absorb reactive P nutrients from the environment for its own use. Thus, it may be controlled within a certain range by some related enzymes [23,24].

Table 3. Comparisons of TP and P diffusion fluxes between the reservoir and other areas.

Site	TP (mg kg^{-1})	PO$_4^{3-}$–P Diffusion Fluxes (mg m^{-2} d^{-1})	Reference
Hongze Lake, China	76.6–932.2	0.172–0.793	[25]
Dianchi Lake, China	1537–4695	1.00–4.36	[26]
Taihu Lake, China	213.7–724.4	0.76–4.57	[27,28]
Three Gorges Reservoir, China	415.5–1047.9	−0.003–0.013	[24,29]
Yuecheng Reservoir, China	1576.3–2172.6	0.10–0.31	This study

3.5. Relationship of Sediment Characteristics and P Fractions

The multiple regression analysis of characteristics and different P forms of the surface sediments are listed in Table 4. A non-parametric test was performed since environmental data do not usually follow a good normal distribution. The pairs of Ca–Pi with TP, with Fe/Al–Pi, and with Po clearly showed an association and implied an external Ca–Pi input to the reservoir (Table 4) [30]. The correlation coefficient of Ca–Pi and TP was 0.95, also indicating that the variance of TP concentrations was mainly related to the Ca–Pi contents. In addition, there was a significant correlation among TN, TOC, Ca–Pi and TP, which reflected that the sediment probability mainly came from the same origin [17].

Table 4. Physical and chemical characteristics of the sediments.

Sample	TP	TN	TOC	C/N	Po	Fe/Al-Pi	Ca-Pi
TP	1						
TN	0.77 **	1					
TOC	0.82 **	0.84 *	1				
C/N	0.64 *	0.49	0.88 **	1			
Po	0.72 *	0.61	0.79 **	0.76 *	1		
Fe/Al-Pi	0.75 *	0.64 *	0.86 **	0.85 **	0.82 **	1	
Ca-Pi	0.95 **	0.71 *	0.68 *	0.46	0.49	0.56	1

** Correlation is significant at the 0.01 level (2-tailed). * Correlation is significant at the 0.05 level (2-tailed).

4. Conclusions

This study showed that different P-forms and PO$_4^{3-}$–P diffusion fluxes of the surface sediments of the Yuecheng Reservoir, which is located in the productive and intensively cultivated agricultural region of North China. The rank order of P-fraction concentrations obtained from sample sites was Ca–Pi > Po > Fe/Al–Pi. Ca–Pi, varying from 72.4% to 81.9% of TP, was the primary driver of variation in TP. The C/N ratios ranged from 12.6 to 15.7, which suggested that Po mainly originated from endogenetic sources and terrestrial plants or materials. The analysis of environmental factors indicated that there was an association among P-forms. Although the Yuecheng Reservoir is located in one of the most productive and intensively cultivated agricultural regions in North China, Po is not a major part of the sediment, which is simply because the construction of the cascade reservoirs in the upper stream blocked a large number of terrestrial plants in the basin. Thus, for water environment management in this basin, although the construction of a large number of water conservancy projects in the upper reaches of the river has resulted in the decrease of inflow runoff, the pollution from terrestrial plants or materials played a key role in the sediment phosphorus fraction, and this should be emphasized in the water environment management of the river basin. Future work on the biological characterization of the sediments would provide more information on the status of this reservoir.

Author Contributions: Conceptualization, Y.C. and C.D.; methodology, C.D. and M.L.; investigation, Z.M., F.Z., C.C., L.Y. and Y.L.; writing—original draft preparation, Y.C. and M.L.

Funding: The study was financially supported by the National Natural Science Foundation of China (U1802241, 51509066 and 51909053), University Science and dTechnology Research Project of Hebei Province (ZD2019005) and Graduate Innovation Foundation of Hebei Province (CXZZBS2020152).

Conflicts of Interest: The authors declare no conflict of interest.

References

1. Correll, D. Phosphorus: A rate limiting nutrient in surface waters. *Poult. Sci.* **1999**, *78*, 674–682. [CrossRef] [PubMed]

2. Reddy, K.; Kadlec, R.; Flaig, E.; Gale, P. Phosphorus retention in streams and wetlands: A review. *Crit. Rev. Environ. Sci. Technol.* **1999**, *29*, 83–146. [CrossRef]

3. Correll, D.L. The role of phosphorus in the eutrophication of receiving waters: A review. *J. Environ. Qual.* **1998**, *27*, 261–266. [CrossRef]

4. Pan, G.; Krom, M.D.; Herut, B. Adsorption—Desorption of phosphate on airborne dust and riverborne particulates in east mediterranean seawater. *Environ. Sci. Technol.* **2002**, *36*, 3519–3524. [CrossRef] [PubMed]

5. Pan, G.; Krom, M.D.; Zhang, M.; Zhang, X.; Wang, L.; Dai, L.; Sheng, Y.; Mortimer, R.J.G. Impact of suspended inorganic particles on phosphorus cycling in the yellow river (China). *Environ. Sci. Technol.* **2013**, *47*, 9685–9692. [CrossRef]

6. Chuai, X.; Ding, W.; Chen, X.; Wang, X.; Miao, A.; Xi, B.; He, L.; Yang, L. Phosphorus release from cyanobacterial blooms in meiliang bay of lake taihu, China. *Ecol. Eng.* **2011**, *37*, 842–849. [CrossRef]

7. Huang, L.; Fu, L.; Jin, C.; Gielen, G.; Lin, X.; Wang, H.; Zhang, Y. Effect of temperature on phosphorus sorption to sediments from shallow eutrophic lakes. *Ecol. Eng.* **2011**, *37*, 1515–1522. [CrossRef]

8. Wang, Y.; Shen, Z.; Niu, J.; Liu, R. Adsorption of phosphorus on sediments from the three-gorges reservoir (China) and the relation with sediment compositions. *J. Hazard. Mater.* **2009**, *162*, 92–98. [CrossRef]

9. Młynarczyk, N.; Bartoszek, M.; Polak, J.; Sułkowski, W. Forms of phosphorus in sediments from the goczałkowice reservoir. *Appl. Geochem.* **2013**, *37*, 87–93. [CrossRef]

10. Qin, L.; Zeng, Q.; Zhang, W.; Li, X.; Steinman, A.D.; Du, X. Estimating internal p loading in a deep water reservoir of northern China using three different methods. *Environ. Sci. Pollut. Res.* **2016**, *23*, 18512–18523. [CrossRef]

11. Elser, J.; Bennett, E. Phosphorus cycle: A broken biogeochemical cycle. *Nature* **2011**, *478*, 29. [CrossRef] [PubMed]

12. Zhang, W.; Jin, X.; Zhu, X.; Shan, B.; Zhao, Y. Phosphorus characteristics, distribution, and relationship with environmental factors in surface sediments of river systems in Eastern China. *Environ. Sci. Pollut. Res.* **2016**, *23*, 19440–19449. [CrossRef] [PubMed]

13. Liu, J.; Li, Y.; Huang, G.; Zeng, X. A dual-interval fixed-mix stochastic programming method for water resources management under uncertainty. *Resour. Conserv. Recycl.* **2014**, *88*, 50–66. [CrossRef]

14. Liu, J.; Li, Y.; Huang, G.; Zhuang, X.; Fu, H. Assessment of uncertainty effects on crop planning and irrigation water supply using a monte carlo simulation based dual-interval stochastic programming method. *J. Clean. Prod.* **2017**, *149*, 945–967. [CrossRef]

15. Bao, Y.; Wang, Y.; Hu, M.; Wang, Q. Phosphorus fractions and its summer flux from sediments of deep reservoirs located at a phosphate-rock watershed, Central China. *Water Sci. Technol. Water Supply* **2017**, *18*, 688–697. [CrossRef]

16. Association, A.P.H.; Association, A.W.W.; Federation, W.P.C.; Federation, W.E. *Standard Methods for the Examination of Water and Wastewater*; American Public Health Association: Washington, DC, USA, 1915; Volume 2.

17. Zhang, W.; Jin, X.; Zhu, X.; Shan, B. Characteristics and distribution of phosphorus in surface sediments of limnetic ecosystem in Eastern China. *PLoS ONE* **2016**, *11*, e0156488. [CrossRef]

18. Berner, R.A. *Early Diagenesis: A Theoretical Approach*; Princeton University Press: Princeton, NJ, USA, 1980.

19. Krom, M.D.; Berner, R.A. The diffusion coefficients of sulfate, ammonium, and phosphate ions in anoxic marine sediments 1. *Limnol. Oceanogr.* **1980**, *25*, 327–337. [CrossRef]

20. Cade-Menun, B.J. Characterizing phosphorus in environmental and agricultural samples by 31p nuclear magnetic resonance spectroscopy. *Talanta* **2005**, *66*, 359–371. [CrossRef]

21. Meyers, P.A.; Lallier-Vergès, E. Lacustrine sedimentary organic matter records of late quaternary paleoclimates. *J. Paleolimnol.* **1999**, *21*, 345–372. [CrossRef]

22. Kaiserli, A.; Voutsa, D.; Samara, C. Phosphorus fractionation in lake sediments—Lakes Volvi and Koronia, N. Greece. *Chemosphere* **2002**, *46*, 1147–1155. [CrossRef]

23. Perez, E.; Sulbaran, M.; Ball, M.M.; Yarzabal, L.A. Isolation and characterization of mineral phosphate-solubilizing bacteria naturally colonizing a limonitic crust in the south-eastern venezuelan region. *Soil Biol. Biochem.* **2007**, *39*, 2905–2914. [CrossRef]

24. Yao, Y.; Wang, P.; Wang, C.; Hou, J.; Miao, L.; Yuan, Y.; Wang, T.; Liu, C. Assessment of mobilization of labile phosphorus and iron across sediment-water interface in a shallow lake (Hongze) based on in situ high-resolution measurement. *Environ. Pollut.* **2016**, *219*, 873–882. [CrossRef] [PubMed]

25. Mao, J.; Wang, Y.; Zhao, Q.; Wu, X. Preliminary study on phosphorus release of internal load in dianchi lake sediment. *J. China Inst. Water Resour. Hydropower Res.* **2005**, *3*, 229–233.

26. Zhang, R.-Y.; Wang, L.-Y.; Wu, F.-C.; Zhu, Y.-R. Distribution patterns of phosphorus forms in sediments interstitial water of lake taihu and the effects of sediment-water phosphorus release in spring. *Chin. J. Ecol.* **2012**, *31*, 902–907.

27. Qiu, H.; Geng, J.; Ren, H.; Xu, Z. Phosphite flux at the sediment–water interface in northern lake taihu. *Sci. Total Environ.* **2016**, *543*, 67–74. [CrossRef]

28. Wang, Y.; Niu, F.; Xiao, S.; Liu, D.; Chen, W.; Wang, L.; Yang, Z.; Ji, D.; Li, G.; Guo, H. Phosphorus fractions and its summer's release flux from sediment in the China's three gorges reservoir. *J. Environ. Inform.* **2015**, *25*, 36–45. [CrossRef]

29. Yuan, Y.; Bi, Y.; Zhu, K.; Hu, Z. Alkaline phosphatase activity in the sediments of the three gorges reservoir. *Environ. Sci. Technol.* **2014**, *37*, 60–64.

30. Sudha, V.; Ambujam, N.K. Characterization of bottom sediments and phosphorus fractions in hyper-eutrophic krishnagiri reservoir located in an agricultural watershed in India. *Environ. Dev. Sustain.* **2013**, *15*, 481–496. [CrossRef]

Article

Priority Pollutants in Water and Sediments of a River for Control Basing on Benthic Macroinvertebrate Community Structure

Xiang Liu [1,2,3], **Jin Zhang** [4], **Wenqing Shi** [1,2,*], **Min Wang** [1,2], **Kai Chen** [1,2] **and Li Wang** [1,2]

[1]   State Key Laboratory of Hydrology-Water Resources and Hydraulic Engineering, Nanjing 210029, China; gclx_2007@126.com (X.L.); wm910618@mail.ustc.edu.cn (M.W.); ckai2005@gmail.com (K.C.); syauwangli@126.com (L.W.)

[2]   Center for Eco-Environmental Research, Nanjing Hydraulic Research Institute, Nanjing 210029, China

[3]   State Key Joint Laboratory of Environment Simulation and Pollution Control, School of Environment, Tsinghua University, Beijing 100084, China

[4]   School of Civil Engineering, Yantai University, Yantai 264005, China; zhangjin513@outlook.com

*   Correspondence: shiwenqing2005320@126.com; Tel.: +86-25-85829765

Received: 22 May 2019; Accepted: 14 June 2019; Published: 17 June 2019

Abstract: Understanding the drivers of macroinvertebrate community structure is fundamental for adequately controlling pollutants and managing ecosystems under global change. In this study, the abundance and diversity of benthic macroinvertebrates, as well as their chemical parameters, were investigated quarterly from August 2014 to April 2015 in four reaches of the Huai River basin (HRB). The self-organizing map (SOM) algorithm and canonical correspondence analysis (CCA) were simultaneously applied to identify the main factors structuring the benthic community. The results showed that the benthic community structure was always dominated by gastropoda and insecta over seasons and presented obvious spatial and temporal heterogeneity along different pollution levels. The insects were always the top contributors to number density of the benthic community, except for the summer, and the biomass was mainly characterized by mollusca in all seasons. Statistical analysis indicated that TN and NH_3-N in water, as well as Hg, As, Cd, and Zn in sediments, were the dominant factors structuring the community, which determined the importance of sediment heavy metal concentrations in explaining the benthic community composition in comparison with other factors. These major factors should be given priority in the process of river pollutant control, which might be rated as a promising way to scientifically improve river health management and ecological restoration.

Keywords: macroinvertebrate; metal levels; water quality; sediment; canonical correspondence analysis; self-organizing map

1. Introduction

Rivers are the most essential component of natural ecosystems, which are often seen as the ecological channel of material circulation and energy flow between terrestrial ecosystems and aquatic ecosystems [1,2]. They provide some irreplaceable functions in sustaining human beings, such as water supply, irrigation, food production, and transportation [3,4]. However, in recent decades, most rivers have been gradually affected by different anthropogenic disturbances, such as pollution, habitat deterioration, channelization, and spatial isolation [5,6]. All these pressures have imposed severe threats to human water security and biodiversity loss [7]. Since well-balanced and adaptive communities can only be maintained by a healthy aquatic ecosystem, the biological community structure is the best indicator of the health of aquatic ecosystems. Therefore, an in-depth understanding of community–environment relationships is of great concern for the ecological management of rivers, especially for chemical pollutants in water and sediments.

Up to now, rapid industrialization and urbanization have caused increasing pollution in rivers, through the discharge of runoff, urban sewage, and industrial wastewater [8,9]. A large number of sluices and dams have been built across rivers for flood management and water resource utilization, which have adversely altered the hydrological regimes and indirectly interfere with both the migration and transformation of pollutants [10,11]. The nutrients (N, P) and organic matter in water directly lead to the environmental problems of eutrophication and make water black and smelly. Consequently, rivers are undergoing a perennial challenge linking the aquatic organisms to water environment degradation, due to a high level of pollutant accumulation.

It is well known that sediments act as sinks, and may in turn act as sources of pollution [12]. A large quantity of hazardous chemicals (e.g., heavy metals and persistent organic pollutants) can be absorbed by suspended solids and then accumulate in sediments. In particular, heavy metals may be transformed into persistent metallic compounds with high toxicity, that first bioaccumulate in organisms, subsequently magnify in the food chain, and ultimately threaten human health [13,14]. However, they are easily released into water columns worldwide under altered pH and redox potential, hydrodynamic disturbances, and movement of benthic biota [11,15], which have posed severe threats to the aquatic flora and fauna due to their toxicity, ubiquity, and persistence.

In 2000, the European Union passed the Water Framework Directive (WFD, European Parliament Council, 2000), mandating the use of different organismal groups to monitor the ecological status of surface waters [16]. Therefore, comprehensive evaluations, including the aquatic community structure and the chemicals in water or sediments, have been widely conducted [9,17]. In the aquatic food webs, benthic macroinvertebrates—sediment-dwelling organisms—are ubiquitous in freshwater ecosystems, which are regarded as the primary material exchangers across the sediment–water interface [18,19]. In addition, they also play important roles in trophic dynamics by cycling nutrients and providing food for higher trophic levels [20]. They are among the most diverse and abundant organisms in freshwater ecosystems that are vital for ecological functions. At present, the integrity of their community structure is successfully used to detect the evolution of water ecology on the temporal and spatial scale [7,21,22]. Thus, there is an increasing trend to use benthic macroinvertebrate communities in rivers as indicators for the environmental quality or diagnosing which pollutants should be controlled preferentially, according to the response of organisms.

Here, the objectives of this study were to: (1) analyze the characteristics of the environment and macroinvertebrate community from the time scale in a river basin; (2) establish an approach to reveal community–environment relationships and identify priority pollutants for control from the perspective of water ecological health protection. The study was also expected to provide new knowledge on biomonitoring and river basin management.

2. Material and Methods

2.1. Study Area

The study was conducted in the Huai River basin (HRB, 30°55′–36°36′ N, 111°55′–121°25′ E), China (Figure 1), which is located between the Yangtze River and the Yellow River, with a drainage area of 270,000 km^2, and provides some services, including water supply, agricultural irrigation, flood control, shipping, and aquaculture. The mainstream originates from Tongbai Mountain in Henan Province and flows eastward for approximately 1000 km through four provinces (Henan, Hubei, Anhui, and Jiangsu), before joining the Yangtze River in Sanjiangying in Jiangsu Province. The Shaying River is the largest tributary, which is 557 km long, and Hongze Lake is the largest lake of the HRB.

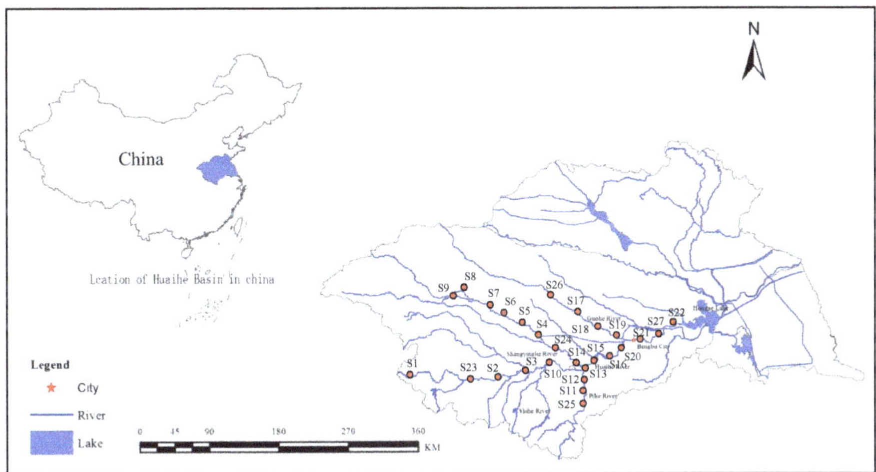

Figure 1. Location of study area, distribution of the sampling sites for the benthic community and environmental variables in the mainstream and other three important tributaries of the Huai River basin (HRB).

The HRB is an important agriculture production area and the most crowded region in terms of population and water projects. By 2005, the population had accounted for 13.1% of the national total [23]. There were about 11,000 dams and sluices built by the end of 2010, accounting for approximately half of those in China [24]. Furthermore, severe channelization, sand mining, and excessive pollutant discharge have directly disturbed the river ecosystem [25], and the HRB is facing serious risks of water quality deterioration and ecosystem degradation.

Previous studies have shown that approximately 80% of the reaches of the Huai River were contaminated by high concentrations of chemical oxygen demand (COD_{Mn}) and ammonia nitrogen (NH_3-N) [26,27], and the water quality ranked between Class IV and V of the Environmental Quality Standard for Surface Water (GB 3838-2002). Meanwhile, the middle parts of the Shaying River and Guo River were highly polluted by organic pollutants [26]. In addition, 98.15% of the sediments in the Huai River, particularly the reaches within Anhui Province, were highly enriched with heavy metals due to anthropogenic activities [27].

2.2. Field Sampling and Data Collection

To better diagnose the riverine ecosystem health of key regions and identify main pollutant structuring communities, water, sediments, and benthic macroinvertebrates were sampled at 27 sites (Figure 1). Four field surveys were conducted in summer (August), autumn (November), winter (January), and spring (April) from 2014 to 2015.

Water samples for chemical analysis were collected at about 50 cm below the river surface, and were conserved in 500 mL polyethylene bottle with the addition of 1 mL concentrated sulfuric acid (analytical reagent) to inhibit microbial activity. All water samples were placed in an ice chest at 4 °C, and were analyzed immediately after the samples arrived at the laboratory. Water variables such as the suspended solids (SS), chemical oxygen demand (COD_{Mn}), total nitrogen (TN), ammonia nitrogen (NH_3-N), nitrate nitrogen (NO_3^--N), nitrite nitrogen (NO_2^--N), and total phosphorus (TP) were measured based on procedures from the National Environment Protection Agency (NEPA, 2012). Besides this, dissolved oxygen (DO) was measured in situ at all sampling sites with multiparameter water quality sonde (YSI EXO2, SonTech, OH, USA). For the measurement of NH_3-N, NO_2^-N, and NO_3^--N, water samples were filtrated using 0.45 μm nitrocellulose filters. Heavy metals, including nickel (Ni), zinc (Zn), copper (Cu), chromium (Cr), cadmium (Cd), lead (Pb), arsenic (As), and mercury (Hg), were measured via inductively coupled plasma mass spectrometry (ICP-MS, XSERIES 2).

Sediments were collected using grabs at a depth of 0–10 cm and were sealed immediately in polyethylene airtight bags for heavy metal analysis. Sediment samples were first dried at room temperature. Subsequently, they were all ground with a mortar and sieved through a 100-mesh nylon, prior to analysis. Sediment samples (0.2000 g each) were digested with an HCl-HNO_3-HF-$HClO_4$ mixture and then measured by instruments three times. Heavy metals, including Ni, Zn, Cu, Cr, Cd, and Pb were also analyzed by ICP-MS (XSERIES 2), except for As and Hg, which were digested using HCl-HNO_3 and measured by the reduction gasification-atomic fluorescence spectrophotometer (AFS-230E) method.

Benthic macroinvertebrates were collected simultaneously with water and sediment samples. Different replicates of samples were collected with a 30 cm wide D-frame kick net of 500 μm mesh at each site, according to the occurrence of different habitats. All samples from the randomly selected sampling sites were field rinsed on a sieve (500 μm mesh size) to remove silt and detritus. Then, the D-net was inspected for macroinvertebrates adhering to the mesh. The macroinvertebrates were put into sealed plastic bags and preserved with 75% ethanol for later classification in the laboratory. Organisms were mainly identified to the species or genus level with a stereoscopic dissection microscope (LEICA MZ 95). Some groups were only identified to the higher taxonomic level, limited by their identification ability. For instance, Ephemeroptera, Plecoptera, and Trichoptera larvae were identified to the lowest taxonomic level, mostly to the genus level, while Odonata, Hemiptera, and Diptera were only identified to the family level, and the non-insect taxa to the order level. Identification of macroinvertebrates followed the key of Kawai (1985), Kang (1993), Morse et al. (1994), and Merritt et al. (2008) [28–31].

2.3. Data Analysis

The self-organizing map (SOM) is an adaptive unsupervised learning algorithm [32] and is often used to visualize and explore linear and non-linear relationships in high-dimensional datasets [33]. Up to now, this method has been widely used to find out which environmental variables had major influences on the presence of organisms [34]. SOM consists of two layers of input and output connected with computational weights [35]. The input layer acquires information from a data matrix (environmental variables or taxa abundance), whereas the output layer consists of a two-dimensional network of nodes arranged in a hexagonal lattice. The number of output neurons in an SOM can be selected using the heuristic rule suggested by Vesanto et al. (2000) [36] and applied in Park et al. (2006) [37]. The optimal number of map units is close to $5\sqrt{n}$, where n is the number of training samples. An alternative to the calculation of eigenvalues is to consider quantization error (QE) and topographic error (TE) [38]. Here, the SOM is trained with different map sizes, and the optimum size is selected based on minimum values for QE and TE. In general, the more complex the structure of the neural network, the stronger the ability to deal with non-linear problems, but the training time will be prolonged [39]. This calculation procedure was realized in the SOM toolbox package (Ver. 2.0, Laboratory of Information and Computer Science, Espoo, Helsinki, Finland) for Matlab R 2013b.

Detrended correspondence analysis (DCA) was applied using CANOCO (4.5, Centre for Biometry, Wageningen, Gelderland, the Netherlands) to examine whether redundancy analysis (RDA) or canonical correspondence analysis (CCA) would be appropriate to analyze the data [40]. If the gradient lengths analyzed by DCA were higher than three standard deviations, CCA was used to analyze the relationships between organisms and environmental variables. Prior to the multivariate statistical analysis, macroinvertebrate abundances were log $(x + 1)$ transformed to obtain homogeneity of variances [41]. After the forward selection and Monte-Carlo permutation test ($p < 0.05$), ten environmental variables were selected as independent factors. All data were used together to construct the plots. The nonmetric multi-dimensional scale (NMDS) was applied to perform the benthic community structure, using PC-ORD for Windows 4.0 (MjM Software Design, Gleneden Beach, OR, USA) based on the Bray–Curtis distance measure.

3. Results

3.1. Environmental Variables

Variations of environmental variables, including DO, SS, COD_{Mn}, TN, NH_3-N, NO_2^--N, NO_3^--N, and TP in water columns and Ni, Zn, Cu, Cr, Cd, Pb, As, and Hg in sediments in all seasons across the HRB are summarized in Figure 2. As a whole, HRB was suffering from a severe contamination in comparison to other sub-basins, especially for the pollution of TN, Cd, and Pb.

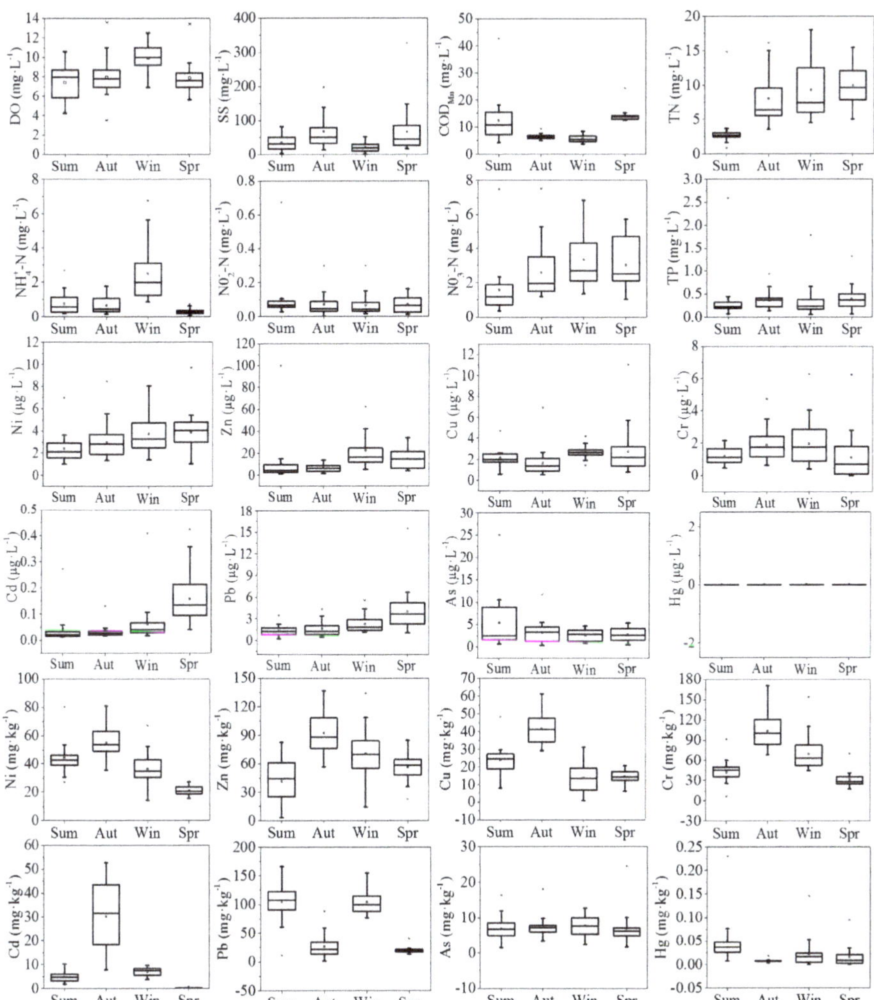

Figure 2. The temporal variations of environmental variables, including water quality and heavy metals pollution, in sediments at 27 monitoring stations of Huai River basin in the year 2014–2015.

In water, the DO in winter, with an average value of 9.94 mg/L, was significantly higher than that in the other seasons (ANOVA: Dunnett T3, $p < 0.001$). The SS concentrations were lower in summer and winter, whereas relatively high concentrations occurred in autumn and spring. The mean COD_{Mn} concentration varied between seasons, with the high mean value in spring (13.86 mg/L). The most severe pollutant was TN, with a water quality level worse than Grade V through the year, and a concentration in summer significantly lower than that in the other seasons (ANOVA: Dunnett T3, $p <$ 0.001). NO_3^--N had similar trends to TN, with the lowest concentration in summer, but the highest concentration of NO_3^--N occurred in winter, which was slightly different to the change in TN. NH_3-N also varied seasonally, with the highest concentration occurring in winter. NO_2^--N and TP levels had no significant difference among the four periods. Concentrations of heavy metals in water were relatively low, all of which were detected at the microgram level. In particular, Hg was not detected at any sampling sites over the seasons.

In sediments, the concentrations of Ni, Zn, Cu, Cr, and Cd had similar trends, with the high concentrations occurring in autumn. As and Hg levels were relatively stable during the four seasons. In particular, the average concentration of Cd was highest in autumn and achieved approximately 400 times the background concentration (0.079 mg/kg). The average concentration of Pb was higher in summer and winter than that in the other two seasons, which were about 4.5 times the background concentration (23.50 mg/kg) on the whole.

3.2. Macroinvertebrate Community Structure

A total of 10,722 individuals belonging to 3 phyla, 6 classes, 18 orders, 42 families, and 61 genera were collected from the 103 samples distributed in the typical reaches of HRB in the year 2014–2015. Nonparametric multidimensional scaling (NMDS) ordination was performed and is shown in Figure 3. The benthic community structure changed with the seasons, and its community in summer and winter presented obvious differences, whereas a certain degree of similarity was observed in autumn and spring. As shown in Figure 4, in particular, the composition of the benthic community based on the number of taxa showed a similar trend over seasons, including Gastropoda, Lamellibranchia, Hirudinea, Oligochaeta, Crustacea, and Insecta. Among these, Insecta and Gastropoda were the main components of the genera collected in the basin, ranging from 51.11% to 59.46% and 18.18% to 31.11%, respectively.

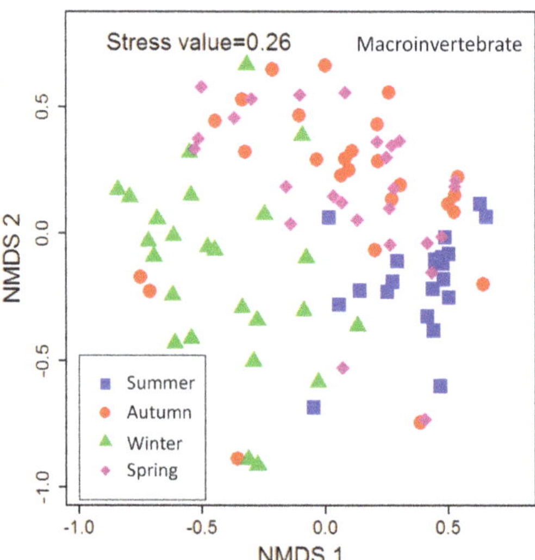

Figure 3. Nonparametric multidimensional scaling (NMDS) ordination of sampling sites based on the relative abundance data of benthic macroinvertebrates across different seasons.

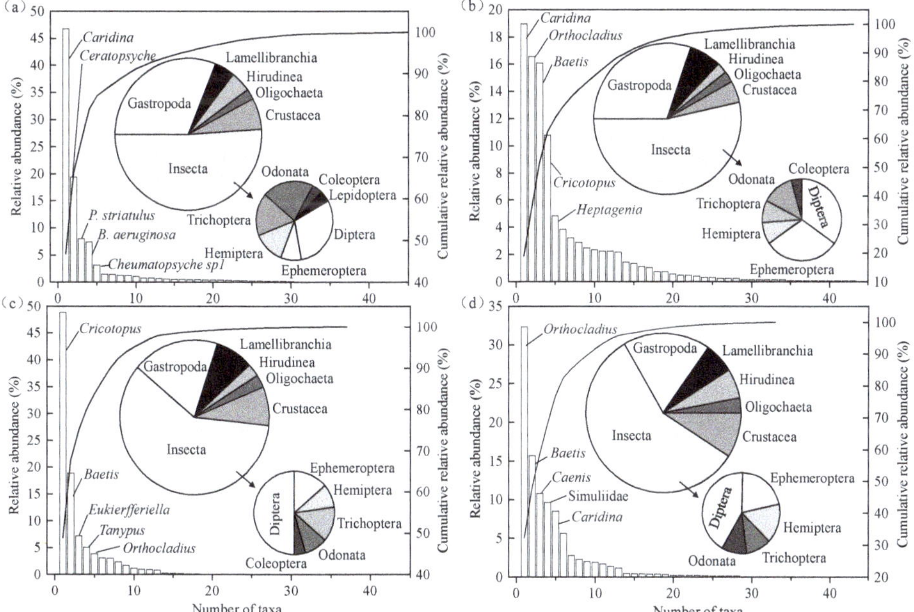

Figure 4. Macroinvertebrate communities of the Huai River basin sampled on summer (**a**), autumn (**b**), winter (**c**), and spring (**d**) in the year 2014–2015. Vertical bars represent relative abundances of each taxon from all sites in the study region, with the five most abundant taxa labeled. Solid lines depict the asymptotic nature of cumulative relative abundance in these communities (right-hand axis). Pie graphs show the composition of macroinvertebrates among classes based on the number of taxa. The small pie graphs show the detailed composition of Insecta among orders.

With regard to relative abundance, the five most-abundant taxa of genera also varied between seasons (Figure 4). In summer (Figure 4a), the most abundant species was in the genus *Caridina*, representing 46.72% of total abundance, and the other four dominant species were within the genera *Ceratopsyche*, *P. striatulus*, *B. aeruginosa*, and *Cheumatopsyche sp*1. In autumn (Figure 4b), the most abundant species was still in the genus *Caridina*, representing 18.96% of total abundance, and the other four dominant species were within the genera *Orthocladius*, *Baetis*, *Cricotopus*, and *Heptagenia*. In winter (Figure 4c), the most abundant species was in the genus *Cricotopus*, representing 48.90% of total abundance, and the other four dominant species were within the genera *Baetis*, *Eukierfferiella*, *Tanypus*, and *Orthocladius*. In spring (Figure 4d), the most abundant species in the genus *Orthocladius*, representing 32.36% of total abundance, and the other four dominant species were within the genera *Baetis*, *Caenis*, Simuliidae, and *Caridina*.

The density and biomass of macroinvertebrates fluctuated sharply throughout seasons (Figure 5a). The highest density was recorded in winter (85.20 ind./m^2) and the lowest value was observed in autumn (27.48 ind./m^2). Insecta was an important contributor, which accounted for more than 60% of the total number density, except that crustacea exhibited a relatively higher percentage than other species in summer (Figure 5b). In contrast, the biomass decreased from summer to winter, and then increased over time, showing a maximum value of 7.86 g/m^2 in summer and a minimum value of 0.57 g/m^2 in winter (Figure 5a). The biomass of mollusca was remarkably higher than other species, whereas annelida accounted for less than 1% of the total biomass in all seasons (Figure 5c).

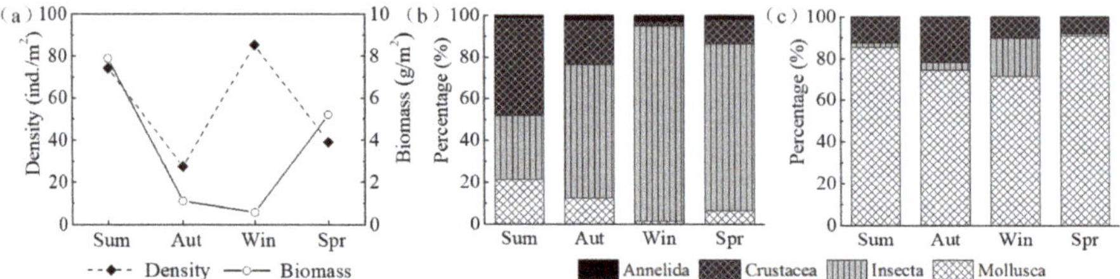

Figure 5. Seasonal changes of macroinvertebrate density and biomass (**a**), the percentage of different components in density (**b**), and biomass (**c**) in the typical reaches of the Huai River basin. The ind. represents individuals.

3.3. Macroinvertebrate Community Structure in Relation to Environmental Variables

The macroinvertebrate communities were patterned according to the similarity of community compositions through training the SOM ($54 = 9 \times 6$) (Figure 6). The sampling sites were reasonably well distributed in the SOM (Figure 6a), and the final values of QE and TE were 1.1499 and 0, respectively, indicating a good fit for the SOM training. According to the cluster analysis of k-means, seven main clusters were formed under the minimum principle of the Davies–Bouldin index (DBI) (Figure 6a). Six dominant species were associated with different clusters under a variety of pollution gradients (Figure 6b). Clusters V and VI were two larger groups located on the left of the SOM map, which were dominated by *Bellamya aeruginosa*. Clusters I, II and III were located in the bottom fourth of the SOM map, and were dominated by *Caridina*, *Orthocladius*, *Cricotopus*, and *Baetis*. Among them, the sampling sites in Cluster I were distributed in the upstream of the Huai river mainstream and Shaying River. The sampling sites in Cluster II were only distributed in the headwater region of Shaying River, and the sampling sites in Cluster III were only distributed in the headwater region of Huai river mainstream. Clusters IV and VII were located on the right of the SOM map, and were dominated by *Eukiefferiella thienemann*.

Twenty-four environmental variables involving water and sediments were observed at the sites and clustered in the SOM (Figure 7). Through the comparison and analysis of Figures 6 and 7, the qualitative influence of each pollutant on the benthic community structure can be clearly seen. In particular, none of the six dominant species adapted to the high concentration of SS. However, the influence of TN on the benthic community structure was not an absolutely negative effect. For example, *E. thienemann* could largely survive under a high concentration of TN, whereas the abundance of *B. aeruginosa* could be highly affected by TN. Meanwhile, *E. thienemann* was still positive with the high concentration of NH_3-N, and the influence caused by NO_3^--N presented a similar trend with TN. The concentrations of COD_{Mn}, NO_2^-N, and TP had no significant influence on the benthic community. As for the heavy metals in water with relatively low concentrations, they posed no regular influence on the benthic community, and even some positive effects were observed, like Cd on *Cricotopus* and *Baetis*, and As on *Caridina*. However, for the heavy metals in sediments, the abundance of *E. thienemann* showed a negative effect on the Hg concentration, and the pollution of Cu, Cd, and As generally posed negative effects on the benthic community.

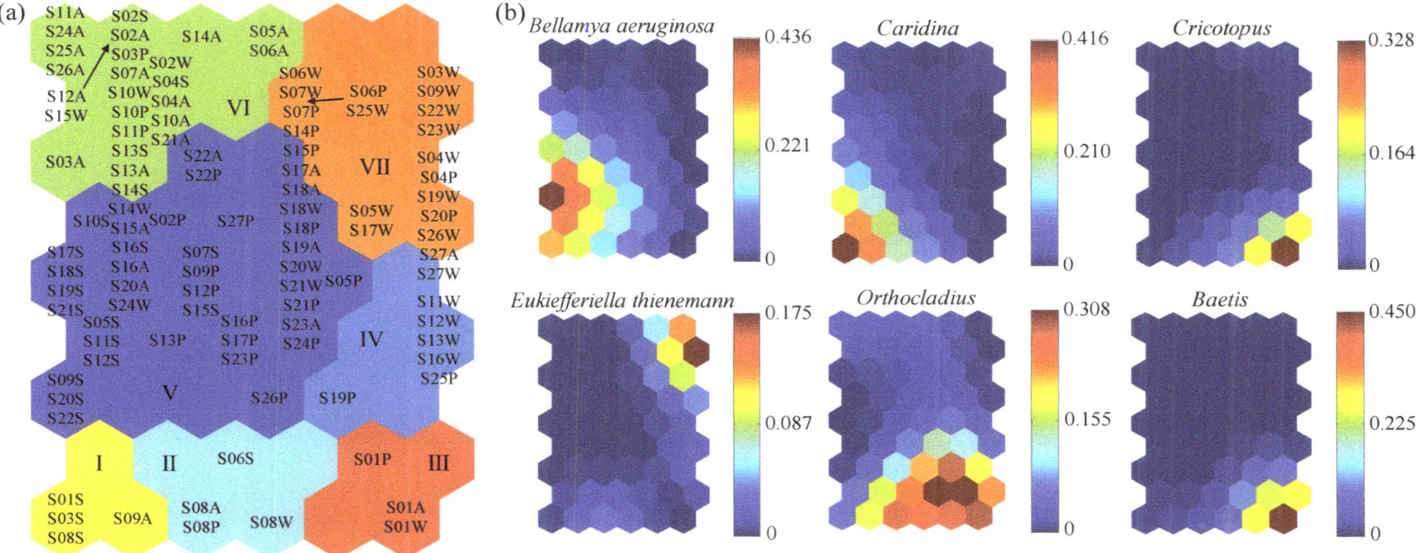

Figure 6. (**a**): distribution and clustering of samples on the self-organizing map (SOM) according to the abundance of 72 macroinvertebrate taxa. Codes within each hexagon correspond to individual samples (S: summer, A: autumn, W: winter, P: spring, 01–27 sampling stations) which were derived from the *k*-means algorithm applied to the weights of the 72 taxa in the 60 output neurons of the SOM; (**b**): profile of abundance of the prevalent taxa matched to clusters based on the trained SOM. The values in the vertical bar indicate densities (individuals/m^2).

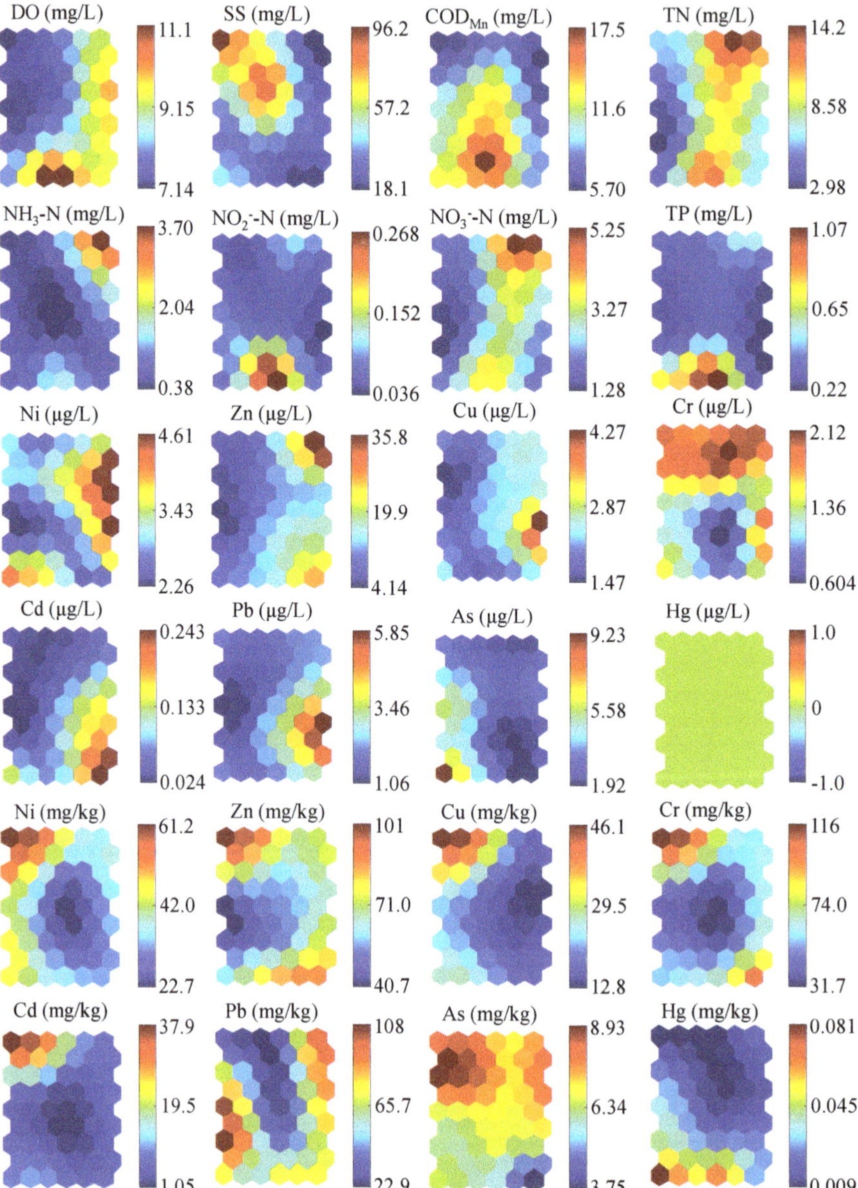

Figure 7. Visualization of environmental variables in waters and sediments. The mean value for each variable was calculated in each output neuron of the SOM previously trained with macroinvertebrate data.

Results obtained from CCA showed strong relationships between the macroinvertebrate community and environmental variables (Figure 8). After the forward selection and Monte Carlo permutation test ($p < 0.05$), 11 environmental variables, mainly including TN, TP, NH$_3$-N, and Cd in water and Ni, Zn, Cu, Cr, Cd, Hg, and As in sediments, were selected and could interpret 69.60% of data variability in the first two axes. The species–environment correlations of CCA axes 1 and

2 were 0.908 and 0.892, respectively, and the first two axes accounted for 41.40% of the variance of species–environment relationship.

Of these, Hg, As, Cd, and Zn were the most important representative factors in sediments, and TN and NH$_3$-N were the most significant representative factors in water. Hg pollution was the first important factor, and most of the species, such as *R. swinhoei*, *Orthocladius*, *Oligochaeta*, and *E. thienemann*, showed a negative correlation with Hg level, whereas *P. striatulus*, *Caenis*, and *Baetis* presented their preferences to the relatively higher level of Hg. In addition, *E. thienemann* and Oligochaeta were strongly positive with TN. *Orthocladius* and *R. swinhoei* showed preferences to As and *B. aeruginosa*. Moreover, *Caridina* was positive to the levels of Cu and Ni, but negatively related to the levels of Cd and Zn.

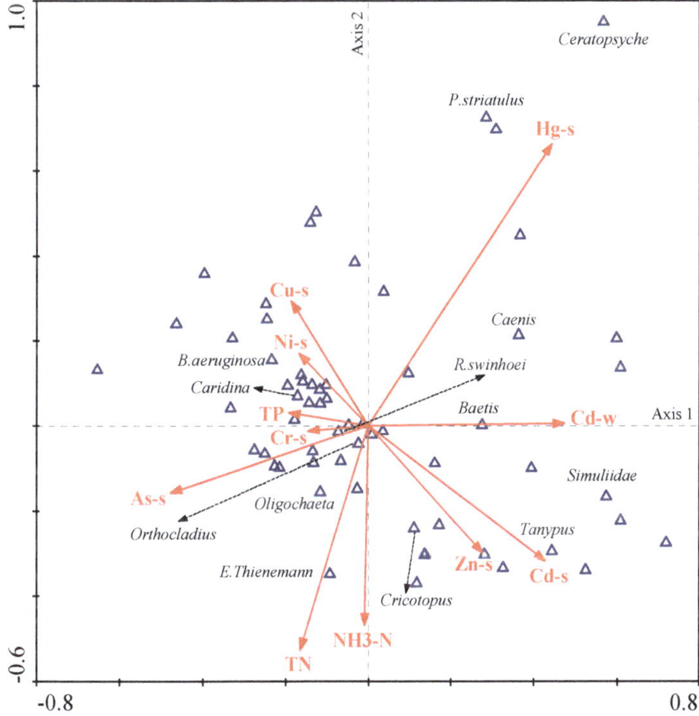

Figure 8. Ordination diagram of canonical correspondence analysis (CCA) on species abundances and selected environmental variables in the typical reaches of the Huai River basin (some high abundance species are marked on the diagram).

4. Discussion

With the rapid development of the economy, water quality deterioration has become a prominent issue threatening river ecological health and water security throughout the world [42]. HRB, as the sixth largest basin in China, is faced with the most severe pollution, both in water and sediments, which is closely related to the industrial structure, environmental management, and the surrounding land use. Due to the construction of wastewater treatment plants and pipeline networks, point source pollution has been effectively controlled, and urban domestic sewage for COD has decreased by 44.57% from 1993 to 2005 [26]. Of course, some wastewater randomly discharged by small enterprises with no permission could be another important factor for water quality degradation in the HRB, which is usually accompanied by the occurrence of sudden pollution events. Additionally, non-point source pollution arising from the daily lives of village residents and framing might be an underlying and increasing driving factor for the deterioration of water quality. Besides this, various types of land use also have significant influences on the degradation of water quality. In the HRB, as the first leading industry and the contribution of agricultural non-point source pollution, agriculture could account for almost 70% of the total pollutants during the flood season, especially for the SS and nutrients [5,43]. In this study, the high concentrations of nutrients (N, P) observed in the typical reaches of the HRB were probably due to the use of too much fertilizer in agricultural activities. Water quality improvement during the flood season could be further attributed to the dilution function of rivers. However, a large number of tannery and electroplating factories distributed in the HRB might be the main point source contributing to the heavy metal pollution. For instance, Shaying River flows through an area of paper mills and tanneries, where high concentrations of metals were observed. With regard to the spatio–temporal variation in pollutants in surface sediments, this may be attributed to the highly regulated sluices closed for water supply in the non-flood season and opened for flood control in the flood season [26,44].

So far, in addition to a focus on how to reduce the concentration of pollutants, we are more concerned with their ecological effects under the proposal of water ecological civilization construction. Benthic macroinvertebrates, as the middle ecological niche in the aquatic food chains, have a lot of advantages in evaluating the water ecological health when compared to other aquatic organisms [15,20]. They survive perennially in the water–sediment interface and are highly sensitive to the pollutants both in the water and sediment. In recent years, many studies have demonstrated that the benthic community structure is highly correlated with environmental variables, including physical and chemical factors, which have therefore been considered as outstanding indicators for environmental conditions [7,45].

Understanding the drivers of community structure is a fundamental work for adequately controlling pollutants and managing ecosystems. In this study, community–environment relationships were established by SOM and CCA and received reliable results by complementing and verifying each other. The SOM approach can complete the site clustering, based on the similarity of community structure and present visualization effects. Our study suggested that several chemical parameters, such as TN and NH$_3$-N in water, as well as some heavy metals in sediment, such as Hg, As, Cd, and Zn, had the largest contribution to the macroinvertebrate taxonomic variation. Due to the sensitivity difference of species to different environmental pollutants, the influential mechanisms of environmental factors on organisms are various. Some of these factors, particularly nutrient concentrations, have a direct influence on primary production characteristics and food availability, as well as on the dispersal and reproduction of benthic macroinvertebrates [46]. Previous studies showed that nutrients (N, P) can cause eutrophication with a change of benthic algal composition [47], and further affect macroinvertebrate assemblages by bottom-up regulation [48]. On the other hand, they can also easily change the DO concentration and the transparency of water [16]. These two aspects may be the main reasons for the influence of nutrients on the structure of benthic communities.

This study found that heavy metals in water and sediment can have an impact on benthic fauna to different degrees, but the effects of heavy metals in sediments explained a larger part of the variation in the benthic community structure. Zhang et al. (2014) fund that Cu, Zn, and Pb in sediment were the main elements affecting the benthic community structure in the Baiyang, especially for insects [49]. Iwasaki and Ormerod (2012) showed that the total abundance and richness of Trichoptera had strong negative correlations with heavy metals [50]. Thus, we can conclude from the above results that different heavy metals have different effects on the benthic community in different water periods, and the influence of heavy metals on macroinvertebrates is mainly attributed to their genotoxic and neurotoxic effects, which can further affect many physiological and cellular processes in different macroinvertebrates [51,52]. However, the response difference of benthic communities to heavy metals can be modified by chemical conditions, tolerance values, biomass, community composition, and species interactions [53]. For example, it has been found that some Ephemeroptera and Trichoptera taxa were highly tolerant of metals [54]. Thus, facing the community–environment relationships, the more factors are considered, the more accurate the results are.

However, it is also worth mentioning that some limitations still exist in this study. For example, the effects of physical factors, including hydrological conditions, habitat type, and riparian vegetation on the benthic community structure are currently not considered. Secondly, we only focused on the effects of conventional water quality parameters and heavy metals on macroinvertebrates, ignoring the other toxic organic matters, such as pesticides. Additionally, the data analysis method applied in this study generally still stays on the qualitative level. However, omitting these effects at the current stage did not represent an oversimplification. In the next work, more factors will be focused on during the field study and the specific quantitative study on environmental variables on the individual of benthos under controlled laboratory will be soon addressed. In a possible situation, we can further study the toxic-mode-of-action of different chemical classes to macroinvertebrates by molecular biotechnology.

5. Conclusions

The abundance and diversity of benthic macroinvertebrates, as well as chemical parameters, were investigated quarterly in the HRB, and the impacts of environmental variables on macroinvertebrate assemblages were determined clearly. Our results showed that the benthic community structure showed significant spatial and temporal heterogeneity along the different pollutant levels, but it was always dominated by gastropoda and insecta throughout the year. Dominant species in the studied area were *B. aeruginosa*, *Caridina*, *Cricotopus*, *E. thienemann*, *Orthocladius*, and *Baetis*. Community–environment relationships were analyzed by a combination of SOM and CCA, indicating that TN and NH_3-N in water and Hg, As, Cd, and Zn in sediments were the main factors structuring the community. Meanwhile, results also revealed that heavy metals in sediments explained a larger part of the variation in the benthic community structure as compared to other water parameters. Hence, these main factors could be listed as priority pollutants for river ecological management in the HRB, and macroinvertebrate assemblages could also be used as a biomonitoring tool to better assess water quality in the river system.

Author Contributions: Conceptualization, X.L. and W.S.; Methodology, X.L. and K.C.; Software, X.L.; Validation, X.L., K.C. and W.S.; Formal Analysis, X.L.; Investigation, X.L., M.W., K.C. and L.W.; Resources, W.S.; Data Curation, W.S.; Writing-Original Draft Preparation, X.L.; Writing-Review & Editing, J.Z.; Visualization, X.L. and J.Z.; Supervision, W.S.; Project Administration, W.S. and K.C.; Funding Acquisition, W.S.

Funding: This research was funded by National Natural Science Foundation of China (51425902, 91547206) and the National Water Pollution Control and Treatment Major Science and Technology Project (2014ZX07204-006-01) and the Special Fund for Scientific Research in the Public Interest (201501007).

Conflicts of Interest: The authors declare no conflict of interest.

References

1. Fullerton, A.H.; Burnett, K.M.; Steel, E.A.; Flitcroft, R.L.; Pess, G.R.; Feist, B.E.; Torgersen, C.E.; Miller, D.J.; Sanderson, B.L. Hydrological connectivity for riverine fish: Measurement challenges and research opportunities. *Freshw. Biol.* **2010**, *55*, 2215–2237. [CrossRef]

2. Islam, M.S.; Ahmed, M.K.; Raknuzzaman, M.; Habibullah-Al-Mamun, M.; Islam, M.K. Heavy metal pollution in surface water and sediment: A preliminary assessment of an urban river in a developing country. *Environ. Int.* **2015**, *48*, 282–291. [CrossRef]

3. Gordon, L.J.; Finlayson, C.M.; Falkenmark, M. Managing water in agriculture for food production and other ecosystem services. *Agric. Water Manag.* **2010**, *97*, 512–519. [CrossRef]

4. Pan, B.Z.; Wang, Z.Y.; Xu, M.Z.; Xing, L.H. Relation between stream habitat conditions and macroinvertebrate assemblages in three Chinese rivers. *Quat. Int.* **2012**, *282*, 178–183. [CrossRef]

5. Tang, T.; Cai, Q.H.; Liu, R.Q.; Xie, Z.C. Distribution of epilithic algae in the Xiangxi River system and their relationships with environmental factors. *J. Freshw. Ecol.* **2002**, *17*, 345–352. [CrossRef]

6. Yi, Y.J.; Yang, Z.F.; Zhang, S.H. Ecological risk assessment of heavy metals in sediment and human health risk assessment of heavy metals in fishes in the middle and lower reaches of the Yangtze River basin. *Environ. Pollut.* **2011**, *159*, 2575–2585. [CrossRef] [PubMed]

7. Carlson, P.E.; Johnson, R.K.; Mckie, B.G. Optimizing stream bioassessment: Habitat, season, and the impacts of land use on benthic macroinvertebrates. *Hydrobiologia* **2013**, *704*, 363–373. [CrossRef]

8. Hilton, J.; O'Hare, M.; Bowes, M.J. How green is my river? A new paradigm of eutrophication in rivers. *Sci. Total Environ.* **2006**, *365*, 66–83. [CrossRef]

9. Roig, N.; Sierra, J.; Ortiz, J.D.; Merseburger, G.; Schuhmacher, M.; Domingo, J.L.; Nada, M. Integrated study of metal behavior in Mediterranean stream ecosystems: A case-study. *J. Hazard. Mater.* **2013**, *263*, 122–130. [CrossRef]

10. Zhang, M.; Shao, M.L.; Xu, Y.Y.; Cai, Q.H. Effect of hydrological regime on the macroinvertebrate community in Three-Gorges Reservoir, China. *Quat. Int.* **2010**, *226*, 129–135. [CrossRef]

11. Grill, G.; Dallaire, C.O.; Chouinard, E.F.; Sindorf, N.; Lehner, B. Development of new indicators to evaluate river fragmentation and flow regulation at large scales: A case study for the Mekong River Basin. *Ecol. Indic.* **2014**, *45*, 148–159. [CrossRef]

12. Suresh, G.; Ramasamy, V.; Meenakshisundaram, V.; Venkatachalapathy, R.; Ponnusamy, V. Influence of mineralogical and heavy metal composition on natural radionuclide contents in the river sediments. *Appl. Radiat. Isot.* **2011**, *69*, 1466–1474. [CrossRef] [PubMed]

13. Su, L.Y.; Liu, J.L.; Christensen, P. Spatial distribution and ecological risk assessment of metals in sediments of Baiyangdian wetland ecosystem. *Ecotoxicity* **2011**, *20*, 1107–1116. [CrossRef] [PubMed]

14. Li, C.; Song, C.W.; Yin, Y.Y.; Sun, M.H.; Tao, P.; Shao, M.H. Spatial distribution and risk assessment of heavy metals in sediments of Shuangtaizi estuary. *China Mar. Pollut. Bull.* **2015**, *98*, 358–364. [CrossRef] [PubMed]

15. Colas, F.; Vigneron, A.; Felten, V.; Devin, S. The contribution of a niche-based approach to ecological risk assessment: Using macroinvertebrate species under multiple stressors. *Environ. Pollut.* **2014**, *185*, 24–34. [CrossRef] [PubMed]

16. Virtanen, L.K.; Soininen, J. Temporal variation in community–environment relationships and stream classifications in benthic diatoms: Implications for bioassessment. *Limnologica* **2016**, *58*, 11–19. [CrossRef]

17. Fu, J.; Zhao, C.P.; Luo, Y.P.; Liu, C.S.; Kyzas, G.Z.; Luo, Y.; Zhao, D.Y.; An, S.Q.; Zhu, H.L. Heavy metals in surface sediments of the Jialu River, China: Their relations to environmental factors. *J. Hazard. Mater.* **2014**, *270*, 102–109. [CrossRef]

18. Keizer-Vlek, H.E.; Verdonschot, P.F.M.; Verdonschot, R.C.M.; Goedhart, P.W. Quantifying spatial and temporal variability of macroinvertebrate metrics. *Ecol. Indic.* **2012**, *23*, 384–393. [CrossRef]

19. Hussain, Q.A.; Pandit, A.K. Macroinvertebrates in streams: A review of some ecological factors. *Int. J. Fish. Aquacult.* **2012**, *4*, 114–123.

20. Teferi, E.; Uhlenbrook, S.; Bewket, W.; Wenninger, J.; Simane, B. The use of remote sensing to quantify wetland loss in the Choke Mountain range, Upper Blue Nile basin, Ethiopia. *Hydrol. Earth Syst. Sci.* **2010**, *14*, 2415–2428. [CrossRef]

21. Gray, N.F.; Delaney, E. Comparison of benthic macroinvertebrate indices for the assessment of the impact of acid mine drainage on an Irish river below an abandoned Cu-S mine. *Environ. Pollut.* **2008**, *155*, 31–40. [CrossRef]

22. Giorgio, A.; Bonis, S.D.; Guida, M. Macroinvertebrate and diatom communities as indicators for the biological assessment of river Picentino (Campania, Italy). *Ecol. Indic.* **2016**, *64*, 85–91. [CrossRef]

23. Xia, J.; Zhang, Y.Y.; Zhan, C.S.; Ye, A.Z. Water quality management in China: The case of the Huai River Basin. *Int. J. Water Resour. Dev.* **2011**, *27*, 167–180. [CrossRef]

24. Liu, C.M.; Zhao, C.S.; Xia, J.; Sun, C.L.; Wang, R.; Liu, T. An instream ecological flow method for data-scarce regulated rivers. *J. Hydrol.* **2011**, *398*, 17–25. [CrossRef]

25. Kagalou, I.I.; Kosiori, A.; Leonardos, I.D. Assessing the zooplankton community and environmental factors in a Mediterranean wetland. *Environ. Monit. Assess.* **2010**, *170*, 445–455. [CrossRef] [PubMed]

26. Zhai, X.Y.; Xia, J.; Zhang, Y.Y. Water quality variation in the highly disturbed Huai River Basin, China from 1994 to 2005 by multi-statistical analyses. *Sci. Total Environ.* **2014**, *496*, 594–606. [CrossRef]

27. Wang, J.; Liu, G.J.; Lu, L.L. Geochemical normalization and assessment of heavy metals (Cu, Pb, Zn, and Ni) in sediments from the Huaihe River, Anhui, China. *Catena* **2015**, *129*, 30–38. [CrossRef]

28. Kawai, T. *An Illustrated Book of Aquatic Insects of Japan*; Tokai Univ. Press: Tokyo, Japan, 1985.

29. Kang, S.C. *Ephemeroptera of Taiwan (Excluding Baetidae)*; National Chung Hsing University: Taichung, Taiwan, 1993.

30. Morse, J.C.; Yang, L.F.; Tian, L.X. (Eds.) *Aquatic Insects of China Useful for Monitoring Water Quality*; Hohai University Press: Nanjing, China, 1994.

31. Merritt, R.W.; Cummins, K.W.; Berg, M.B. *An Introduction to the Aquatic Insects of North America*, 4th ed.; Kendall/Hunt Publishing Company: Dubuque, IA, USA, 2008.

32. Kohonen, T. Self-organized formation of topologically correct feature maps. *Biol. Cybern.* **1982**, *43*, 59–69. [CrossRef]

33. Chon, T.S. Self-Organizing Maps applied to ecological sciences. *Ecol. Inf.* **2011**, *6*, 50–61. [CrossRef]

34. Park, Y.S.; Chon, T.S.; Kwak, I.S.; Lek, S. Hierarchical community classification and assessment of aquatic ecosystems using artificial neural networks. *Sci. Total Environ.* **2004**, *327*, 105–122. [CrossRef] [PubMed]

35. Li, F.Q.; Cai, Q.H.; Qu, X.D. Characterizing macroinvertebrate communities across China: Large-scale implementation of a self-organizing map. *Ecol. Indic.* **2012**, *23*, 394–401. [CrossRef]

36. Vesanto, J.; Himberg, J.; Alhoniemi, E.; Parhankangas, J. *SOM Toolbox for Matlab 5. Technical Report A57*; Neural Networks Research Centre, Helsinki University of Technology: Helsinki, Finland, 2000.

37. Park, Y.S.; Tison, J.; Lek, S.; Giraudel, J.L.; Coste, M.; Delmas, F. Application of a self-organizing map to select representative species in multivariate analysis: A case study determining diatom distribution patterns across France. *Ecol. Inf.* **2006**, *1*, 247–257. [CrossRef]

38. Park, Y.S.; Céréghino, R.; Compin, A.; Lek, S. Applications of artificial neural networks for patterning and predicting aquatic insect species richness in running waters. *Ecol. Model.* **2003**, *60*, 265–280. [CrossRef]

39. Song, M.Y.; Hwang, H.J.; Kwak, I.S. Self-organizing mapping of benthic macroinvertebrate communities implemented to community assessment and water quality evaluation. *Ecol. Model.* **2007**, *203*, 18–25. [CrossRef]

40. ter Braak, C.J.F.; Smilauer, P. *CANOCO Reference Manual and CanocoDraw for Windows User's Guide: Software for Canonical Community Ordination (version 4.5)*; Centre for Biometry: Wageningen, the Netherlands, 2002.

41. Mereta, S.T.; Boets, P.; Bayih, A.A. Analysis of environmental factors determining the abundance and diversity of macroinvertebrate taxa in natural wetlands of Southwest Ethiopia. *Ecol. Inf.* **2012**, *7*, 52–61. [CrossRef]

42. Khadse, G.K.; Patni, P.M.; Kelkar, P.S.; Devotta, S. Qualitative evaluation of Kanhan river and its tributaries flowing over central Indian plateau. *Environ. Monit. Assess.* **2008**, *147*, 83–92. [CrossRef]

43. Lehner, B.; Reidy Liermann, C.; Revenga, C.; Vörösmarty, C.; Fekete, B.; Crouzet, P.; Döll, P.; Endejan, M.; Frenken, K.; Magome, J.; et al. High-resolution mapping of the world's reservoirs and dams for sustainable river-flow management. *Front. Ecol. Environ.* **2011**, *9*, 494–502. [CrossRef]

44. Suthar, S.; Nema, A.K.; Chabukdhare, M.; Gupta, S.K. Assessment of metals in water and sediments of Hindon River, India: Impact of industrial and urban discharge. *J. Hazard. Mater.* **2009**, *171*, 1088–1095. [CrossRef]

45. Zamora-Muñoz, C.; Sáinz-Cantero, C.E.; Sánchez-Ortega, A.; Alba-Tercedor, J. Are biological indices BMPW' and ASPT' and their significance regarding water quality seasonally dependent? Factors explaining their variations. *Water Res.* **1995**, *29*, 285–290. [CrossRef]

46. Wolfram, G.; Höss, S.; Orendt, C.; Schmitt, C.; Adámek, Z.; Bandow, N.; Großschartner, M.; Kukkonen, J.V.K.; Leloup, V.; López Doval, J.C.; et al. Assessing the impact of chemical pollution on benthic invertebrates from three different European rivers using a weight-of-evidence approach. *Sci. Total Environ.* **2012**, *438*, 498–509. [CrossRef]

47. Leland, H.V.; Porter, S.D. Distribution of benthic algae in the upper Illinois River basin in relation to geology and land use. *Freshw. Biol.* **2000**, *44*, 279–301. [CrossRef]

48. Wang, X.; Tan, X. Macroinvertebrate community in relation to water quality and riparian land use in a substropical mountain stream, China. *Environ. Sci. Pollut. Res. Int.* **2017**, *3*, 1–8. [CrossRef] [PubMed]

49. Zhang, L.L.; Liu, J.L. Relationships between ecological risk indices for metals and benthic communities metrics in a macrophyte-dominated lake. *Ecol. Indic.* **2014**, *40*, 162–174. [CrossRef]

50. Iwasaki, Y.; Ormerod, S.J. Estimating safe concentrations of trace metals from inter-continental field data on river macroinvertebrates. *Environ. Pollut.* **2012**, *166*, 182–186. [CrossRef]

51. Martins, R.J.E.; Pardob, R.; Boaventura, R.A.R. Cadmium (II) and zinc (II) adsorption by the aquatic moss *Fontinalis antipyretica*: Effect of temperature, pH and water hardness. *Water Res.* **2004**, *38*, 693–699. [CrossRef] [PubMed]

52. Michailova, P.; Warchalowska-Śliwa, E.; Szarek-Gwiazda, E.; Kownacki, A. Does biodiversity of macroinvertebrates and genome response of *Chironomidae larvae* (Diptera) reflect heavy metal pollution in a small pond? *Environ. Monit. Assess.* **2012**, *184*, 1–14. [CrossRef] [PubMed]

53. Bian, B.; Zhou, Y.; Fang, B.B. Distribution of heavy metals and benthic macroinvertebrates: Impacts from typical inflow river sediments in the Taihu Basin, China. *Ecol. Indic.* **2016**, *69*, 348–359. [CrossRef]

54. Bere, T.; Dalu, T.; Mwedzi, T. Detecting the impact of heavy metal contaminated sediment on benthic macroinvertebrate communities in tropical streams. *Sci. Total Environ.* **2016**, *572*, 147–156. [CrossRef]

Article

First-Principles Study on the Migration of Heavy Metal Ions in Ice-Water Medium from Ulansuhai Lake

Chi Sun [1], Changyou Li [1,*], Jianjun Liu [2], Xiaohong Shi [1], Shengnan Zhao [1], Yong Wu [3] and Weidong Tian [1]

[1] College of Water Conservancy and Civil Engineering, Inner Mongolia Agricultural University, Hohhot 010018, China; sunchi0505@163.com (C.S.); imaushixiaohong@163.com (X.S.); zhaoshengnan2005@163.com (S.Z.); tianweidong1122@163.com (W.T.)
[2] Shanghai Institute of Ceramics, Chinese Academy of Sciences, Shanghai 200000, China; jliu@mail.sic.ac.cn
[3] College of Resources and Environment, Henan Agricultural University, Zhengzhou 450002, China; wuyong526@126.com
* Correspondence: nndlichangyou@163.com; Tel.: +86-0471-4307990

Received: 19 July 2018; Accepted: 24 August 2018; Published: 28 August 2018

Abstract: Energy is a fundamental driver that causes material movement. It is important to discover changes in energy by studying the internal mechanism of pollutant migration between system components during the freezing process. To explore the migration mechanism of heavy metal ions (HMIs) from ice to water in a lake, we carried out a laboratory freezing experiment and simulated the distribution and migration of HMIs (Fe, Cu, Mn, Zn, Pb, Cd, and Hg) under different conditions. Then, we analyzed the use of energy by first-principle calculations. The results showed that HMIs are more stable in an aqueous environment than in an ice environment. For the same HMI, the binding energy in water is smaller than that in ice. Hence, the HMIs migrated from ice to water as the lake was freezing. The ability of different kinds of heavy metals to migrate from ice to water is related to their binding energy in ice. The concentrations of HMIs in ice are positively correlated with their binding energies. This study investigated the migration characteristics and mechanisms of HMIs in the process of lake freezing.

Keywords: heavy metal ions; ice-water medium; migration mechanism; first-principles; binding energy

---

1. Introduction

In aqueous ecosystems, substances migrate, transform, and are metabolized. At present, there is much research at home and abroad on the processes and mechanisms of material transfer and transformation, but most of these focus on the liquid water phase [1,2]. The increase in ice media due to freezing has an important influence on the migration of chemical substances. Research on the ice-water media involved in changes in lake environments and pollutants in multi-media environments has drawn much attention [3]. Lakes of the world that are located in high latitudes undergo a freeze-thaw cycle every year, and their environmental characteristics are obviously different from those of other lakes. Ice bodies play an important role in the environmental impact on the lakes during ice growth. Therefore, it is very important to study the distribution and migration mechanism of pollutants in the ice-water environment of cold regions.

The solute release into liquid water during freezing in an artificial environment has been extensively studied in various fields, such as technology, chemical analysis, food, medicine [4–7], and pollutant removal. In particular, freezing enrichment technology is used in industry to purify solutions [8]. In natural water, salt is redistributed from ice to water when the ocean freezes [9]. In recent years, there have been many investigations into the migration of pollutants from ice to water in lakes. Wang [10] found that the concentrations of Cr, Hg, and As varied widely between ice and water. In some parts of the lake, the amount of As in the water under the ice was more than ten times higher than in the ice. By simulating the migration of heavy metals in the ice-water medium, Lv [11] determined that the migration coefficients of heavy metals in ice-water systems were 0.12–0.84, also showing that heavy metals migrated from ice to water during freezing. Liu et al. [12] concluded that during the ice growth period in Ulansuhai Lake, Hg concentrations increased in the water first during the freezing process, then the dynamic equilibrium of the Hg between the water and sediments became upset, and finally a portion of the Hg migrated to the sediments, as a result of the difference in the equilibrium gradient. Based on the above studies, the migration and distribution of pollutants from ice to water in ice-water systems were obtained. The question is why do pollutants exhibit such a pattern of behavior? We all know that migration is a form of movement and that energy is the fundamental cause of material movement. Therefore, the freezing process must be accompanied by changes in energy. Finding the change of energy in this process is important to study the internal mechanism of pollutant migration in the lake ice-water media during the freezing period.

Lake water is a natural electrolyte solution. After freezing in a lake, there will be both ice and liquid water. In this study, we think of ice and water as two materials. These two materials are calculated from the perspective of solid state physics and quantum chemistry and analyzed to investigate the energy changes of pollutants in ice and water systems. This change of energy and the determination of the stability of contaminants in ice and water systems will involve a first-principles study. First-principles is usually associated with computation, when there are no other experimental, empirical, or semi-empirical parameters in the calculations that inform the program of which atoms are in use and where they are located. As the basis for evaluating things, first principles and empirical parameters represent the two extremes. First-principles can simulate the periodic structure, and it is independent of any empirical parameters that can solve the system of a ground state electronic structure and properties, while it is also a way to design materials. The research focus of first-principles is on the prediction of alloying effects in the processing of metals, the choice of atom positions, and the effects of electronic structures and properties on surfaces and interface layers [13–17]. Therefore, it is necessary to introduce first-principles calculations in the study of the transport mechanism of pollutants in lake ice-water media.

Inevitably, there is a difference in energy between lake pollutants that have migrated to water from ice. It is necessary to obtain the pollutant energy values in the most stable state in each media using the first-principles to explore the energy change of the pollutants in the freezing process. First of all, the morphology of pollutants in ice and water media should be determined using first-principles to calculate the energy. Water molecules in the liquid state, do not exist in the form of individual molecules, but in the form of a dimer, trimer, or tetrahedron [18]. There are at least 13 types of ice that have been characterized. Ice, such as frost and snow, which are exposed to daily life, are termed ice-I_h [19,20], while pure ice is a crystalline structure containing a tetrahedron [21]. Therefore, the structure of water and ice is important in establishing the microstructural model.

Lake freezing in natural conditions is a macroscopic and complex process. The complexity is mainly reflected in three aspects: (1) types of pollutants; (2) forms of contaminants; and (3) external environmental conditions. These factors have resulted in hardly any studies of the mechanism of pollutant migration from the perspective of energy. In this study, simulated lake freezing experiments conducted in the laboratory can control the environmental conditions and the types and forms of pollutants into individual variables, such that the migration mechanism of heavy metal ions can be explored in-depth. In this work, we focused on the migration of HMIs from ice to water in the course of the lake freezing using simulation experiments. First-principles were applied to calculate the binding energy in the different media in the ice-water system. This topic is important to better understand the mechanisms of lake ecosystems in cold regions, especially during the course of climate change.

2. Materials and Methods

2.1. Study Site

Ulansuhai Lake (40°36′–41°03′ N, 108°43′–108°57′ E) is on the Inner Mongolian Plateau, in the interior of Inner Mongolia, China. Its surface area is 283 km^2. It usually enters the freezing period from November to April of the next year. The average thickness of the ice over many years is 0.63 m. It is the only drainage channel for local farmland, industrial waste water, and domestic sewage [22]. Every year, about 5 × 10^8 m^3 of water is abstracted from the lake for farm irrigation, while 2 × 10^8 m^3 of industrial waste water and sanitary waste with lead, mercury, arsenic, chromium, cadmium, and other heavy metals are discharged into it [23]. During the ice-covered period, the concentrations of heavy metals in the water are greater than that of the open water period. In particular, the concentrations of Hg in water during winter exceed the standard for III-type water quality, and the concentrations in the lake inlet region exceed the standard for V-type water quality [11]. For example, in January 2017, concentrations of heavy metals (Cu, Fe, Mn, Zn, Pb, Cd, and Hg) in the lake water were 1–3 (mean = 1.67) times higher than those in November 2016.

2.2. Experiment

2.2.1. Simulation Experiment Device

In order to simulate the top-down freezing process of water in natural lakes, a one-dimensional, unidirectional ice simulator was used (Figure 1). The experimental apparatus consisted of one cylinder within another, with the intervening space filled with insulating material. The inner cylinder holds the experimental sample. The insulation is meant to prevent heat loss from the side and bottom of the water column, ensuring that the cooling starts at the top (identical to freezing in natural lakes). The column body can be independently removed, not only for cleaning but also for easy removal of the formed ice. A piston was provided near the bottom to buffer pressure from expansion during the freezing, which otherwise could cause damage to the experimental instrument. To simulate the cold weather during which the lake freezes over, the experiment was carried out in a low-temperature storage.

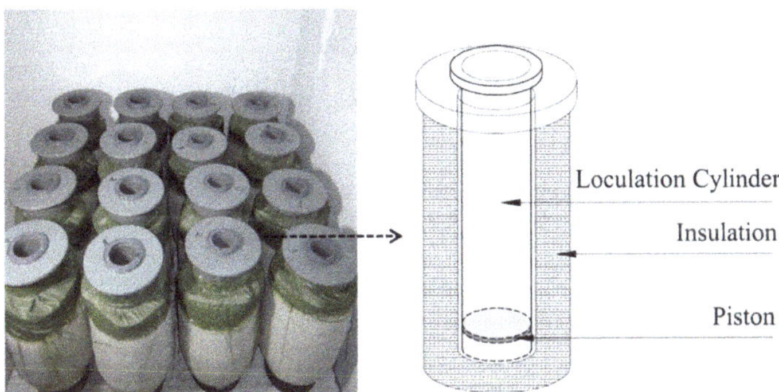

Figure 1. Device for unidirectionally simulating the freezing of lake water.

2.2.2. Experimental Design and Methods

The freezing conditions of the indoor experimental simulation were designed based on the data for Ulansuhai Lake during the freezing period over many years. Water samples from this lake were not used during the field experiment work. Because the interaction of various pollutants and ions in the natural lake is complicated, it is hard to control the conditions for the types and concentrations of each ion. Therefore the characteristics of lake water were only used as a guide for the design of the experiment. Heavy metals exist in various forms in the lake water environment as simple ions and metal complexes [24]. Different forms exhibit unique chemical behaviors and play different roles in the aquatic environment of lakes. Zhao [25] used the geochemical software (PHREEQC, U.S. Geological Survey, Reston, VA, USA) to study the existence of heavy metals in Ulansuhai Lake. Simple ion heavy metals are the dominant form in Ulansuhai Lake. Therefore, the experimental design for this study only considered simple ion heavy metals in the microscopic simulation. The common heavy metals in Ulansuhai Lake (Cu, Fe, Mn, Zn, Pb, Cd, and Hg) were selected to prepare the water samples, using ultrapure water (resistivity: 18.2 MΩ·cm) and primary standards for each heavy metal (Table 1). Cl^- is the dominant anion in the water of Ulansuhai Lake. The average concentration of the whole lake is 750.6 mg/L. The hydrochemical type of the lake is [Cl]NaIII and the TDS is more than 1000 mg/L. It belongs to high salinity brackish water [26]. In our experiment, water of V type and 10,000 µg/L was prepared according to the standard materials (Sample number in Table 1). The concentration of standard materials is 1000 µg/mL with the medium of $C(HNO_3)$ = 1.0 mol/L. We stepwise diluted it using ultrapure water by the standard materials. Water quality is a V-type in Ulansuhai Lake based on measured data over many years. Hence, we prepared the experimental samples according to V type. In order to study the migration of heavy metal ions under the same initial concentration conditions, and easily and accurately detect the ions concentration, we set the initial concentration to 10,000 µg/L. Each water sample of 2 L was prepared in the simulator. The freezing temperatures were set as −15 °C and −25 °C in the laboratory. According to the statistical results of the temperature data of Ulansuhai Lake from 1977–2015, the average low temperature is about −22.2 °C during the lake freezing period, so the experiment was designed with a freezing temperature of −25 °C; the average temperature is about −13.9 °C, so the experiment was designed with a freezing temperature of −15 °C. According to the 'Environmental quality standards of surface water' [27], Ulansuhai Lake has a V-type water quality [28] based on data measured over many years. To reduce the experimental error, the initial concentrations of these heavy metals in the water samples were all set as the maximum values allowed for V-type water, while the other initial concentrations were 10,000 µg/L (Table 1). The pH values of 5, 7, and 11 were chosen to represent acidic, neutral, and alkaline conditions, respectively. The limits for iron and manganese are in the supplementary items of surface water sources for the centralized drinking water standard in China. The devices were assembled, placed in the freezer, and taken out when the ice was about 1/3 of the total volume (which matches the ice: water ratio in Ulansuhai Lake during the freezing period; i.e., 1/3 to 1/2, [29]). After measuring the ice thickness, the ice was placed in separate containers to melt. For all water samples, the heavy metal concentrations were tested in triplicate according to the 'Water and Wastewater monitoring analysis method' [30]. The samples were first digested and underwent other appropriate pre-treatments. Then, the concentrations of Fe, Cu, Zn, Pb, Cd, and Mn were measured in a graphite oven-atomic absorption spectrometer, while those of Hg were measured with a double-channel atomic fluorescence spectrometer. The simulation experiments were completed in the laboratory of the National Position Observation Research Station of the Ulansuhai Wetland System in Inner Mongolia, China. In this study, water samples from this lake were not used during the experiment. But the setting up of simulated experimental conditions was based on the water environment characteristics of Ulansuhai Lake. Although the water environmental conditions of Ulansuhai Lake were used as a reference in the experiment design, the results can be applied to other lakes with a long freezing period. The lakes in high altitude areas of China have common characteristics of 4–7 months freezing period, and the ice thickness about 0.5–1.5 m, and environmental average temperature below −20 °C during the freezing period, which is the same as other lakes in North Europe [31].

Table 1. Concentration values of the standard materials and the initial concentrations (mg/L).

Standard Materials	Sample Number	C_1	C_2
Cu	GSB04-1725-2004	1.0	10.0
Fe	GSB04-1726-2004	3.0	10.0
Mn	GSB04-1736-2004	0.1	10.0
Zn	GSB04-1761-2004	0.1	10.0
Pb	GSB04-1742-2004	0.1	10.0
Cd	GSB04-1721-2004	0.01	10.0
Hg	GSB04-1729-2004	0.001	10.0

The heavy metal concentrations were measured in each layer of ice and the underlying water. To study the changes in the distribution of heavy metals in the ice-water system under different conditions, the distribution coefficient (K) was calculated individually for each heavy metal as the ratio between the mean concentration in the three ice layers (C_S) and that of the underlying water (C_L).

$$K = C_S/C_L \tag{1}$$

2.3. Calculation

2.3.1. Microstructure Model

Choosing Cu^{2+} is just an example for introducing model setup. The models of other divalent heavy metal ions are basically similar to Cu^{2+}. The only difference is the change of hydration number of ions in the water environment. Water system model: HMIs are surrounded by a number of water molecules [32]. The hydrated metal ions are formed because the strong action of the electric field near the ions causes the partial solvent water molecules to be arranged around them. Water molecules surround the metal ions with a hydration layer in a certain sequence around the metal ions according to Frank and Wen [33]. The water molecules and ions are bounded in firm combinations. The water molecules in the first hydration layer lose a translational degree of freedom, often accompanied by ions moving together. The number of molecules does not vary relative to temperature. The hydration numbers of a set of HMIs were determined by the ^{1}H nuclear magnetic resonance method according to Swift and Sayre [34]. Therefore, when establishing the water system model in the case of Cu^{2+}, firstly, two OH^- were simulated near Cu^{2+} in order to balance positive and negative charges in a liquid system and optimize the cluster structure, and then five H_2O were simulated around the Cu^{2+} in the first hydration layer. In addition, two H_2O were simulated to surround and fix the two OH^-. Figure 2a illustrates a model in the liquid phase, where the H_2O outside the first hydration layer were not considered in the model because the H_2O of the second hydration layer were acting weakly and were in a disordered structure.

Ice system model: Compared to liquid water, whose microstructural arrangement is disordered, solid ice has an organized order. The structure of ice is a tetrahedral formation with hexagonal rings. Each hydrogen atom is connected with two oxygen atoms, which are bounded by covalent bonds and hydrogen bonds, respectively [35]. Ice exhibits different configurations at different temperatures and pressures. There are 13 configurations discovered up to the present. According to the temperature and pressure conditions, lake ice occurs as ice-I_h [20,36]. A single crystal cell of ice-I_h is expanded into a $2 \times 2 \times 2$ super pure ice cell structure. Depending on the state of most of the contaminants in the ice, they occur in bubbles or channels [3]. Thus, when we established the model, a hole was set up in the ice. In order to determine the structure of HMIs in ice with the lowest energy, the cluster structure of HMIs and two OH^- were placed in the middle of the two H_2O layers, and replacing the positions of one, two or three H_2O in one layer. After optimizing these four structures, the energy of each whole system was calculated. Finally, the lowest energy was found to be the structure that replaced the position of one H_2O in the center of the super pure ice cell. Figure 2b is the optimizing model for Cu^{2+}

in ice. Based on the above established model of Cu^{2+} in the water and ice environments, several other heavy metal models were found to be similar.

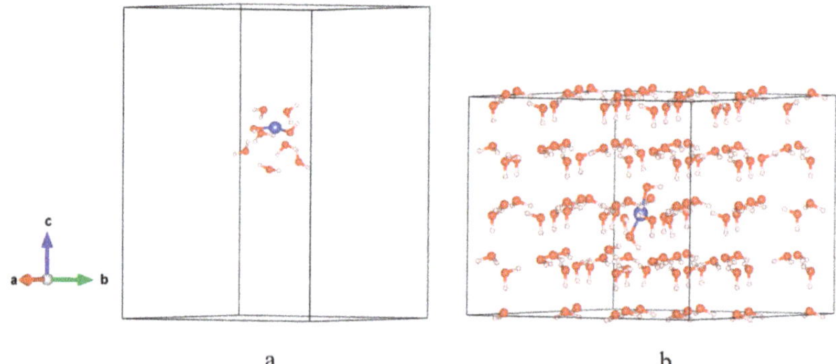

a   b

Figure 2. (**a**) The model of Cu^{2+} in the aqueous environment. (**b**) The model of Cu^{2+} in the ice environment (a represents x-axis, b represents y-axis, c represents z-axis; the blue atom is Cu, the red atoms are O, the grey atoms are H).

2.3.2. Binding Energy Calculation Method

The definition of binding energy is the energy released by a single atom (system) into the corresponding material with which it has become bound. The more energy released, the more stable the material, and vice versa. The binding energies of HMIs in an ice body and a water body can be calculated. The stability of HMIs in the two media can be assessed by the size of the binding energies, which can reveal the mechanism of migration of HMIs in ice and water media. The first-principles calculations were performed at the Institute of Integrated Computational Materials Research Centre, Shanghai Institute of Ceramics, Chinese Academy of Sciences.

The DFT calculations were performed using the Vienna Ab initio Simulation Package (VASP) [37, 38] with the Perdew-Burk-Ernzerh parameterization of the generalized gradient approximation (GGA) adopted for the exchange correlation potential [39]. An energy cut off of 400 eV was consistently used in our calculations. The atomic positions are fully relaxed with the conjugate gradient procedure until the residual forces vanished within 0.05 eV/A. The standard of energy convergence is 10^{-5} eV/A $2 \times 2 \times 1$ Monkhorst–Pack k-point was used to sample the surface Brillouin zone.

To compare the binding energy of HMIs in ice-water systems, two models of ice and water were established. The calculation formula for the binding energies was:

$$E = [E(M) + E(H)] - E(M + H) \tag{2}$$

In the calculation formula of the liquid (or solid) state system, $E(M + H)$ is the energy of the equilibrium geometry when the contaminant molecule M and the liquid water (or solid state) molecule H form the interaction system, and $E(M)$ and $E(H)$ are the equilibrium geometry energies when M and H molecules exist individually.

3. Results

3.1. The Distribution and Transfer of HMIs in Ice and Water

The concentrations of the different HMIs in ice and water were obtained based on the lake freezing simulation experiments. If the temperature is $-25\,°C$ and $-15\,°C$, the other two conditions are kept same, which are the pH is 7 and the initial concentration is 10,000 µg/L. If the pH is 5 or 11, the other two conditions are kept the same; temperature is $-25\,°C$ and the initial concentration is 10,000 µg/L. If the initial concentration is V-type or 10,000 µg/L, the other two conditions are the same; the temperature is −25 °C and the pH is 7. Figure 3 shows the concentrations of HMIs in the ice medium. At the same condition, the concentrations of HMIs in ice were in the following order: Fe > Cu > Mn > Zn > Cd > Hg > Pb. The results were calculated according to Equation (1) and the distribution coefficients (K) are given in Table 2. Regardless of whether the temperature is −25 °C or −15 °C, the pH is 5 or 11, or the initial concentration is V-type or 10,000 µg/L, the concentration in ice is less than the initial concentration of the solution before freezing, and significantly less than the concentration in the water under ice. This indicates that HMIs migrate from ice to water when the lake is freezing, and are released during the freezing. At the same condition, the concentrations of HMIs in ice were in the following order: Fe > Cu > Mn > Zn > Cd > Hg > Pb. This result is consistent with the result of Lv Hongzhou, who did indoor simulation of Ulansuhai Lake [11,12]. From Figure 3, the concentration in ice under acidic condition is greater than that under alkaline. This result is consistent with the result of Wang Shuang, that the concentration of heavy metals in ice is acidic > neutral > alkaline according to the simulation of frozen Ulansuhai Lake [10]. Under the specific conditions of high concentration of heavy metals and high pH, the Ulansuhai Lake is easier to concentrate in surface sediments.

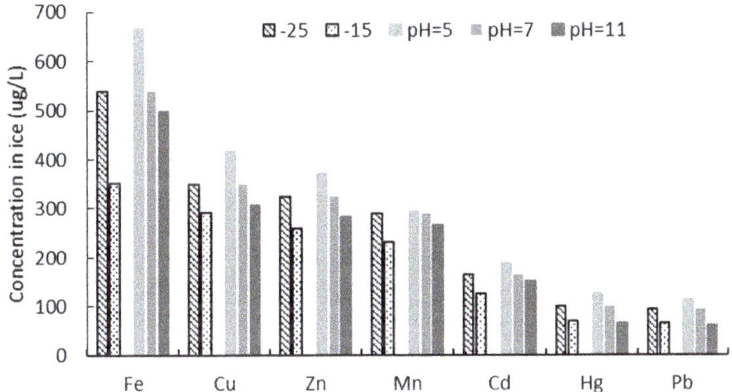

Figure 3. The heavy mental concentrations in ice under different conditions.

Table 2. The distribution coefficients (*K*) of heavy metals under different conditions.

HMIs	T = −25 °C	T = −15 °C	pH = 5	pH = 11	V-Type	10,000 µg/L
Fe	0.037	0.024	0.045	0.034	0.061	0.037
Cu	0.024	0.020	0.028	0.021	0.105	0.024
Zn	0.022	0.017	0.025	0.019	0.017	0.022
Mn	0.019	0.016	0.020	0.018	0.013	0.019
Cd	0.011	0.008	0.013	0.010	0.118	0.011
Hg	0.007	0.005	0.008	0.004	0.055	0.007
Pb	0.006	0.004	0.007	0.004	0.010	0.006

3.2. Energy of HMIs in Ice and Water

The results were calculated according to Equation (2), and the binding energy calculation process is shown in Table 3. The binding energies of HMIs were also calculated, as shown in Figure 4. It can be seen that the binding energies of the different heavy metals in water are greater than those in ice. In order to verify the results, we calculated the hydrogen bond energy of Ih-ice according to the first principle, and obtained the hydrogen bond energy of 0.276 eV, which is similar to the result

(6.87 kcal/mol = 0.268 eV) of Liu Yi by the AIM method [40]. Therefore, the binding energy calculated based on this method is reliable.

Table 3. Binding energy calculation process (eV).

Heavy Metals	E(M)	Calculation in Ice		Calculation in Water	
		E (M + H)	E (H)	E (M + H)	E (H)
Fe	−28.09	−1268.80	−1237.85	−118.12	−87.15
Cu	−16.10	−1256.49	−1237.85	−105.95	−87.15
Zn	−15.41	−1255.76	−1237.85	−105.27	−87.15
Mn	−18.98	−1259.33	−1237.85	−108.89	−87.15
Cd	−14.10	−1254.42	−1237.85	−103.99	−87.15
Hg	−13.50	−1253.64	−1237.85	−103.24	−87.15
Pb	−17.63	−1257.73	−1237.85	−107.46	−87.15

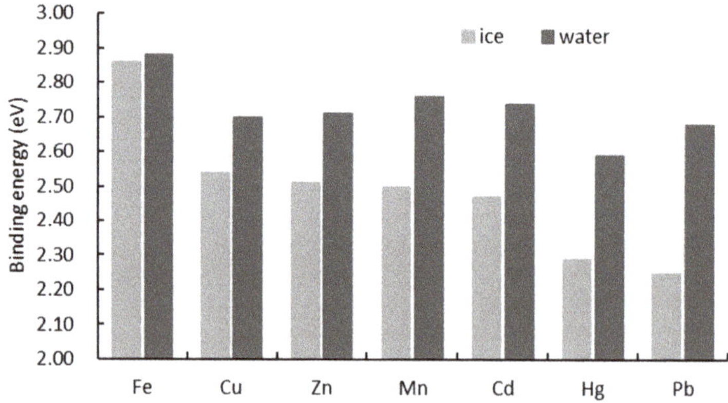

Figure 4. Binding energies of HMIs in ice and water.

4. Discussion

It can be seen from the experimental results of simulating lake freezing that the concentration in water under ice is much greater than that in ice, no matter whether the temperature of freezing is different or the initial concentration is different. Therefore, lake pollutants migrate to the water under ice during freezing. The movement of this migration results from changes in energy, so we discuss the mechanism from three aspects: structural stability, binding energy, and ionic radius.

4.1. Mechanism of HMIs Migration in Ice and Water

4.1.1. Stability of HMIs in Ice and Water

Based on the liquid phase model and solid phase model, we find that the HMIs in water are tightly surrounded by water molecules. Those water molecules are in a relatively free state. That means the binding energy of HMIs that combine with H_2O is relatively high. However, HMIs remain in the ice without enough surrounding free H_2O when the solution begins freezing. At this time, Cu^{2+} will interact with H_2O and form an opening in the regular ice structure, as the binding energies of HMIs are relatively lower. Therefore, the micro-model also shows that HMIs are more stable in the water media than in the ice media. On the other hand, based on the calculated binding energies, the binding energy of the same HMI in water is larger than that in ice.

4.1.2. Binding Energy of HMIs in Ice and Water

Lake freezing is the process of changing from the single medium of water to the multiple media of ice-water. The process described from the molecular viewpoint is a complex process. However, this study captured two states, the state of HMIs in the aqueous medium and in the ice medium. According to the results shown in Figure 3, the binding energy in water was greater than that in ice for the same HMI. From an energy point of view, when the lake is freezing, the ice medium increases. For the two-media system, HMIs are more likely to exist in the relatively more stable aqueous environment. Therefore, the concentrations of HMIs in water are greater than those in ice. Therefore, HMIs migrate from ice to water when freezing.

4.2. Mechanism of Different HMIs in Ice

4.2.1. Effect of Ionic Radius on Concentration in Ice

Under the same temperature and initial concentration, the heavy metals have different behaviors in ice. For ions of the same charge, the larger the ionic radius, the weaker its influence on the water structure. Similarly, among ions with similar ionic radii, those with higher charges affect the water structure less. Taken together, the higher the charge density of the ion, the greater its influence on the water structure [41]. The heavy metal ions selected in this study all promote molecular association (i.e., stronger hydration and lower mobility of solute in the solution; [42]). Based on the relationship between concentration in ice and ionic radius, as shown in Figure 5, we find that when a single environmental factor changes, the concentrations of different heavy metals in the ice body are negatively correlated with their ionic radius. With an increase in ionic radius, the charge density decreases, and the influence on the structure of water is weakened. The effect of hydration on promoting the association of water molecules is also weakened, and therefore the solute mobility is lower. In contrast, under the same conditions, the concentrations of ionic pollutants in the ice gradually decreases with increasing ionic radius.

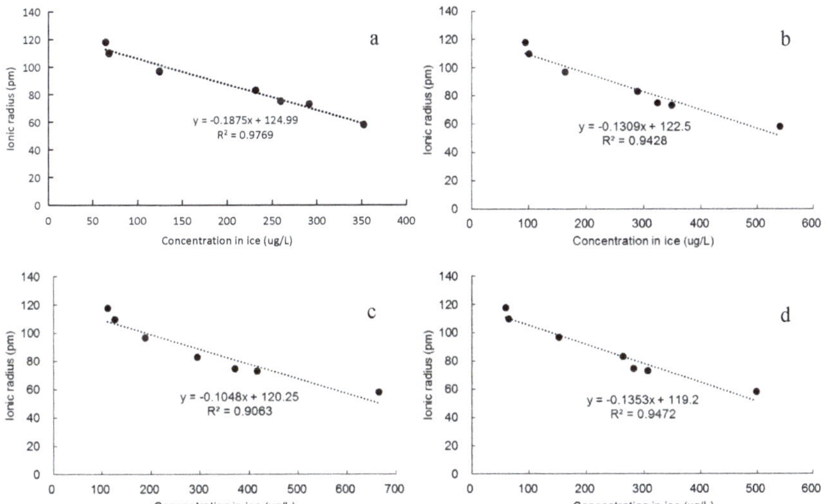

Figure 5. The relationship between concentration in ice and ionic radius ((**a**) T = −25 °C, pH = 7, initial concentration is 10,000 µg/L; (**b**) T = −15 °C, pH = 7, initial concentration is 10,000 µg/L; (**c**) pH = 5, T = −25 °C, initial concentration is 10,000 µg/L; (**d**) pH = 11, T = −25 °C, initial concentration is 10,000 µg/L).

4.2.2. Effect of Binding Energy on Concentration in Ice

According to the results of the experimental freezing under different conditions, the concentration trends of HMIs in ice were the same regardless of the different external conditions. It can be seen that the migration capacities of HMIs from ice to water were independent of changes in the external conditions. In order to explore the reasons for the concentration distribution of HMIs in ice, the binding energies of the different HMIs in ice were calculated. The results showed that the trends in the concentrations of the HMIs in ice were the same. The differences among the concentrations of different HMIs in ice results from the different binding capacity of the HMIs. As shown in Figures 3 and 4, Fe^{3+} has the highest binding energy in ice among the heavy metals, such that the concentrations of Fe^{3+} in the ice were higher than those of the other HMIs. The binding energy of Pb^{2+} in ice was the lowest and its concentrations were also less. The binding energy of heavy metal ions in ice is related to the characteristic of ions. Fe loses three electrons to form Fe^{3+}, so there are three OH^{-}, forming a neutral cluster around Fe^{3+}. The cluster is held more firmly by H_2O in the ice structure, so the binding energy between Fe^{3+} and ice is the largest in these seven kinds of heavy metal ions. Therefore, ion concentration remains larger in ice. Compared with other divalent heavy metal ions, the atoms with larger radii have a longer interaction distance with H_2O and a weaker interaction force, so the binding energy with ice is smaller. The concentration of H_2O in ice is relatively smaller when freezing. Figure 6 shows the relationship between concentration in ice and binding energy in ice, and under these four conditions the concentration of different heavy metals in the ice body are positively correlated with their binding energies.

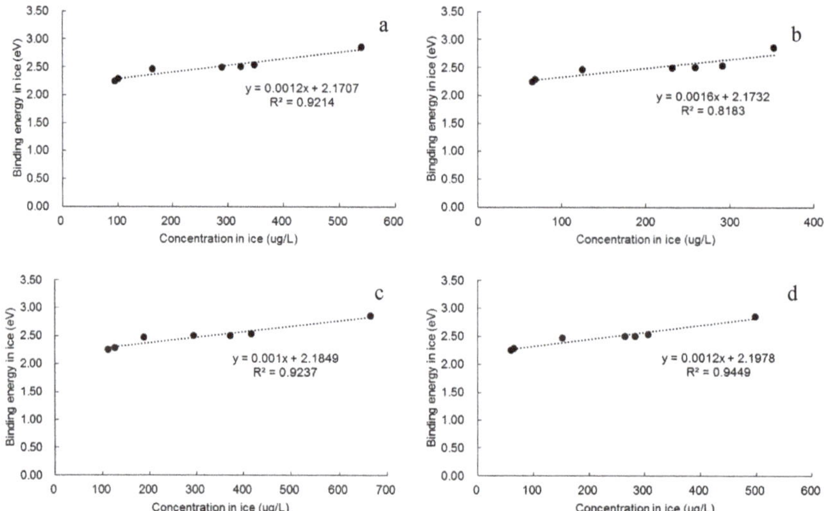

Figure 6. The relationship between concentration in ice and binding energy in ice ((**a**) T = $-25\,^{\circ}$C, pH = 7, initial concentration is 10,000 µg/L; (**b**) T = $-15\,^{\circ}$C, pH = 7, initial concentration is 10,000 µg/L; (**c**) pH = 5, T = $-25\,^{\circ}$C, initial concentration is 10,000 µg/L; (**d**) pH = 11, T = $-25\,^{\circ}$C, initial concentration is 10,000 µg/L).

The molecular model of heavy metal ions in an ice and water body was established based on first principles, which allowed the binding energy for each ion to be determined. The phenomenon of heavy metal pollutants concentrating in the lower bottom water during the frozen period was fully explained from the energy viewpoint, which is the main driver of the migration. The results reflect the interdisciplinary nature of the research methods, and provide a theoretical basis and ability to predict changes in the behavior of heavy metals in lakes during periods of freezing.

5. Conclusions

Above all, we aim to clarify the migration mechanism of HMIs from ice to water, when lakes are freezing from the perspective of energy change. HMIs are stable in a lower energy state, similar to all materials. They are more stable in an aqueous environment than in an ice environment. For the same HMI, the binding energy in water is smaller than that in ice. Thus HMIs migrate from ice to water when the lake is freezing. The ability of different HMIs to migrate from ice to water is related to the binding energies in ice. The concentrations of HMIs in ice are positively correlated with their binding energies.

6. Implications and Future Research

In our future research, we intend to study the relationship of the binding energies and concentration values of HMIs. Based on these relationships, we can predict the contents of some heavy metals in ice bodies and water bodies during periods of freezing. Using an equation to describe the relationships, we can not only reduce the field workload during freezing periods, but we can also check the agreement between the measured and predicted values and test the theoretical reliability of this experimental study.

Author Contributions: Data curation, C.S., X.S., S.Z. and W.T.; Formal analysis, Y.W.; Methodology, C.S., S.Z. and Y.W.; Software, C.S. and J.L.; Writing-original draft, C.S.; Writing-review & editing, C.L. and X.S.

Funding: This research was funded by the National Natural Science Foundation of China (Grant Nos.s 51339002, 51669022, 51509133, 51569019).

Acknowledgments: We are grateful for assistance with the Institute of Integrated Computational Materials Research Centre, Shanghai Institute of Ceramics, Chinese Academy of Sciences and the laboratory of the National Position Observation Research Station of the Ulansuhai Wetland System in Inner Mongolia, China.

Conflicts of Interest: The authors declare no conflicts of interest.

References

1. Chen, A.; Zeng, G.; Zhang, J.; Lu, L.; Chen, M.; Xu, P. Global landscape of total organic carbon, nitrogen and phosphorus in lake water. *Sci. Rep.* **2015**, *5*, 15043. [CrossRef] [PubMed]

2. Yu, J.; Fan, C.; Zhong, J.; Zhang, Y.; Wang, C.; Lei, Z. Evaluation of in situ simulated dredging to reduce internal nitrogen flux across the sediment-water interface in lake Taihu, China. *Environ. Pollut.* **2016**, *214*, 866–877. [CrossRef] [PubMed]

3. Zhang, Y.; Li, C.Y.; Shi, X.H.; Chao, L.I. The migration of total dissolved solids during natural freezing process in Ulansuhai Lake. *J. Arid. Land* **2012**, *4*, 85–94. [CrossRef]

4. Bengtsson, L. Mixing in ice-covered lakes. *Hydrobiologia* **1996**, *322*, 91–97. [CrossRef]

5. Fedotov, V.I.; Cherepanov, N.V.; Tyshko, K.P. Some features of the growth, structure and metamorphism of East Antarctic landfast sea ice. In *Antarctic Sea Ice: Physical Processes, Interactions and Variability*; Jeffries, M.O., Ed.; American Geophysical Union: Washington, DC, USA, 2013; pp. 343–354.

6. Baker, R.A. Trace organic contaminant concentration by freezing-IV: Ionic effects. *Water Res.* **1970**, *4*, 559–573. [CrossRef]

7. Huige, N.J.J.; Thijssen, H.A.C. Production of large crystals by continuous ripening in a stirrer tank. *J. Cryst. Growths* **1972**, *13–14*, 483–487. [CrossRef]

8. Muller, M.; Sekoulov, I. Waste water reuse by freeze concentration with falling film reactor. *Water Sci. Technol.* **1992**, *26*, 1475–1482. [CrossRef]

9. Weeks, W. Sea ice: The potential of remote sensing. *Oceanus* **1981**, *24*, 39–48.

10. Wang, S. *Study on Distribution Law and Laboratory Simulation Experiment of Heavy Metals during the Icebound Season in Wulansuhai Lake*; Inner Mongolia Agriculture University: Inner Mongolia, China, 2012. (In Chinese)

11. Lv, H.Z. *Studies on Pollutant Distribution in Ice-Water System of Ulansuhai Lake*; Inner Mongolia Agriculture University: Inner Mongolia, China, 2015. (In Chinese)

12. Liu, Y.; Li, C.Y.; Anderson, B.; Zhang, S.; Shi, X.H.; Zhao, S.N. A modified QWASI model for fate and transport modeling of mercury between the water-ice-sediment in Lake Ulansuhai. *Chemosphere* **2017**, *176*, 117–124. [CrossRef] [PubMed]

13. Zhao, L.; Sun, Y.; Li, Y.G.; Meng, X.F.; Song, Q.L. First-principles studies of the effects of microalloy elements on Fe/Al interface. *J. At. Mol. Phys.* **2007**, *24*, 853–857. (In Chinese)

14. Hu, X.L.; Zhang, Y.; Lu, G.H.; Wang, T.M.; Xiao, P.H.; Yin, P.G.; Xu, H.B. Effect of impurity on structure and mechanical properties of NiAl intermetallics: A first-principles study. *Intermetallics* **2009**, *17*, 358–364. [CrossRef]

15. Xu, Q.C.; Ven, A.V. First-principles investigation of migration barriers and point defect complexes in B_2-NiAl. *Intermetallics* **2009**, *17*, 319–329. [CrossRef]

16. Wei, H.; Liang, J.J.; Sun, B.Z.; Zheng, Q.; Sun, X.F.; Peng, P.; Yao, X.; Dargusch, M.S. Site preference of Re in NiAl and valence band structure of NiAl containing Re: First-principles study and photoelectron spectrum. *Appl. Phys. Lett.* **2009**, *94*, 1731–1756.

17. Kogita, T.; Kohyama, M.; Kido, Y. Structure and dynamics of NiAl(110) studied by high-resolution ion scattering combined with density functional calculations. *Phys. Rev. B* **2009**, *80*, 308–310. [CrossRef]

18. Tu, T.S.; Fang, H.P. The microstructure of liquid water. *J. Phys. Chem.* **2010**, *39*, 79–84. (In Chinese)

19. Zhou, P. *A Theoretical Calculated Investigation on Hydrogen Bonding Clusters of Water-Ammonia Molecules*; Shandong Normal University: Shandong, China, 2008. (In Chinese)

20. Michalarias, I.; Beta, I.; Ford, R.; Ruffle, S.; Li, J.C. Inelastic neutron scattering studies of water in DNA. *Appl. Phys. A* **2002**, *74*, s1242–s1244. [CrossRef]

21. Isakov, S.V.; Moessner, R.; Sondhi, S.L.; Tennant, D.A. Analytical theory for proton correlations in common water ice I_h. *Phys. Rev. B* **2015**, *91*, 245152. [CrossRef]

22. Wu, Y.; Li, C.Y.; Zhang, C.F.; Shi, X.H.; Bourque, C.P.A.; Zhao, S.N. Evaluation of the applicability of the SWAT model in an arid piedmont plain oasis. *Water Sci. Technol.* **2016**, *73*, 1341–1348. [CrossRef] [PubMed]

23. Zhao, S.N.; Shi, X.H.; Li, C.Y.; Zhang, S.; Sun, B.; Wu, Y.; Zhao, S.X. Diffusion flux of phosphorus nutrients at the sediment-water interface of the Ulansuhai Lake in northern China. *Water Sci. Technol.* **2017**, *75*, 1455–1465. [CrossRef] [PubMed]

24. Ye, S.Y.; Sun, J.C.; Jiang, C.Y. Current situation and advances in hydrogeochemical researches. *Acta Geosci. Sin.* **2002**, *23*, 477–482. (In Chinese)

25. Zhao, S.N. *Enviromental Geochemistry of Heavy Metal and Modelling Analysis of Their Speciation for Ulansuhai Lake in Inner Mongolia*; Inner Mongolia Agriculture University: Inner Mongolia, China, 2013. (In Chinese)

26. Cui, F.L.; Li, C.Y.; Shi, X.H.; Shi, Y.Q.; Fu, X.J. Seasonal changing characteristics of the major ions in Ulansuhai Lake. *J. Arid. Land Resour. Environ.* **2013**, *8*, 137–142.

27. *Environment Quality Standards for Surface Water*; GB3838-2002; State Environmental Protection Administration of the People's Republic of China: Beijing, China, 2002. (In Chinese)

28. Li, W.P.; Xu, J.; Yu, L.H.; Han, P.J.; Zhao, Z.; Jing, J. Distribution characteristics of nutrients and phytoplankton in Ulansuhai Lake during the icebound season. *Ecol. Environ. Sci.* **2014**, *23*, 1007–1013. (In Chinese)

29. Yang, F.; Li, C.Y.; Leppäranta, M.; Shi, X.H.; Zhao, S.N.; Zhang, C.F. Notable increases in nutrient concentrations in a shallow lake during seasonal ice growth. *Water Sci. Technol.* **2016**, *74*, 2773–2783.

30. The State Environmental Protection Administration the Water and Wastewater Monitoring Analysis Method Editorial Board. *Water and Wastewater Monitoring Analysis Method*, 4th ed.; China Environmental Science Press: Beijing, China, 2002; pp. 38–47. (In Chinese)

31. Sergey, K.; Mikhail, N. Lake ladoga ice phenology: Mean condition and extremes during the last 65 years. *Hydrol. Process.* **2011**, *25*, 2859–2867.

32. Wang, W.H.; Zhao, L.; Yan, B. Effects of ions on structure of liquid water. *Chemistry* **2010**, *73*, 491–498. (In Chinese)

33. Frank, H.S.; Wen, W.Y. Ion-solvent interaction. Structural aspects of ion-solvent interaction in aqueous solutions: A suggested picture of water structure. *Discuss. Faraday Soc.* **1957**, *24*, 133–140. [CrossRef]

34. Swift, T.J.; Sayre, W.G. Determination of hydration numbers of cations in aqueous solution by means of proton NMR. *J. Chem. Phys.* **1966**, *44*, 3567–3574. [CrossRef]

35. Goto, A.; Hondoh, T.; Mae, S. The electron density distribution in ice I_h determined by single-crystal X-ray diffractometry. *J. Chem. Phys.* **1990**, *93*, 1412–1417. [CrossRef]

36. Petrenko, V.F.; Whitworth, R.W. *Physics of Ice*; Oxford University Press: Oxford, UK, 1999.

37. Kresse, G.; Hafner, J. Ab initio molecular dynamics for liquid metals. *Phys. Rev. B* **1993**, *47*, 558–561. [CrossRef]

38. Blöchl, P.E. Projector augmented-wave method. *Phys. Rev. B Condens. Matter* **1994**, *50*, 17953–17979. [CrossRef] [PubMed]

39. Perdew, J.P.; Burke, K.; Ernzerhof, M. Generalized gradient approximation made simple. *Phys. Rev. Lett.* **1998**, *77*, 3865–3868. [CrossRef] [PubMed]

40. Liu, Y. *Atoms-in-Molecules Molecular-Dynamics Study of Ion-Water Bonding of K^+ Solvation in Water*; North University of China: Taiyuan, China, 2010. (In Chinese)

41. Wang, W.H.; Zhao, L.; Yan, B.; Tan, X.; Qi, Y.; He, B. Effects of concentration and freeze-thaw on the first hydration shell structure of Zn^{2+} ions. *J. Tianjin Univ.* **2011**, *17*, 381–385. (In Chinese) [CrossRef]

42. Marcus, Y. Effect of ions on the structure of water: Structure making and breaking. *Chem. Rev.* **2009**, *109*, 1346–1370. [CrossRef] [PubMed]

Article

The Optimal Width and Mechanism of Riparian Buffers for Storm Water Nutrient Removal in the Chinese Eutrophic Lake Chaohu Watershed

Xiuyun Cao [1,†], Chunlei Song [1,*,†], Jian Xiao [1,2,†] and Yiyong Zhou [1,†]

[1] Key Laboratory of Algal Biology, State Key Laboratory of Freshwater Ecology and Biotechnology, Institute of Hydrobiology, Chinese Academy of Sciences, Wuhan 430072, China; caoxy@ihb.ac.cn (X.C.); alexxiao811@gmail.com (J.X.); zhouyy@ihb.ac.cn (Y.Z.)

[2] University of Chinese Academy of Sciences, Beijing 100039, China

* Correspondence: clsong@ihb.ac.cn; Tel.: +86-27-6878-0709

† All authors contributed equally to the paper.

Received: 28 September 2018; Accepted: 17 October 2018; Published: 22 October 2018

Abstract: Riparian buffers play an important role in intercepting nutrients entering lakes from non-point runoffs. In spite of its ecological significance, little is known regarding the underlying mechanisms of riparian buffers or their optimal width. In this study, we examined nutrient removal efficiency, including the quantity of nutrients and water quality, in the littoral zone of different types of riparian buffers in the watershed around eutrophic Lake Chaohu (China), and estimated the optimal width for different types of riparian buffers for effective nutrient removal. In general, a weak phosphorus (P) adsorption ability and nitrification-denitrification potential in soil resulted in a far greater riparian buffer demand than before in Lake Chaohu, which may be attributed to the soil degradation and simplification of cover vegetation. In detail, the width was at least 23 m (grass/forest) and 130 m (grass) for total P (TP) and total nitrogen (TN) to reach 50% removal efficiency, respectively, indicating a significantly greater demand for TN removal than that for TP. Additionally, wetland and grass/forest riparian buffers were more effective for TP removal, which was attributed to a high P sorption maximum (Q_{max}) and a low equilibrium P concentration (EPC_0), respectively. The high potential nitrification rate (PNR) and potential denitrification rate (PDR) were responsible for the more effective TN removal efficiencies in grass riparian buffers. The nutrient removal efficiency of different types of riparian buffers was closely related with nutrient level in adjacent littoral zones around Lake Chaohu.

Keywords: riparian buffer; nitrification-denitrification; phosphorus adsorption; Lake Chaohu

1. Introduction

Transport of nutrients to lakes and the resulting acceleration of eutrophication is a serious concern around the world. Runoff from land surfaces such as grass, forest, farmland, and wetlands has been found to be an important nutrient source for aquatic ecosystems [1]. Riparian buffers have been considered as an effective and sustainable means of buffering aquatic ecosystems against nutrient stressors, such as N and P from runoff [2], and thus they have been considered as the last line of defense for nutrient removal. Riparian buffers attenuate nutrients through soil adsorption, microbial immobilization, plant uptake, and coupled nitrification-denitrification [3–5].

Riparian buffer width is an important reference for managers and planners of watershed in designing the restoration of riparian buffers or protection objectives and scope. The determination of riparian buffer width range depends on the physiochemical and microbial properties of soil, slope, vegetation cover, the form and content of nutrients, and so on [6]. Sandy soil shows higher retention efficiency for P than silty clay [7]. Combinations of the dense, stiff, and native warm-season grass and woody vegetation were proved to significantly improve the removal effectiveness for non-point source pollutants [8]. Buffers composed of trees generally remove more N from runoffs [6], which should be attributed to the strong coupled nitrification-denitrification potential mediated by rhizosphere microorganisms. However, the riparian buffer width demand studied by many researchers indeed varied greatly between 5 m and 300 m [9–17], and little research has been focused on the mechanisms and explanations, which has lacked sufficient evidence. In addition, the actual effect of nutrient removal through riparian buffers on adjacent aquatic ecosystems was rarely involved.

In this study, nutrient levels (TN and TP) from storm water runoffs in different types of riparian buffer and surface water in littoral and deep zones of Lake Chaohu were investigated to reveal the nutrient removal efficiency and width demand of riparian buffers and the effects on aquatic ecosystems. The P adsorption ability and nitrification-denitrification potential of soil from riparian buffers were also analyzed to explain the corresponding mechanisms. The following points will be addressed: (1) comparison of riparian buffer width demand for nutrient removal in Chinese eutrophic lake watershed to European and American studies; (2) the optimal riparian buffer types for P and N removal as well as their mechanisms; (3) whether riparian buffers can make a crucial effect on nutrient levels in adjacent aquatic ecosystems, especially in littoral zones.

2. Materials and Methods

2.1. Study Site

Lake Chaohu, one of the five largest freshwater lakes in China, is located in the Yangtze-Huaihe region, central of Anhui Province. Its surface area is about 780 km^2, drainage total area is 13,486 km^2, and the shoreline length around the whole lake is 184.66 km. The climate of Lake Chaohu belongs to sub-tropical monsoon climates, and the annual average rainfall is 1120 mm [18]. In the last decades, due to the over-development of land, over-fertilization of agriculture, construction of highway around the lake that is too close to the shoreline, large-scale artificial plant restoration, vegetation and microorganism biodiversity as well as the properties of soil in the riparian buffer has noticeably declined, resulting in weaker nutrient removal efficiency and more serious lake eutrophication (see the Supplementary Materials Figures S1 and S2).

Different types of riparian buffer around Lake Chaohu was chosen depending on vegetation cover type, including grass (37 sites), forest (27 sites), wetlands (33 sites), grass/forest (27 sites), S and farmland (20 sites).

2.2. Storm Water Collection and Sampling

The storm water sampler received runoff in riparian buffers of different widths around Lake Chaohu during June to September in 2013 and 2014 through a 50 cm piece of 2 cm diameter PVC pipe placed flush with the surface of the ground on a selected sloping surface of an angle of about 10–30 degrees with the horizon line. Runoff entered the pipe through a slit cut along the entire length of pipe. The pipe was wrapped with a fiberglass screen to prevent insects and large debris from entering. Wooden clothespins with small pieces of nylon rope held the pipes in place. Water from the pipe flowed into a sampler through a notched cap with silicon tubing, which ran from the end of the PVC pipes to the sample bottle. The sampler was a 1 L glass bottle placed in a 10 cm diameter protective PVC sleeve. The original storm water was collected directly from the sky as the initial untreated storm water samples. All the sample bottles were taken back to the laboratory for chemical analysis as soon as possible after runoff had stopped, and the water volume in each bottle and the riparian width from the top to the collection site was recorded [19]. Two samplers for each site were designed and installed in the upper and lower sites of the riparian buffer, respectively, in order to calculate the efficiency of nutrient removal through the specific width.

Storm water was collected at each site of different widths for two to three rain events. Soil samples under the surface of the vegetation 10–20 cm at each site were collected for the analysis of P adsorption and potential nitrification-denitrification rate. Surface waters of littoral (22 sites) and deep (11 sites) zones in Lake Chaohu were also obtained with a Friedinger sampler. The positions of the sampling sites in littoral (L) and deep (C) zones as well as the types of riparian buffers are shown in Figures 1 and 2.

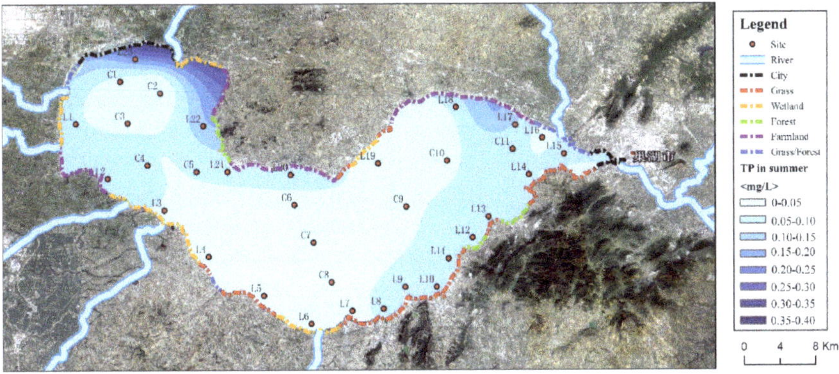

Figure 1. A map showing TP distribution in the littoral and deep zones of Lake Chaohu, including the distribution of riparian buffer type.

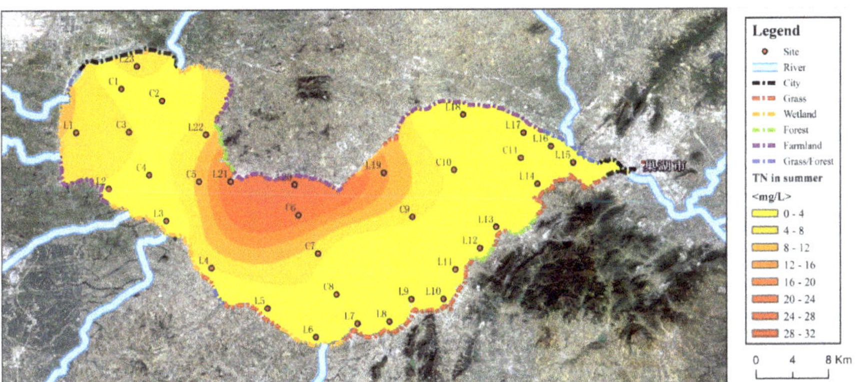

Figure 2. A map showing TN distribution in the littoral and deep zones of Lake Chaohu, including the distribution of riparian buffer types.

2.3. Chemical Analysis

TN and TP from water samples were measured following national standards [20].

Batch P sorption isotherm experiments were conducted in triplicate for soil samples under reducing conditions, using 0.01 mol L^{-1} KCl solutions containing 0, 0.1, 0.2, 1, 2, 5, 8, 10, 15, 20, 30, 40, 50, and 100 mg P L^{-1} KH_2PO_4 as sorption solution matrices [21]. The sealed centrifuge tubes with the mixed solution were shaken on a reciprocal shaker at a speed of 200 cycles min^{-1} for 24 h at 25 ± 3 °C. The suspension was centrifuged at 3500 rpm for 20 min. The supernatants were filtered through a 0.22 μm mixed cellulose ester membrane and determined for soluble reactive phosphorus. P sorption parameters of soils were simulated by the Langmuir and Freundlich isothermal model, and equilibrium P concentration (EPC$_0$) was determined according to the Freundlich equation. P sorption maximum (Q$_{max}$) was determined according to the Langmuir equation [22].

The soil samples were washed by phosphate buffer to extracted biofilm [23] which was used to extract DNA and measure potential PNR and PDR. The PNR was measured according to the shaken-slurry method [24]. Slurry containing 3 g of biofilm pellet, 100 mL of phosphate buffer (1 mM, pH 7.4), and 0.5 mL of $(NH_4)_2SO_4$ (0.25 M) was incubated on an orbital shaker (180 rpm) at 25 °C for 24 h. Subsamples (5 mL) of the slurry were taken at 1, 4, 10, 16, and 24 h after the start of the incubation. The subsamples were used for the analyses of NO_2^--N and NO_3^--N. The PNR was calculated as NO_2^--N and NO_3^--N production per unit time.

Meanwhile, the PDR was measured based on the Denitrifying Enzyme Activity (DEA) Assay [25]. Briefly, 5 g of biofilm pellet was moved to a special tailor-made borosilicate glass media bottle with 20 mL DEA solution (7 mM KNO_3, 3 mM glucose and 5 mM chloramphenicol), and oxygen was purged out from each bottle by continuously pumping helium and acetylene which was added to reach a final concentration of 10%. Sample bottles were placed on an orbital shaker and shaken (125 r/min) in dark at 25 °C. Gas samples were taken out at 0, 0.5, 1, 1.5, and 2 h for N_2O measurement using gas chromatography. The PDR was calculated as the slope of a curve of best fit in a plot of N_2O concentration against time.

2.4. Statistical Analysis

Pearson correlation coefficients and log-normal regression analyses were performed using SigmaPlot 2000 and SPSS 13.0 for Windows. Pearson's test was performed using the SPSS 18.0 package (SPSS, Chicago, IL, USA), with a selected value of 0.05 for significance.

3. Results

The relationships between riparian buffer width and nutrient removal efficiency are shown in Figure 3 (TP) and Figure 4 (TN), and exhibited log-normal distributions (except the forest type in TP profile). The farmland data are not shown due to heavy nutrient loss (nutrient levels got higher through the riparian buffer). Based on the non-linear regression model, the buffer width was calculated under specific removal efficiencies (50%, 75%, and 90%). On the condition of the same percentage removal efficiency, the needed width for TN removal was longer than for TP (Figures 3 and 4, and Table 1). In detail, the wetland and grass/forest riparian buffers were more effective for TP removal, and their widths were 44 m, 204 m, and 512 m and 23 m, 203 m, and 755 m to achieve a 50%, 75%, and 90% removal efficiency, respectively. The grass riparian buffer was less effective in TP removal, which needed more width to reach removal efficiency. The removal effectiveness of TP in the forest riparian buffer had a weak linear relationship with buffer width, showing a longer width than other riparian buffers (Table 1, Figure 3). The best TN removal efficiency was observed in the grass riparian buffer, which was 130 m, 749 m, and 2143 m in width for 50%, 75%, and 90% removal efficiency, respectively, followed by grass/forest and wetland riparian buffers. The forest riparian buffer could function for TN removal at considerably large widths, which was difficult to construct (Table 1).

Table 1. The relationship between TN and TP removal efficiency and riparian buffer width. Based on a given percent effectiveness (50%, 75%, and 90%), buffer widths are approximate values predicted by the non-linear model, $y = a \times \ln(x) + b$.

Riparian Buffer Type	N	Relationship between TP Removal Efficiency and Buffer Width		Approximate Buffer Width (m) by Predicted TP Effectiveness			Relationship between TN Removal Efficiency and Buffer Width		Approximate Buffer Width (m) by Predicted TN Effectiveness		
		Model	R^2	50%	75%	90%	Model	R^2	50%	75%	90%
Grass	37	$y = 0.1155\ln(x) - 0.0367$	0.7124	104	908	3325	$y = 0.1426\ln(x) - 0.1938$	0.9043	130	749	2143
Forest	27	$y = 3.6864 + 0.0436x$	0.4493	1062	1636	1980	$y = 0.1108\ln(x) - 0.2841$	0.6577	1184	11305	43775
Wetland	33	$y = 0.1628\ln(x) - 0.1157$	0.8073	44	204	512	$y = 0.1327\ln(x) - 0.2678$	0.8336	325	2142	6636
Grass/Forest	27	$y = 0.1142\ln(x) + 0.1432$	0.7411	23	203	755	$y = 0.1586\ln(x) - 0.3401$	0.8513	200	966	2488

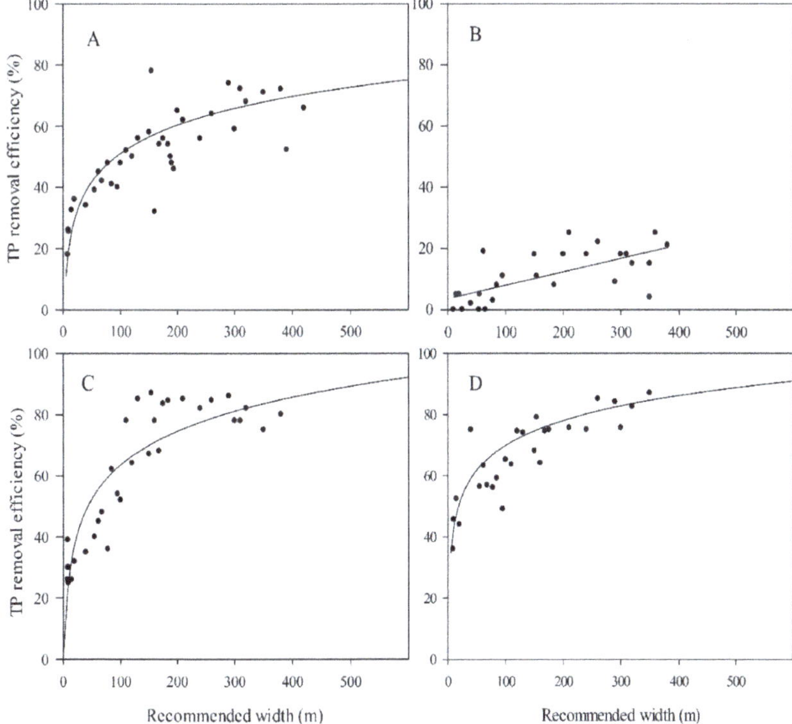

Figure 3. Relationships between TP removal efficiency for storm water and the width of different types of riparian buffer in the Lake Chaohu watershed. (**A**) Grass; (**B**) Forest; (**C**) Wetland; (**D**) Grass/Forest.

Compared to previous research, significantly higher EPC_0 (0.24 mg L^{-1} on average) and lower Q_{max} (352.5 mg kg^{-1} on average) were recorded in the riparian buffers of Lake Chaohu (Figure 5). In horizontal profile, the highest Q_{max} value in the wetland riparian buffer and the lowest EPC_0 value in the grass/forest riparian buffer, as well as the lower Q_{max} value and the higher EPC_0 value in farmland and forest riparian buffers, are shown in Figure 5. Similarly, considerably low PNR and PDR levels were found in this study (Figure 6). In horizontal profile, the higher PNR and PDR were observed in grass and forest riparian buffers and the lowest PNR and PDR were observed in the farmland riparian buffer (Figure 6).

The low TP concentration was distributed in the littoral zone of the southeast area (mainly wetland riparian buffer from Site L2 to L7) and the northeast area (mainly grass/forest riparian buffer from Site L15 to L17), as well as Site L19 in Lake Chaohu. The high TP concentration was distributed in the littoral zone of the north area (mainly farmland riparian buffer in Site L1, L2, L17, L18, L20, L21, and L22) and the southeast area (mainly grass riparian buffer from Site L7 to L15) in Lake Chaohu (Figure 1).

The TN showed similar tendencies to the TP. In the southeast area, the grass riparian buffer contributed mainly to the low TN value in littoral zones, and the farmland riparian buffer, with nutrient leaching, was responsible for the high TN level in the littoral zone of the north area of Lake Chaohu (Figure 2).

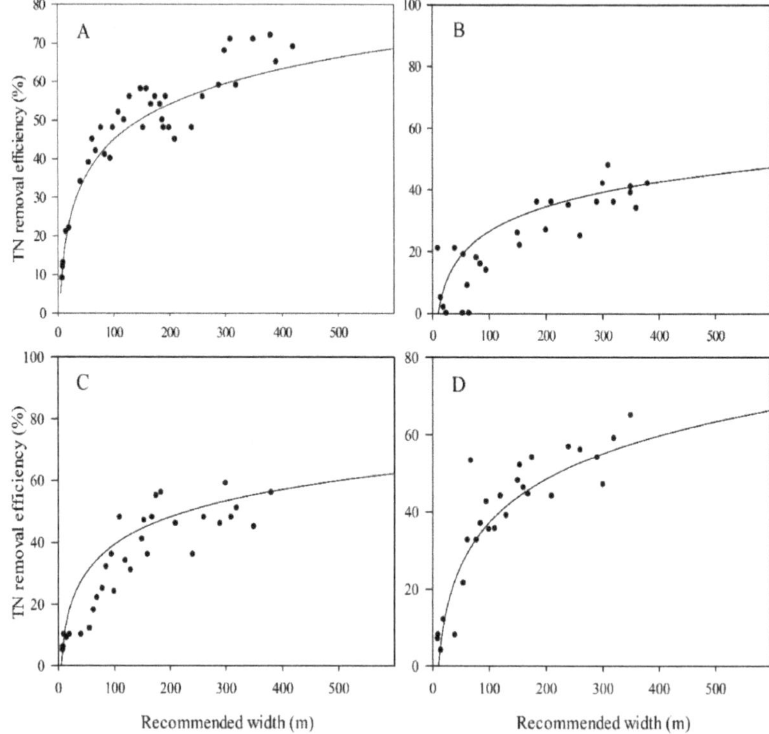

Figure 4. Relationships between TN removal efficiency for storm water and the width of different types of riparian buffer in the Lake Chaohu watershed. (**A**) Grass; (**B**) Forest; (**C**) Wetland; (**D**) Grass/Forest.

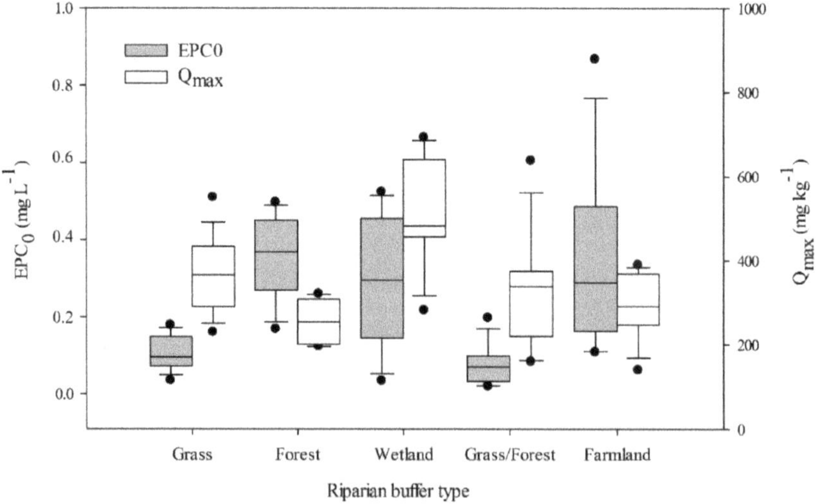

Figure 5. Comparison of phosphorus adsorption ability (Q_{max} and EPC_0) in different types of riparian buffer soil.

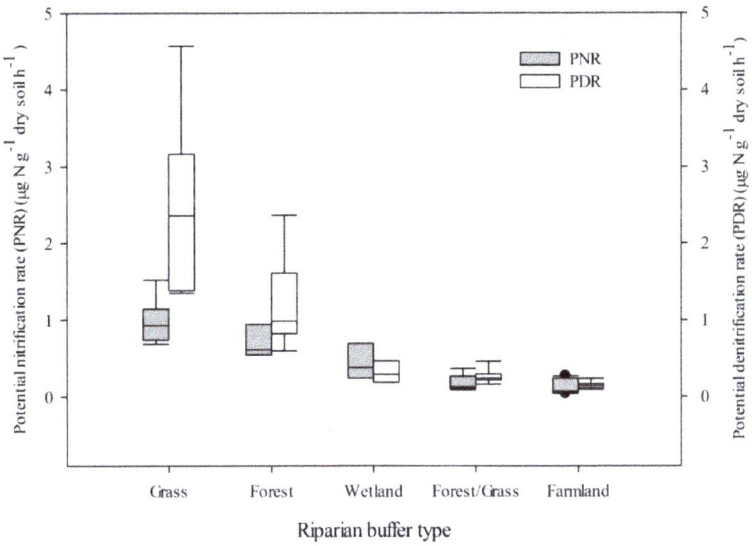

Figure 6. Comparison of PNR and PDR in different types of riparian buffer soil.

4. Discussion

The log-normal relationships between riparian buffer width and nutrient removal efficiency for TP and TN are in agreement with previous reports [6,16]. The longer width demand for TN compared to TP at specific nutrient removal efficiencies can be explained by nutrient retention capabilities. Strong retention of inorganic P and weaker retention of inorganic N by soil indicates greater contribution to the increased export of nitrogen [26]. Through the comparison of riparian buffer width demand between this study and previous research, it is not hard to find that the width demand in this study is greater than that of previous research. In rural temperate watersheds of Brazil, riparian buffer zones with 60 m width and composed of woody soils were more effective in phosphorus (99.9%) and nitrogen (99.9%) removal when compared to shrub (66.4% and 83.9%, respectively) or grass vegetation (52.9% and 61.6%, respectively) areas [27]. In Europe and the United States, riparian width (smaller than 10 m) provided limited protection against nitrate contamination [13,14]. During rainfall simulation, the switch-grass buffer removed 64 and 72% of the incoming TN and TP, respectively. The switch grass-woody buffer removed 80 and 93% of the incoming TN and TP, respectively [8]. A meta-analysis of literature has shown that more than 96% of the buffers studied for N and P removal were less than 20 m wide, and buffers composed of trees had higher N and P removal efficacy than buffers composed of grasses or mixtures of grasses and trees [6]. The huge width demand of riparian buffers in this study should be attributed to the decline of nutrient removal ability due to soil degradation and low vegetation diversity. A close relationship exists between land use, soil properties, and nutrient removal [28]. On the other hand, aggressive non-native plant species have colonized and dominated, the diversity of native plants has declined, and the plant community structure has been simplified [29], leading to less efficiency of riparian buffers as sinks for nutrients. Structurally diverse riparian buffers, i.e., those that contain a mix of trees, shrubs, and grass, are much more effective at capturing a wide range of nutrients than a riparian buffer that is solely of trees or grass [30]. Hence, the over-use and over-fertilization of soil and human intervention on vegetation restoration in riparian buffers around Lake Chaohu watershed has directly resulted in the decline of soil physicochemical characteristics and rhizospheric microbial activities as well as community composition diversity [31,32], eventually accelerating the nutrient leaching of soil and the decline of nutrient removal efficiency. All these factors have jointly caused the enormous demand on riparian buffer width, which facilitated the administrative organization in reserving or restoring more riparian buffers for effective nutrient removal.

EPC_0 is the soluble reactive phosphorus (SRP) concentration in equilibrium with soil at which neither net sorption nor desorption occurs. In short, soil with a high EPC_0 value will have a greater tendency to desorb (release) SRP. Lower EPCo values would suggest that this particular soil is inclined to adsorb P from runoff [33,34]. Q_{max} represents the P adsorption capability. In this study, significantly higher EPC_0 and lower Q_{max} indicated weak P adsorption ability in riparian buffers of Lake Chaohu (Figure 3) compared to previous reports that exhibited a lower and higher value by an order of magnitude for EPC_0 and for Q_{max}, respectively [35,36]. In horizontal profile, the higher TP removal efficiency in wetland and forest/grass riparian buffers in this study can be explained by the highest Q_{max} value in the wetland riparian buffer and the lowest EPC_0 value in the grass/forest riparian buffer, indicating high P adsorption capacity in wetland riparian buffers and strong P retention ability in grass/forest riparian buffers. At the same time, the lower Q_{max} value and the higher EPC_0 value in farmland and forest riparian buffers manifested the weak P adsorption ability, which was in parallel, with no relationship between riparian buffer width and TP removal efficiency in the forest riparian buffer and P leaching from the farmland riparian buffer. All these facts indicates the close relationship between riparian buffer width demand for P removal and P adsorption ability.

TN removal efficiency depended on the PNR and PDR. Similarly, compared to previous reports on soil reviewed by Xu et al. [37], considerably low PNR and PDR levels in this study directly caused the decrease of N removal efficiency and the increase of riparian buffer width demand. In horizontal profile, the highest TN removal efficiency in the grass riparian buffer was found to be coupled with the highest PNR and PDR, indicating strong N removal through coupled nitrification-denitrification. Although the higher PNR and PDR were found in the forest riparian buffer, the amount of N leaching from the forest probably exceeded that of the N removal, resulting in low TN removal efficiency. In the wetland riparian buffer, the lower PNR restricted the denitrification process and TN removal efficiency. The difference of PNR and PDR in the different types of riparian buffers should be attributed to the quality and quantity of organic carbon, aerobic and anaerobic conditions, and the activity and community composition of rhizobacteria [38]. Taken together, the demand of riparian buffer width for P and N removal was closely related with P adsorption ability and coupled nitrification-denitrification ability in soil, respectively.

The nutrient removal efficiency of the riparian buffers made a crucial effect on the nutrient level of the water column in the adjacent aquatic ecosystem, especially in littoral zones. In this study, high TP removal efficiency in wetland and grass/forest buffers was responsible for low TP distribution in the littoral zone of the southeast area (mainly wetland riparian buffer) and the northeast area (mainly grass/forest riparian buffer) in Lake Chaohu. Weak TP removal efficiency in the other buffers also correspond to high TP levels in littoral zones of the north area (mainly farmland riparian buffer) and the southeast area (mainly grass riparian buffer) of Lake Chaohu. In the southeast area, the grass buffer with high TN removal efficiency contributed mainly to the low TN value in littoral zones, and the farmland buffer with nutrient leaching was responsible for the high TN level in littoral zones of the north area in Lake Chaohu. All these facts mentioned above validated the prediction that nutrient removal of riparian buffers had an important contribution to lake eutrophication.

5. Conclusions

The vegetation cover type of riparian buffers can determine the physiochemical features of riparian buffer soil, which further influences the storm water nutrient removal efficiency, resulting in the change of nutrient levels in littoral zones of lake. Construction of wetland or grass/forest and grass riparian buffer strips is strongly recommended for effective TP and TN removal, respectively.

Supplementary Materials: The following are available online at http://www.mdpi.com/2073-4441/10/10/1489/s1, Figure S1: the distribution of different land use types in the littoral zone around Lake Chaohu, Figure S2: the distribution of different land use types in the riparian buffer zone with a width of 10 km from the shoreline around Lake Chaohu.

Author Contributions: All authors contributed equally to the paper.

Funding: This research was funded by National Natural Science Foundation of China (41877381; 41573110), the Major Science and Technology Program for Water Pollution Control and Treatment (2017ZX07603), the State Key Laboratory of Freshwater Ecology and Biotechnology (2016FBZ07). The APC was funded by National Natural Science Foundation of China (41573110).

Acknowledgments: We thank Siyang Wang, Zijun Zhou and Yao Zhang for sampling helps.

Conflicts of Interest: The authors declare no conflict of interest. The founding sponsors had no role in the design of the study; in the collection, analyses, or interpretation of data; in the writing of the manuscript, and in the decision to publish the results.

References

1. Klapproth, J.C.; Johnson, J.E. Understanding the science behind riparian forest buffers: Effects on water quality. *Va. Polytech. Inst. State Univ. Publ.* **2009**, *22*. Available online: https://vtechworks.lib.vt.edu/bitstream/handle/10919/48062/420-151_pdf.pdf?sequence=1&isAllowed=y (accessed on 1 May 2005).

2. Phillips, J.D. An evaluation of the factors determining the effectiveness of water quality buffer zones. *J. Hydrol.* **1989**, *107*, 133–145. [CrossRef]

3. Syversen, N. Effect of a cold-climate buffer zone on minimizing diffuse pollution from agriculture. *Water Sci. Technol.* **2002**, *45*, 69–76. [CrossRef] [PubMed]

4. Syversen, N. Effect and design of buffer zones in the Nordic climate: The influence of width, amount of surface runoff, seasonal variation and vegetation type on retention efficiency for nutrient and particle runoff. *Ecol. Eng.* **2005**, *24*, 483–490. [CrossRef]

5. Stutter, M.I.; Langan, S.J.; Lunsdom, D.G. Vegetated buffer strips can lead toincreased release of phosphorus to waters: A biogeochemical assessment of themechanisms. *Environ. Sci. Technol.* **2009**, *43*, 1858–1863. [CrossRef] [PubMed]

6. Zhang, X.; Liu, X.; Zhang, M.; Dahlgren, R.A.; Eitzel, M. Review of vegetated buffers and a meta-analysis of their mitigation efficacy in reducing nonpoint source pollution. *J. Environ. Qual.* **2010**, *39*, 76–84. [CrossRef] [PubMed]

7. Schwer, C.B.; Clausen, J.C. Vegetative filter treatment of dairy milkhouse wastewater. *J. Environ. Qual.* **1989**, *18*, 446–451. [CrossRef]

8. Kyehan, L.; Isenhart, T.M.; Schultz, R.C.; Mickelson, S.K. Multispecies riparian buffers trap sediment and nutrients during rainfall simulations. *J. Environ. Qual.* **2000**, *29*, 1200–1205.

9. Lowrance, R.R.; Todd, R.L.; Asmussen, L.E. Nutrient cycling in an agricultural watershed—I: Phreatic movement. *J. Environ. Qual.* **1984**, *13*, 22–27. [CrossRef]

10. Lee, P.; Smyth, C.; Boutin, S. Quantitative review of riparian buffer width guidelines from Canada and the United States. *J. Environ. Manag.* **2004**, *70*, 165–180. [CrossRef]

11. Spruill, T.B. Effectiveness of riparian buffers in controlling ground-water discharge of nitrate to streams in selected hydrogeological settings of the North Carolina Coastal Plain. *Water Sci. Technol.* **2004**, *49*, 63–70. [CrossRef] [PubMed]

12. Vidon, P.G.F.; Hill, A.R. Landscape controls on nitrate removal in stream riparian zones. *Water Resour. Res.* **2004**, *40*, 114–125. [CrossRef]

13. Borin, M.; Vianello, M.; Zanin, G. Effectiveness of buffer strips in removing pollutants in runoff from a cultivated field in North-East Italy. *Agric. Ecosyst. Environ.* **2005**, *105*, 101–114. [CrossRef]

14. Hefting, M.M.; Clement, J.; Bienkowski, P.; Dowrick, D.; Guenat, C.; Butturini, A.; Topa, S.; Pinay, G.; Verhoeven, J.T.A. The role of vegetation and litter in the nitrogen dynamics of riparian buffer zones in Europe. *Ecol. Eng.* **2005**, *24*, 465–482. [CrossRef]

15. Weissteiner, C.J.; Bouraoui, F.; Aloe, A. Reduction of nitrogen and phosphorus loads to european rivers by riparian buffer zones. *Knowl. Manag. Aquat. Ecosyst.* **2013**, *483*, 1–7. [CrossRef]

16. Sweeney, B.W.; Newbold, J.D. Streamside forest buffer width needed to protect stream water quality, habitat, and organisms: A literature review. *J. Am. Water Resour. Assoc.* **2014**, *50*, 560–584. [CrossRef]

17. Miller, J.J.; Curtis, T.; Chanasyk, D.S.; Reedyk, S. Influence of mowing and narrow grass buffer widths on reductions in se. *Can. J. Soil Sci.* **2015**, *95*, 139–151. [CrossRef]

18. Zheng, L.G.; Liu, G.J.; Kang, Y.; Yang, R.K. Some potential hazardous trace elements contamination and their ecological risk in sediments of western Chaohu Lake, China. *Environ. Monit. Assess.* **2010**, *166*, 379–386. [CrossRef] [PubMed]

19. Ward, J.R.; Harr, C.A. (Eds.) *Methods for Collection and Processing of Surface-Water and Bed-Material Samples for Physical and Chemical Analyses*; U.S. Geological Survey Open-File Report; U.S. Geological Survey: Reston, VA, USA, 1990; Volume 71, pp. 90–140.

20. Rice, E.W.; Baird, R.B.; Eaton, A.D.; Clesceri, L.S. *Standard Methods for the Examination of Water and Wastewater*, 22nd ed.; American Public Health Association: Washington, DC, USA, 2012.

21. Istvanovics, V. Fractional composition, adsorption and release of sediment phosphorus in the Kis-Balaton reservoir. *Water Res.* **1994**, *28*, 717–726. [CrossRef]

22. Xiao, W.J.; Song, C.L.; Cao, X.Y.; Zhou, Y.Y. Effects of air-drying on phosphorus sorption in shallow lake sediment, China. *Fresenius Environ. Bull.* **2012**, *21*, 672–678.

23. Huang, X.; Liu, C.X.; Wang, Z.; Gao, C.F.; Zhu, G.F.; Liu, L. The effects of different substrates on ammonium removal in constructed wetlands: A comparison of their physicochemical characteristics and ammonium-oxidizing prokaryotic communities. *Clean Soil Air Water* **2013**, *41*, 283–290. [CrossRef]

24. Fan, F.L.; Zhang, F.S.; Lu, Y.H. Linking plant identity and interspecific competition to soil nitrogen cycling through ammonia oxidizer communities. *Soil Biol. Biochem.* **2011**, *43*, 46–54. [CrossRef]

25. Jha, P.K.; Minagawa, M. Assessment of denitrification process in lower Ishikari River system, Japan. *Chemosphere* **2013**, *93*, 1726–1733. [CrossRef] [PubMed]

26. Harms, T.K.; Ludwig, S.M. Retention and removal of nitrogen and phosphorus in saturated soils of arctic hillslopes. *Biogeochemistry* **2016**, *127*, 291–304. [CrossRef]

27. Aguiar, T.R.; Rasera, K.; Parron, L.M.; Brito, A.G.; Ferreira, M.T. Nutrient removal effectiveness by riparian buffer zones in rural temperate watersheds: The impact of no-till crops practices. *Agric. Water Manag.* **2015**, *149*, 74–80. [CrossRef]

28. Castelle, A.J.; Johnson, A.W.; Conolly, C. Wetland and stream buffer size requirements—A review. *J. Environ. Qual.* **1994**, *23*, 878–882. [CrossRef]

29. Gardner, P.A.; Stevens, R.; Howe, F.P. *A Handbook of Riparian Restoration and Revegetation for the Conservation of Land Birds in Utah with Emphasis on Habitat Types in Middle and Lower Elevations*; Utah Division of Wildlife Resources: Salt Lake City, UT, USA, 1999; Volume 48, pp. 1–48.

30. Jontos, R. Vegetative buffers for water quality protection: An introduction and guidance document. *Connecticut Association of Wetland Scientists White Paper on Vegetative Buffers*, Draft Version; 2004, Volume 1. Available online: http://www.ctwetlands.org/downloads/Draft%20Buffer%20Paper%20Version%201.0.doc (accessed on 21 October 2018).

31. Zhang, C.; Liu, G.; Xue, S.; Wang, G. Changes in rhizospheric microbial community structure and function during the natural recovery of abandoned cropland on the Loess Plateau, China. *Ecol. Eng.* **2015**, *75*, 161–171. [CrossRef]

32. Kuramae, E.E.; Etienne, Y.; Wong, L.C.; Pijl, A.S.; Veen, J.A.; Kowalchuk, G.A. Soil characteristics more strongly influence soil bacterial, communities than land-use type. *FEMS Microbiol. Ecol.* **2012**, *79*, 12–24. [CrossRef] [PubMed]

33. Phillips, I.R. Nitrogen and phosphorus transport in soil using simulated waterlogged conditions. *Commun. Soil Sci. Plan.* **2001**, *32*, 821–842. [CrossRef]

34. Litaor, M.I.; Reichmann, O.; Haim, A.; Auerswald, K.; Shenker, M. Sorption characteristics of phosphorus in peat soils of a semiarid altered wetland. *Soil Sci. Soc. Am. J.* **2005**, *69*, 1658–1665. [CrossRef]

35. Heiberg, L.; Pedersen, T.V.; Jensen, H.S.; Kjaergaard, C.; Hansen, H.C.B. A comparative study of phosphate sorption in lowland soils under oxic and anoxic conditions. *J. Environ. Qual.* **2010**, *39*, 734–743. [CrossRef] [PubMed]

36. Janardhanan, L.; Daroub, S.H. Phosphorus sorption in organic soils in south Florida. *Soil Sci. Soc. Am. J.* **2010**, *74*, 1597–1606. [CrossRef]

37. Xu, Y.; Xu, Z.; Cai, Z.; Reverchon, F. Review of denitrification in tropical and subtropical soils of terrestrial ecosystems. *J. Soil. Sediments* **2013**, *13*, 699–710. [CrossRef]
38. Lowrance, R.; Vellidis, G.; Hubbard, R.K. Denitrification in a restored riparian forest wetland. *J. Environ. Qual.* **1995**, *24*, 808–815. [CrossRef]

Article

Effects of Dredging Season on Sediment Properties and Nutrient Fluxes across the Sediment–Water Interface in Meiliang Bay of Lake Taihu, China

Ji-Cheng Zhong [1,*], Ju-Hua Yu [1], Xiao-Lan Zheng [1], Shuai-Long Wen [1], De-Hong Liu [2] and Cheng-Xin Fan [1]

[1] State Key Laboratory of Lake Science and Environment, Nanjing Institute of Geography and Limnology, Chinese Academy of Sciences, Nanjing 210008, China; yujuhua0000@163.com (J.-H.Y.); zxlan7182@163.com (X.-L.Z.); wensl313@sina.com (S.-L.W.); cxfan@niglas.ac.cn (C.-X.F)
[2] School of Geography Science, Nanjing Normal University, Nanjing 210023, China; l66dehong@126.com
* Correspondence: jczhong@niglas.ac.cn; Tel.: +86-025-8688-2212; Fax: +86-025-5771-4759

Received: 5 September 2018; Accepted: 6 November 2018; Published: 8 November 2018

Abstract: The influence of dredging season on sediment properties and nutrient fluxes across the sediment–water interface remains unknown. This study collected sediment cores from two sites with different pollution levels in Meiliang Bay, Taihu Lake (China). The samples were used in simulation experiments designed to elucidated the effects of dredging on internal loading in different seasons. The results showed that dredging the upper 30 cm of sediment could effectively reduce the contents of organic matter, total nitrogen, and total phosphorus in the sediments. Total biological activity in the dredged sediment was weaker ($p < 0.05$) than in the undredged sediment in all seasons for both the Inner Bay and Outer Bay, but the effect of 30-cm dredging on sediment oxygen demand was negligible. Dredging had a significant controlling effect on phosphorus release in both the Inner Bay and Outer Bay, and soluble reactive phosphorus (SRP) fluxes from the dredged cores were generally lower ($p < 0.05$) than from the undredged cores. In contrast, NH_4^+-N fluxes from the dredged cores were significantly higher ($p < 0.05$) than from the undredged cores in all seasons for both sites, this indicates short-term risk of NH_4^+-N release after dredging, and this risk is greatest in seasons with higher temperatures, especially for the Inner Bay. Dredging had a limited effect on NO_2^--N and NO_3^--N fluxes at both sites. These results suggest that dredging could be a useful approach for decreasing internal loading in Taihu Lake, and that the seasons with low temperature (non-growing season) are suitable for performing dredging projects.

Keywords: dredging season; inorganic nitrogen; soluble reactive phosphorus; benthic fluxes; Taihu Lake

---

1. Introduction

Over the past two decades, many lakes around the world have experienced eutrophication, which has caused serious environmental problems such as frequent algal bloom outbreaks, black and foul water, and pollution [1–4]. Depending on the situation, sediment can serve as a source or sink of pollutants, when sediment acts as a source, sediments are considered important sources of nutrient release to the water column, constantly replenishing nutrients and promoting the formation of algal blooms [5,6]. Studies have shown, even under the conditions of control of external sources, that eutrophication could be maintained for many years because of internal loading [7]. Many techniques are used to control internal nutrient loading, such as in situ capping [8], algae salvage [9], ecological remediation [10], and sediment dredging [11]. As a common ecoengineering technology, sediment dredging is used in many lakes to control internal nutrient loading and to rebuild the natural ecological system [12–15].

Dredging is one of the few options available for improving the ecological balance of lakes by removing contaminated sediments. However, based on available evidence, there are questions regarding the efficacy of the approach and the degree to which dredging results in reduced risk to human health and the environment [16]. Many studies have confirmed the positive effects of dredging on internal nutrient control [17–20]. However, sediment dredging is not always successful in controlling sediment nutrient release [12,21]. Thus, dredging success remains variable [14], and the mechanism underlying the observed effects on nutrient cycling remains puzzling. Previous studies have shown that the controlling effect of sediment dredging on nutrients is affected by many factors, such as the dredging technique, dredging depth, external control, and environmental conditions of the lake itself [17,21–24]. Recently, Chen et al. [25] reported that different dredging seasons have different effects on the prevention of black and odorous formations in the water column, and that dredging to prevent the formation of algae-induced black blooms could be more effective in winter than other seasons. However, the influence of dredging season on internal nutrient control has rarely been reported.

As a physical engineering measure, sediment dredging destroys the original and stable sediment–water interface, and it exposes buried sediment. This newly exposed sediment forms a new and unbalanced sediment–water interface [14,17,18]. Sediment dredging changes the physical, chemical and biological properties of sediments considerably, affecting both the cycling process of nutrients in the sediments and the nutrient flux at the sediment–water interface [19,20,26]. Additionally, nutrient cycling in a lake ecosystem has obvious seasonal characteristics [27–29], which include changes both in the nutrients in porewater and/or sediments and in nutrient fluxes across the sediment–water interface [30–33]. Currently, the selection of dredging season in China, as in many other countries, is random, and the influence of seasons on the effects and effectiveness of dredging is not considered.

In this study, we hypothesized that nutrient fluxes at the sediment–water interface would be different before and after dredging, and we hypothesized further that dredging season would have different effects on internal loading control. We collected sediment cores from two sites with different pollution levels in Meiliang Bay Taihu Lake (China). We used these samples in simulation experiments designed to elucidate the effects of dredging on internal loading in different seasons. The study had three primary objectives: (1) to understand the influence of sediment dredging on sediment properties; (2) to evaluate the effect of dredging season on nitrogen and phosphorus release and recommend the optimum season for dredging with minimal environmental risk; and (3) to study the effectiveness of dredging on internal loading at different pollution levels in sediment. The results could be used to evaluate potential dredging plans and to improve the decision-making process associated with dredging engineering.

2. Materials and Methods

2.1. Site Description and Sampling

Taihu Lake is a typical large shallow lake with a surface area of 2338 km^2 [34]. Taihu Lake is the third-largest freshwater lakes in China, and it serves as an important source of drinking water for surrounding cities, for example, Wuxi and Suzhou. Meiliang Bay is one of the most eutrophic bays in the northern part of Taihu Lake, and in the past two decades, cyanobacterial blooms have occurred there during the warmer seasons [1,35]. The water supply crisis in Wuxi in May 2007 received wide attention both in China and in other countries [3]. In attempting to reconstruct the lake ecosystem, the Chinese government has adopted many different measures since 2007, for example, external controls, ecological water diversion, salvaging blue-green algae, ecological restoration, and sediment dredging [9,36]. The latter is considered an important measure with which to control the internal loading in Taihu Lake. The area of Taihu Lake that was dredged was about 93.65 km^2, and Meiliang Bay was pinpointed as a key area in the dredging plan of Taihu Lake [9].

Two sites (Inner bay and Outer Bay) with different pollution levels in Meiliang Bay (Figure 1) were selected in this study. Inner Bay is close to the city of Wuxi, Previously, untreated sewage was discharged into Meiliang Bay through the Liangxi River, a main river running through Wuxi, Furthermore, the surrounding land is a scenic area, and thus, Inner Bay has been subject to intensive use and it has suffered because of human activities. Outer Bay is located in the inlet of Meiliang Bay, the water and sediment quality at this location are affected primarily by Taihu Lake. The nearby land use is mainly orchards and grassland, and vegetation coverage is high. Therefore, Outer Bay is affected primarily by nonpoint source pollution [37].

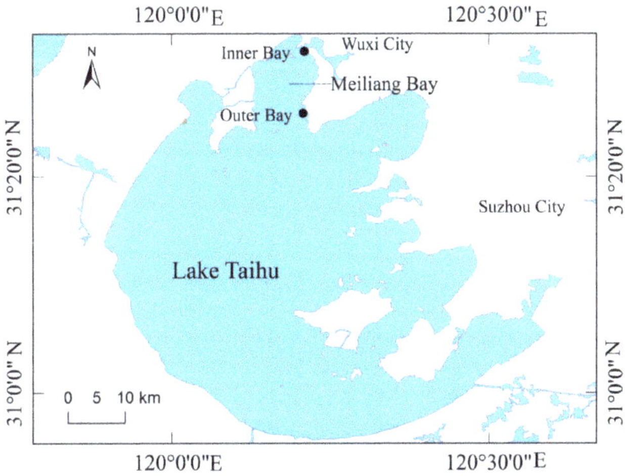

Figure 1. Sampling sites in Taihu Lake, China.

Sampling was conducted at both sites in January (winter), April (spring), July (summer), and October (autumn) through 2013. Sampling was conducted usually in the middle of each month, and 36 sediment cores was collected using a gravity corer (9 cm diameter, 70 cm length) from the two sites in each season. These sediment cores were used to test nutrient flux, sediment oxygen demand and total microbial activity at the sediment-water interface. In January (winter) 2013, one sediment core was collected at each site for analysis of basic physical and chemical properties of the sediments. All sediment cores collected in this study were >50-cm long, and both the top and the bottom of each core was plugged with a rubber stopper. In the sampling and transportation processes, care was taken to avoid disturbance and destruction of the sediment structure. Near-bottom water was sampled in situ at each site and stored in acid-cleaned polyethylene buckets. The water temperature at each site was measured using a multiparameter water quality instrument (YSI, Yellow Springs, OH, USA). All the sediment cores and water samples were then returned to the laboratory for processing.

2.2. Laboratory Microcosm Experiment

For dredging simulation and the laboratory microcosm experiment, we have adopted the procedure described by Kleeberg et al. [17] and Reddy et al. [18]. In the laboratory, overlying water was siphoned off from each core. To simulate dredging, the uppermost 30 cm of sediments in six undisturbed cores from the two sites was skimmed using a suction pump. The remaining sediment cores (9 cm diameter, 20 cm length) were then transferred to cleaned plexiglass tubes to serve as the dredged cores (Supplementary Figure S1). Six additional undisturbed cores from the two sites were adjusted to lengths of 20 cm to serve as the undredged (control) cores, specifically, the excess sediment layers at the bottom were removed, the remaining upper 20 cm sediment layers were then transferred to cleaned plexiglass tubes to serve as the undredged cores (Supplementary Figure S1). There were three replicates for both undredged and dredged treatments. Filtered water from in situ water samples was injected into both the dredged and the undredged cores using the siphon method, and the depth of the overlying water was maintained at 20 cm. Subsequently, the sediment of the cores was wrapped in silver paper to avoid light. The undredged and dredged cores were then transferred to a water bath and incubated at the in situ water temperature ($\pm 2\,^\circ$C). Water samples (50 mL), taken from the incubated cores at designated intervals (0, 12, 24, 36, 48, 60, and 72 h), were filtered through 0.45-μm syringe filters and analyzed for soluble reactive phosphorous (SRP) and inorganic nitrogen. Filtered site water (50 mL) was added to each core immediately after each sampling to maintain water quantity.

The exchange rates of nutrients across the water–sediment interface were calculated according to the following equation [38]:

$$F_i = [V \times (C_n - C_0) + \sum_{j-1}^{n} V_{j-1} \times (C_{j-1} - C_a)]/(S \times T) \tag{1}$$

where F_i represents the exchange rate during the three days (mg m^{-2}d^{-1}); V represents the volume of overlying water in the sediment column (L); C_0, C_n and C_{j-1} represent the nutrient concentration (mg L^{-1}) at the beginning, on day n and on day $j - 1$, respectively; C_a represents the nutrient concentration of the water used to compensate for the sampled water (mg L^{-1}), V_{j-1} represents the water volume sampled each time (L); S represents the area of the sediment column (m^2) and T represents the incubation time (d).

2.3. In Situ Porewater Sampling

Porewater equilibrators, also referred to as peepers [39], were used to obtain the porewater profiles at both sites in different seasons. Each peeper had 36 vertically disposed dialysis cells machined to 1 cm resolution (Supplementary Figure S2). The structure and operating principle of a peeper has been described in detail by Jacobs [40]. When performing interstitial water sampling, the peepers were placed at the sediment-water interface by a self-made device and fixed on a steel shelf by rope at each site, part of the peeper inserted into the sediment, leaving a part of the peeper in the overlying water above the sediment, so that a complete sediment-water interface can be obtained. The peepers were retrieved after 15 days when the interstitial water was considered to have reached equilibrium with the peepers [41]. After retrieval of the peepers, the water of the peepers was sampled immediately, injected into vials containing appropriate fixative agents, and then stored on ice. The nutrient and Fe^{2+} contents of the porewater were analyzed within 3 h following transportation to the laboratory. Unlike the successful recovery of peepers in January 2013, peeper recovery failed in the other seasons. In order to support the results of this study, we also used unpublished porewater data of different seasons obtained at the same two sites in 2007–2008.

2.4. Analyses of Sediment and Water Characteristics

Sediment cores collected from both sites in January 2013 were used for the analysis of sediment physicochemical characteristics. Sediment cores were sliced at 2 cm intervals under anaerobic conditions (N$_2$ atmosphere). Sediment water content was determined by weight loss after drying the sediment samples at 105 $^\circ$C for >24 h. Porosity and bulk density were measured using a cutting ring [26]. Loss on ignition (LOI) was determined after ignition at 550 $^\circ$C for 6 h. Total nitrogen (TN) and organic carbon (OC) content of the sediment were analyzed using CHN elemental analyzer (CE-440, EAI, North Chelmsford, MA, USA). Total phosphorus (TP) in the sediment was analyzed by following the procedure described by Ruban et al. [42].

A further two sets of six cores (9 cm diameter) were used to investigate the effects of on sediment oxygen demand and total microbial activity. Dredging simulations were conducted according to the protocol above. There were three replicates for both undredged and dredging treatments. For the analysis of total microbial activity in the surface sediment, the surface sediments (0–2 cm layers) were sliced from both the undredged and the dredged cores immediately after dredging. The total microbial activity of the sediment was measured using fluorescein diacetate [43]. Sediment oxygen demand of the undredged and dredged cores was determined following the procedure described by Fisher and Reddy [44]. The dissolved oxygen concentrations in the water were recorded at designated intervals using an oxygen microsensor (Presens, Regensburg, Germany). The sediment oxygen demand (*SOD*) rates were calculated using the formula described by Malecki et al. [30].

$$SOD = \left\{ \left[\left(DO_i - DO_f \right) / (T \times S) \right] \times L \right\} \tag{2}$$

where *SOD* is sediment oxygen demand (mg cm^{-2} h^{-1}), DO_i is initial dissolved oxygen (*DO*) concentrations (mg L^{-1}), DO_f is final *DO* concentration (mg L^{-1}), *T* is time (h), *S* is surface area of core (cm^2) and *L* is reflood water volume (L).

Nutrient concentrations in the water samples were analyzed using a flow-injection autoanalyzer (Skalar Sanplus, Breda, The Netherlands). Fe^{2+} in the porewater was determined using the phenanthroline method with a spectrophotometer (UV-2550, Shimadzu, Kyoto, Japan) [45].

2.5. Statistical Analysis

One-way analysis of variance (ANOVA) was used to analyze the differences of nutrient in the porewater and overlying water between the four seasons and two sites. Two-way analysis of variance (ANOVA) was used to analyze the differences of nutrient fluxes, total microbial activity and SOD between undredged and dredged treatments in the four seasons. An independent *t*-test was performed to compare the differences of sediment properties between two sites and different depths. Untransformed data in all cases satisfied assumptions of normality and homoscedasticity. Statistical analysis was performed using the SPSS 20.0 statistical package, and the level of significance used was $p < 0.05$ for all tests.

3. Results and Discussion

3.1. Sediment Characterization

The porosity of sediments generally decreased with depth, whereas the contents of LOI, OC, TN, and TP decreased in the first tested depth and then increased and decreased, showing a clear peak at depths ranging from 10 to 20 cm (Figure 2). The contents of LOI, OC, and TN showed no significant differences ($p > 0.05$) between the Inner Bay and the Outer Bay samples, while the contents of TP differed ($p < 0.05$) because of the different land use types.

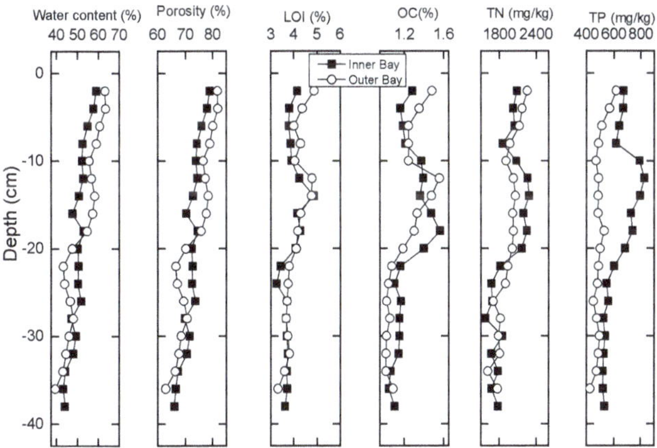

Figure 2. Vertical profiles of selected physicochemical characteristics of sediments.

In this study, we simulated a dredging depth of 30 cm. Figure 2 shows that 30 cm dredging would have considerable impact on the physicochemical properties of the sediments. The water content and porosity of the dredged sediments (buried sediment) were lower ($p < 0.05$) than the undredged sediment (surface sediment) because of the increasing compaction and dehydration of sediment with depth. Furthermore, the contents of LOI, OC, TN, and TP in the dredged sediments were significantly lower ($p < 0.05$) in comparison with the same layer in the undredged sediment. Thus, 30-cm dredging would effectively reduce the loadings of organic matter, nitrogen, and phosphorus in the sediments at both sites.

3.2. Temporal and Spatial Variations of Porewater Chemistry

Peepers are passive porewater samplers with diffusion chambers [38]. Previous studies have demonstrated that their use has good reproducibility in obtaining quality porewater profiles [40]. In this study, porewater profiles in different seasons were obtained from Inner Bay and Outer Bay in order to identify spatial and seasonal trends in biogeochemical processes. The profiles of porewater chemistry obtained by peepers are presented in Figures 3 and 4. The concentrations of NH_4^+-N, SRP and Fe^{2+} in the porewater increased with depth (Figure 3), but the concentrations of NO_3^--N and NO_2^--N decreased with depth. In general, the concentrations of NH_4^+-N and SRP in the porewater of lower sediment layer of Inner Bay were much higher ($p < 0.05$) than in Outer Bay (Figure 3). These porewater results, which are in accord with those of the sediments, show that the pollution status of Inner bay was more serious than Outer Bay.

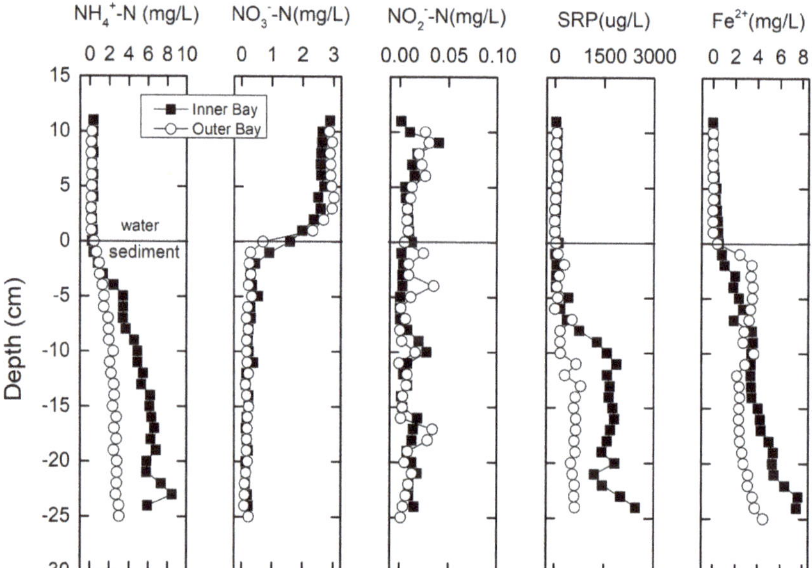

Figure 3. Depth profiles of selected chemical characteristics of porewater in Inner Bay and Outer Bay, samples were collected in January 2013.

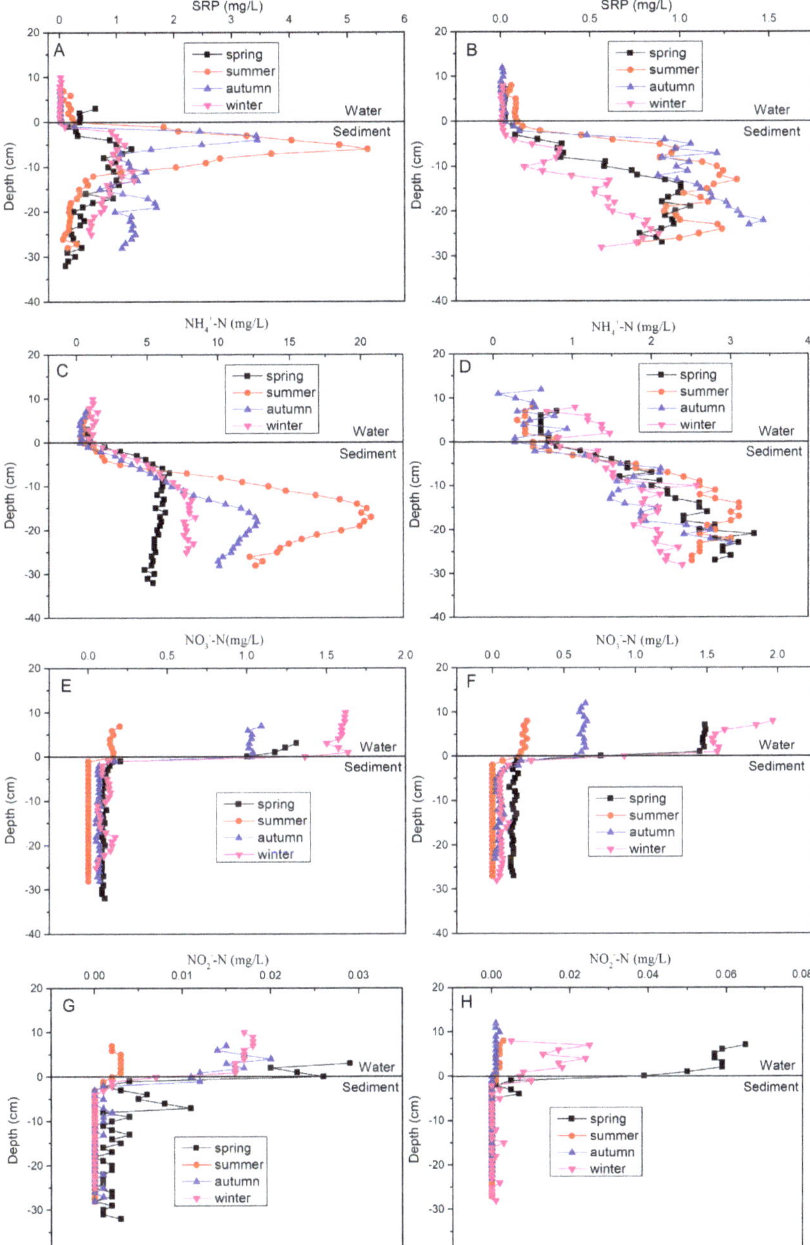

Figure 4. Depth profiles of soluble reactive phosphorus (SRP) and inorganic nitrogen in sediment porewater in different seasons from Inner Bay (**A,C,E,G**) and Outer Bay (**B,D,F,H**), samples were collected in 2007–2008.

The porewater profiles of SRP showed obvious spatial-temporal variations (Figure 4A,B). The concentrations of SRP in the upper 10 cm layer in summer and autumn were generally higher ($p < 0.05$) than in winter and spring, showing seasonal tendency at both sites. This result might be attributable to increased microbial biomass and activity in the sediments during summer and autumn, reflecting increased bioavailable organic matter present on the sediment surface [30]. The SRP profile had a peak between 0 and 15 cm in the Inner Bay, especially in summer and autumn, with high temperatures. The concentration of SRP peaked in sediment layer levels between 0 and 15 cm, which s where the maximum decomposition of organic matter, oxygen consumption, sulfide oxidation, and nutrient reduction occurs [40,46]. The concentration of SRP in porewater was generally higher ($p < 0.05$) than in the overlying water, notably by 40 and 45 times for the Outer Bay and Inner Bay respectively, indicating that sediment is an important source of phosphorus for the overlying water. In addition, the concentrations of SRP in the porewater of the Inner Bay were significantly higher ($p < 0.05$) than in the Outer Bay, especially in the surface 10 cm. These results might be attributable to the different content of TP in the sediments (Figure 2) and different sources of pollution between two sites [37].

All the peepers indicated that NH_4^+-N is the dominant form of inorganic nitrogen in porewater (Figure 4), the proportions of NH_4^+-N to inorganic nitrogen are over 93% for both sites. The profiles of NH_4^+-N showed seasonal characteristics in the Inner Bay, as concentrations of NH_4^+-N were notably higher ($p < 0.05$) in summer and autumn than in winter and spring, and there was a peak between 10- and 25-cm sediment layers. However, the profiles of NH_4^+-N in the Outer Bay had no strong ($p > 0.05$) seasonal variation (Figure 4C,D). In addition, the concentrations of NH_4^+-N in the porewater of the Inner Bay were significantly higher ($p < 0.05$) than in the Outer Bay. These results might be attributable to the different levels of pollution and different sources of pollution between the Inner Bay and Outer Bay [37]. The concentrations of NH_4^+-N in porewater were generally higher ($p < 0.05$) than in the overlying water, suggesting that sediments are an important source of nitrogen for the overlying water at both sites. Unlike NH_4^+-N, the concentrations of NO_3^--N and NO_2^--N in porewater were very low, and they were generally lower ($p < 0.05$) than the overlying water at both sites (Figure 4E,F). Furthermore, the concentrations of NO_3^--N and NO_2^--N in the overlying water showed seasonal tendencies (Figure 4G,H), i.e., the concentrations of NO_3^--N and NO_2^--N in the overlying water in winter and spring were higher ($p < 0.05$) than in summer and autumn. Concentrations of inorganic nitrogen in the water columns showed seasonal trends that are likely influenced by many factors, for example, exogenous inputs, water levels, biological absorption, and nitrogen cycling processes [47–49].

3.3. Effects of Dredging on Total Microbial Activity and Sediment Oxygen Demand

Total microbial activity and SOD are two very important characteristics of sediment. These properties have direct relationships with the decomposition process of organic matter and the transformation process of nutrients in the sediments. In this study, we evaluated the effects of dredging and of dredging season on total microbial activity and SOD. For the undredged treatment, biological activity showed significant seasonal characteristics ($p < 0.05$) in the sediments of the Outer Bay and Inner Bay (Figure 5). The values of biological activity in spring and summer were remarkably higher ($p < 0.05$) than in autumn and winter. For dredged treatment, the microbial activities of the Inner Bay sediments showed significant seasonal differences ($p < 0.05$), while those of the Outer Bay sediments showed no significant seasonal differences ($p > 0.05$). In general, biological activity in the dredged sediments was weaker ($p < 0.05$) than in undredged sediments (Figure 5). Previous research has indicated that the total number of bacteria and hydrolytic activity decrease with sediment depth [50]. In this study, the dredged sediments were buried below 30 cm, and therefore, the dredged sediments had relatively low microbial activity.

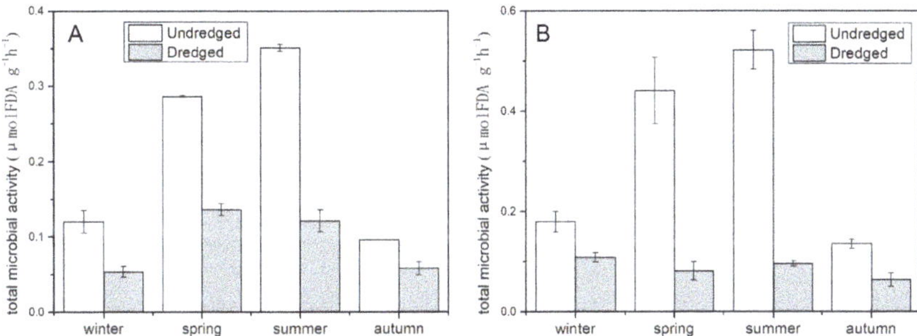

Figure 5. Total microbial activity in undredgedand dredged sediments of the Inner Bay (**A**) and Outer Bay (**B**) in January, April, July, and October 2013, Error bars represent ±1 SE of triplicate samples.

The SOD consumption rates in both the undredged and the dredged cores showed significant seasonal differences ($p < 0.05$); the highest values appeared in summer and, the lowest values appeared in winter (Figure 6). Greater rates of consumption in summer (July) are indicative of increased microbial activity due to increased water column productivity, which results in more bioavailable material than in the winter months. There was no significant difference ($p > 0.05$) in the SOD values between the undredged and dredged cores, except for Inner Bay in spring and for Outer Bay in autumn (Figure 6). This result is different from the total microbial activity in the sediment (Figure 5). Sediment oxygen consumption is attributed to chemical oxidation and microbial-mediated oxidation [30]. In the present study, we simulated a dredging depth of 30 cm, and the deep buried anaerobic sediment was exposed at the sediment–water interface after dredging. We speculate that the amount of chemical aerobic oxidation (e.g., iron oxide) in the dredged sediment was higher than in the undredged sediment. Therefore, despite the undredged sediment having higher microbial activity (Figure 5), indicating higher microbial oxidation than the dredged sediment, in most cases, no significant difference in SOD was found between the dredged and undredged cores.

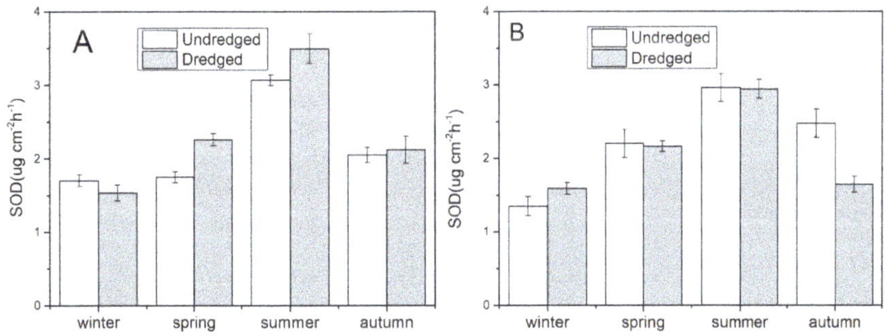

Figure 6. Sediment oxygen demand (SOD) in undredged and dredged sediments of the Inner Bay (**A**) and Outer Bay (**B**) in January, April, July, and October 2013, Error bars represent ±1 SE of triplicate samples.

3.4. Effects of Dredging on Phosphorous Flux

The fluxes of SRP had significant seasonal variation ($p < 0.05$) in the undredged cores from both the Inner Bay and Outer Bay (Figure 7A,B). Phosphorus release fluxes in summer and autumn were higher than in spring and winter; in winter, the SRP fluxes were negative, indicating that the sediment acts as a sink. The SRP fluxes in the dredged cores were negative (i.e., indicating a sink) in all seasons. It was found that the SRP fluxes in the undredged cores were significantly higher ($p < 0.05$) than in the dredged cores from both Inner and Outer Bay (Figure 7A,B). This result indicates that 30 cm dredging could effectively control phosphorus release for both sites in Meiliang Bay.

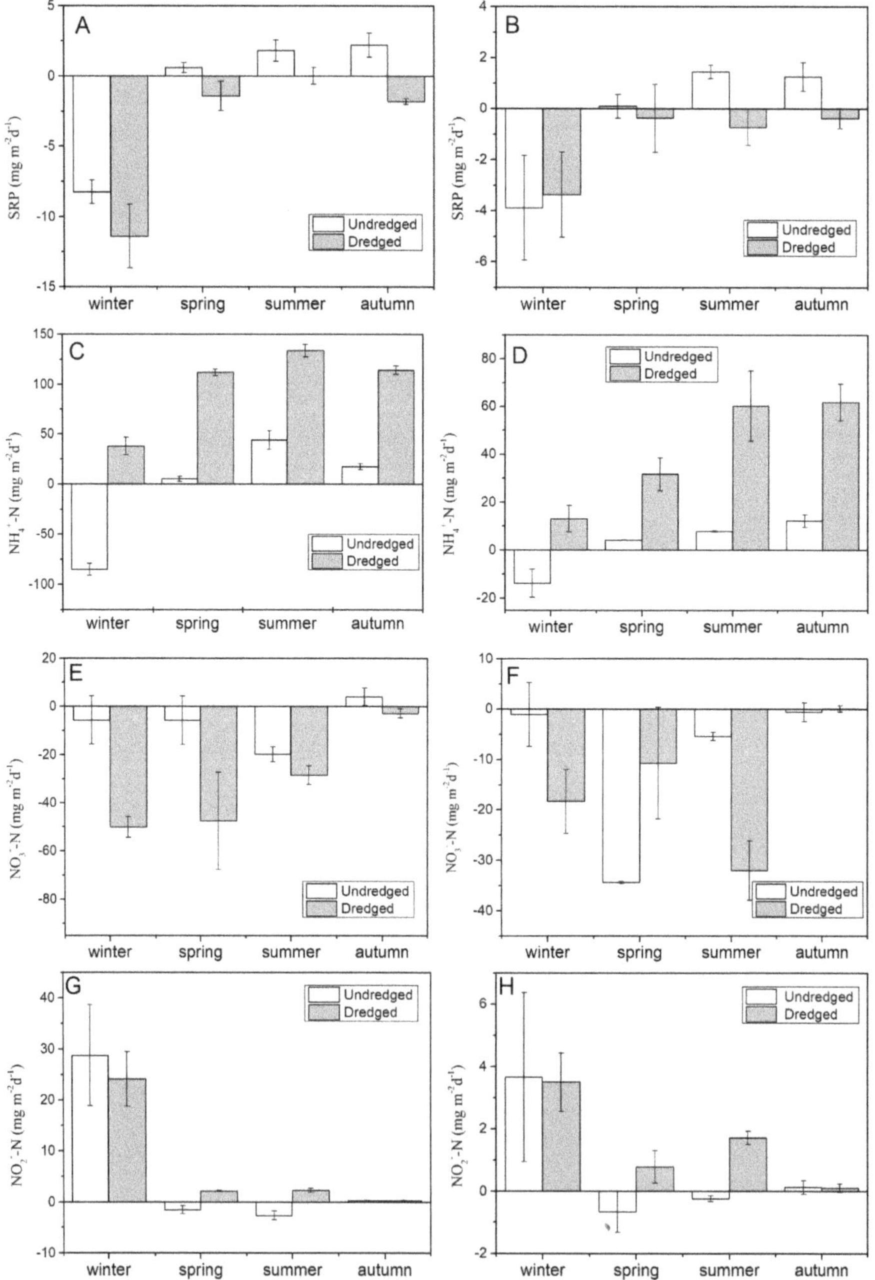

Figure 7. Fluxes of SRP and inorganic nitrogen across the sediment–water interface in undredged and dredged sediments of the Inner Bay (**A,C,E,G**) and Outer Bay (**B,D,F,H**) in January, April, July, and October 2013, Error bars represent ±1 SE of triplicate samples.

Based on simulations and fieldwork, previous studies have proven that sediment dredging can effectively control phosphorus release from sediment [20,23,51]. In the present study, we studied only the controlling effect of dredging on phosphorus release in the short term. Therefore, SRP concentration in the interstitial water, temperature, and sediment characteristics, such as porosity, were the primary factors affecting the fluxes across the sediment–water interface. The reduction in SRP release from the dredged cores in comparison with the undredged cores can be attributed to the following reasons. (1) Sediment porosity decreased considerably (Figure 2) with depth because of the increasing compaction and dehydration of the sediment. Therefore, the porosity of the dredged sediment was lower than that of the undredged cores. The lower sediment porosity of the dredged cores hindered labile phosphorus transfer from the sediments to the water column [17,52]. (2) The SRP concentration in dredged cores also decreased considerably, especially for the Inner Bay (Figure 4A). The peak of SRP in the interstitial water appeared above the 10 cm sediment layer (Figure 4A). Therefore 30 cm dredging could effectively reduce the SRP concentration in the dredged sediment, and the lower SRP concentration in the porewater of the dredged cores might present lower release potential compared with the undredged cores [20]. (3) The soluble Fe^{2+} concentration in the porewater increased considerably with depth (Figure 3). Therefore, the soluble Fe^{2+} concentration in the porewater of dredged cores would be higher than in the undredged cores. The buried and anaerobic sediments exposed at the sediment–water interface after dredging are gradually transformed to aerobic sediments. The formation of an oxic-trapped microlayer formation on the surface sediment of the dredged cores could be attributed to the oxidation of the soluble F^{2+} forms to their insoluble Fe^{3+} oxides and hydroxides [52]. Thus, the phosphorus release at the sediment–water interface of the dredged cores could be inhibited by the oxic-trapped microlayer through adsorption and coprecipitation [6,20].

The temporal effectiveness of sediment dredging in controlling internal loading is a common concern among researchers [11,53]. In this study, 30 cm dredging was shown to be effective in reducing SRP fluxes across the sediment–water interface of dredged cores (Figure 7A,B). Thus, sediment dredging might have a positive effect in reducing internal phosphorus loading just after dredging. The long-term effectiveness of sediment dredging in reducing internal phosphorus loading is affected by many factors, such as external sources, phosphorus regeneration, and the intrinsic nature of the lake itself [18,20,24]. The present study found that 30-cm dredging could reduce the contents of organic matter (LOI) and TP in the dredged sediments (Figure 2). Moreover, pools of readily mobile and labile phosphorus also became reduced [17,18,20]. Additionally, previous research has demonstrated that the lower resupply ability and the higher retention capacity of sediment after dredging could hinder labile phosphorus transfer from the sediments to the overlying water [20]. However, Liu et al. [54] reported that the accumulation of riverine suspended particulate matter in the sediment after dredging could increase internal phosphorus loading and regulate the long-term efficacy of dredging. If external inputs of nutrient loading are not controlled, external phosphorus will lead to the reversion of before dredging levels within a few years after dredging [14,51], which will affect the long-term efficacy of dredging on phosphorus control [54]. Therefore, blocking external phosphorus loading should be a precondition to achieving effective long-term results from dredging [18,51,54].

3.5. Effects of Dredging on Inorganic Nitrogen Flux

The NH_4^+-N fluxes showed significant seasonal variation ($p < 0.05$) in the undredged and dredged cores from both Inner Bay and Outer Bay (Figure 7C,D). The NH_4^+-N fluxes demonstrated lower release rates in winter and spring, and higher release rates in summer and autumn. Unlike SRP fluxes, NH_4^+-N fluxes in the dredged cores were significantly higher ($p < 0.05$) than in the undredged cores for both sites. This result indicates that NH_4^+-N release would be enhanced significantly in the short term after dredging, and this phenomenon should be noted.

In general, NO_3^--N fluxes were mainly negative (Figure 7E,F), that is, the sediment acts as a sink for the overlying water. There were no significant differences ($p > 0.05$) in the NO_3^--N fluxes between the dredged and undregded cores for both Inner Bay and Outer Bay (Figure 7E,F). The NO_3^--N fluxes were generally negative; therefore, the environmental risk posed by NO_3^--N release after dredging can be ignored. Unlike NO_3^--N fluxes, NO_2^--N fluxes had significant seasonal differences ($p < 0.05$) for both the undredged and dredged cores (Figure 7G,H). The NO_2^--N fluxes had higher rates in winter and lower rates in all other seasons. No significant differences ($p > 0.05$) occurred in the NO_2^--N flexes between the dredged and undredged cores for both Inner Bay and Outer Bay (Figure 7G,H). Because of the lower release rates of NO_2^--N compared with NH_4^+-N, the environmental risk posed by NO_2^--N release after dredging is lower.

The release of NH_4^+-N from the sediment–water interface would be enhanced after dredging. This phenomenon has been confirmed by previous studies [23,53,55]. This finding can be attributed to the NH_4^+-N distribution in the vertical profile. The concentration of NH_4^+-N increased with sediment depth (Figure 4C,D). Thus, after 30 cm dredging, buried sediment with higher concentrations of NH_4^+-N would be exposed at the sediment–water interface, forming a new surface layer. Compared with the undredged sediments, the lower porosity of the dredged sediments could prevent the release of NH_4^+-N. However, the higher concentration gradient at the sediment–water interface of the dredged cores would lead to significant increase in NH_4^+-N release [23,55]. Given the more serious pollution in Inner Bay, NH_4^+-N fluxes in the undredged cores from this area were bigger than from Outer Bay (Figure 7C,D). Similarly, after dredging, the NH_4^+-N fluxes in Inner Bay were also bigger than in Outer Bay (Figure 7C,D). Therefore, the environmental risk posed by NH_4^+-N release in Inner Bay is even greater than in Outer Bay.

The present study investigated the effect of dredging on the control of NH_4^+-N release for only a short period after dredging. In our previous study, we evaluated the long-term effectiveness of sediment dredging in reducing internal nitrogen release in Taihu Lake [55]. The results of that work indicated that the release rates of NH_4^+-N could be enhanced for several months after dredging, and that sediment dredging could be considered effective in controlling NH_4^+-N release four months after dredging [55]. In our opinion, the short-term effect of dredging on the release of NH_4^+-N depends on the pool sizes of NH_4^+-N in the interstitial water and labile NH_4^+-N in the sediments just after dredging. However, the long-term effect of dredging on the release of NH_4^+-N is determined by the rate of NH_4^+-N regeneration in the dredged sediment. Sediment dredging can have considerable influence on the physical, chemical, and biological properties of sediment, resulting in significant reduction of nitrogen mineralization and regeneration rates [19,53,56]. Thus, it can demonstrate effective long-term control of nitrogen release [19,23,53].

3.6. Optimum Dredging Season for Minimal Environmental Risk

Currently, in China, as in many other countries, the choice of dredging season is largely random. The implementation of a dredging engineering project depends mainly on government budgets. However, the influence of season on the effects and effectiveness of dredging is not considered in the decision-making process. This study showed that the concentrations of inorganic nitrogen and SRP in the interstitial water and the fluxes across the sediment–water interface have obvious seasonal characteristics, especially for the Inner Bay (Figures 4 and 7). The controlling effects of dredging on the benthic release of SRP and inorganic nitrogen are different in the short term after dredging. For the control of phosphorus release, under the premise of a reasonable depth of dredging, this study found 30-cm dredging in all seasons could effectively control phosphorus release. However, because of the higher SRP concentration in the interstitial water and the higher release rates of SRP in seasons with high temperatures (Figure 4A,B and Figure 7A,B), if the dredging depth were not appropriate, dredging projects conducted in seasons with high temperatures could have greater environmental risks in relation to phosphorus release after dredging. Therefore, in our view, the seasons with low temperature are appropriate for dredging projects intended to control internal phosphorus loading. However, greater attention should be given to determining the optimum dredging depth.

In the present study, we found the risk of release of NO_3^--N and NO_2^--N could be negligible because of low and/or negative fluxes across the sediment–water interface. However, the flux of

NH_4^+-N became enhanced for a short period after dredging because of the high concentration of NH_4^+-N in the porewater, even though the risk posed by NH_4^+-N release could be eliminated when a lake has been dredged for years [24,53], and despite sediment dredging being shown to be effective in controlling NH_4^+-N release from sediments in the long term [55,56], the environmental risks cannot be ignored, especially for the Inner Bay. The NH_4^+-N release rates of dredged cores were lowest in winter for both the Inner Bay and Outer Bay (Figure 7C,D), which means the risk of release of NH_4^+-N was minimized in winter. Moreover, winter is a dry season with shallow water depth, which facilitates the construction of dredging projects. In addition, the implementation of dredging projects in winter avoids the growth season of aquatic organisms, and facilitates the restoration of benthic animals and macrophytes after the implementation of dredging projects. In order to reduce internal nitrogen loading in Taihu Lake (and in other eutrophic lakes), we recommend the seasons with low temperature (non-growing season) as the seasons suitable for dredging engineering projects with minimal NH_4^+-N release risk.

It must be pointed out that we only evaluated the short-term effects of the dredging season on sediment properties and nutrient release at the sediment-water interface in this study. Longer-term studies are more conducive to understanding the environmental effects of dredging season. In our future studies, we will focus on the medium- and long-term effects of the dredging season in order to provide a reference for dredging projects. In addition, future research should also pay attention to the environmental effects of dredging depth, because the depth of dredging is not only related to the control effect of sediment dredging on the internal loading [18,57], but also directly related to the investment of dredging projects.

4. Conclusions

This study evaluated the potential effect of dredging season on sediment properties and nutrient release from the sediments in Meiliang Bay of Taihu Lake, China. The results indicated that 30 cm dredging could effectively reduce organic matter, total nitrogen and total phosphorus loading for both the Inner Bay and Outer Bay. The total biological activity in the dredged sediment was weaker than in the undredged sediment in all seasons for both Inner Bay and Outer Bay, and the effect of 30 cm dredging on sediment oxygen demand was negligible. Dredging in all seasons could effectively control phosphorous release from the sediment at both sites; however, ammonium release might pose a risk to the water body for a short period after dredging. The seasons with low temperature (non-growing season) are considered the most suitable time for dredging engineering projects, because of the low concentrations of nitrogen and phosphorus in the porewater. To increase the efficiency of internal loading control, greater attention should be paid to determining the optimum dredging depth in the decision-making process. The long-term effectiveness of dredging time on the control of internal loading needs further study.

Supplementary Materials: The following are available online at http://www.mdpi.com/2073-4441/10/11/1606/s1, Figure S1: Schematic diagram of fabrication process for the dredged and undredged sediment cores in the microcosm experiment; Figure S2: Schematic and physical photograph of Peeper.

Author Contributions: J.-C.Z. and J.-H.Y. designed the experiments; J.-H.Y. and X.-L.Z. performed the experiment; S.-L.W. and D.-H.L. analyzed the data; C.-X.F. provided laboratory equipment; J.-C.Z. wrote the article.

Funding: This work was supported by the National Natural Science Foundation of China (41771516, 41371457), State Major Project of Water Pollution Control and Management (2013ZX07113001), and Key Project of Chinese Academy of Sciences (KZZD-EW-10-02).

Acknowledgments: We are sincerely grateful to colleagues from the State Key Laboratory of Lake Science and Environment for providing support and help. We do appreciate the anonymous reviewers for their valuable comments and efforts to improve this manuscript.

Conflicts of Interest: The authors declare no conflict of interest.

Abbreviations

The following abbreviations are used in this manuscript:

TP	Total phosphorus
TN	Total nitrogen
OC	Organic carbon
LOI	Loss on ignition
Fe	Iron
SRP	Soluble reactive phosphorus
SOD	Sediment oxygen demand

References

1. Zhang, Y.L.; Qin, B.Q.; Liu, M.L. Tempral-spatial variations of chlorophyll a and primary production in Meiliang Bay, Lake Taihu, China from 1995–2003. *J. Plankton Res.* **2007**, *39*, 707–719. [CrossRef]
2. Song, L.R.; Chen, W.; Peng, L. Distribution and bioaccumulation of microcystins in water columns: A systematic investigation into the environmental fate and the risks associated with microcystins in Meiliang Bay, Lake Taihu. *Water Res.* **2007**, *41*, 2853–2864. [CrossRef] [PubMed]
3. Yang, M.; Yu, J.W.; Li, Z.L.; Guo, Z.H.; Burch, M.; Lin, T.F. Taihu Lake not to blame for Wuxi's woes. *Science* **2008**, *319*, 158. [CrossRef] [PubMed]
4. Lu, G.H.; Ma, Q.; Zhang, J.H. Analysis of black water aggregation in Taihu Lake. *Water Sci. Eng.* **2011**, *4*, 374–385.
5. Schindler, D. Evolution of phosphorus limitation in lakes. *Science* **1977**, *195*, 260–262. [CrossRef] [PubMed]
6. Penn, M.R.; Auer, M.T.; Doerr, S.M.; Driscoll, C.T.; Brooks, C.M.; Effer, S.W. Seasonality in phosphorus release rates from sediments of a hypereutrophic lake under a matrix of pH and redox conditions. *Can. J. Fish. Aquat. Sci.* **2000**, *57*, 1033–1041. [CrossRef]
7. Murphy, T.P.; Lawson, A.; Kumagai, M.; Babin, J. Review of emerging issues in sediment treatment. *Aquat. Ecosyst. Health* **1999**, *2*, 419–434. [CrossRef]
8. Berg, U.; Neumann, T.; Donnert, D.; Nüesch, R.; Stüben, D. Sediment capping in eutrophic lakes-efficiency of undisturbed calcite barriers to immobilized phosphorus. *Appl. Geochem.* **2004**, *19*, 1759–1771. [CrossRef]
9. Wang, M.R.; Zhang, H.Q.; Zhu, X. *Innovation and Practice of Cyanobacteria Control in Taihu Lake*; China Water Resources and Hydropower Publishing Press: Beijing, China, 2012. (In Chinese)
10. Bianchi, V.; Masciandaro, G.; Ceccanti, B.; Doni, S.; Lannelli, R. Phytoremediation and bio-physical conditioning of dredged marine sediments for their re-use in the environment. *Water Air Soil Pollut.* **2010**, *210*, 187–195. [CrossRef]
11. Gustavson, K.E.; Burton, G.A.; Francingues, N.R.; Reible, D.D.; Vorhees, D.J.; Wolfe, J.R. Evaluating the effectiveness of contaminated-sediment dredging. *Environ. Sci. Technol.* **2008**, *42*, 5042–5047. [PubMed]
12. Ryding, S.O. Lake Trehörningen restoration project, Changes in water quality after sediment dredging. *Hydrobiologia* **1982**, *9*, 549–558. [CrossRef]
13. Does, V.D.; Verstraelen, P.; Boers, P.; Roestel, J.V.; Roijackers, R.; Moser, G. Lake restoration with and without dredging of phosphorus-enriched upper sediment layers. *Hydrobiologia* **1992**, *233*, 197–210. [CrossRef]
14. Fan, C.X.; Zhang, L.; Wang, J.J.; Zheng, C.H.; Gao, G.; Wang, S.M. Processes and mechanism of effects of sludge dredging on internal source release in lakes. *Chin. Sci. Bull.* **2004**, *49*, 1853–1859. [CrossRef]
15. Cao, X.Y.; Song, C.L.; Li, Q.M.; Zhou, Y.Y. Dredging effects on P status and phytoplankton density and composition during winter and spring in Lake Taihu, China. *Hydrobiologia* **2007**, *581*, 287–295. [CrossRef]
16. National Research Council. *Sediment Dredging at Superfund Megasites: Assessing the Effectiveness*; The National Academies Press: Washington, DC, USA, 2007.
17. Kleeberg, A.; Kohl, J.G. Assessment of the long-term effectiveness of sediment dredging to reduce benthic phosphorus release in shallow Lake Müggelsee (Germany). *Hydrobiologia* **1999**, *394*, 153–161. [CrossRef]
18. Reddy, K.R.; Fisher, M.M.; Wang, Y.; White, J.R.; James, R.T. Potential effects of sediment dredging on internal phosphorus loading in a shallow, subtropical lake. *Lake Reserv. Manag.* **2007**, *23*, 27–38. [CrossRef]
19. Yu, J.H.; Fan, C.X.; Zhong, J.C.; Zhang, L.; Zhang, L.; Wang, C.H.; Yao, X.L. Effects of sediment dredging on nitrogen cycling in Lake Taihu, China: Insight from mass balance based on a 2-year field study. *Environ. Sci. Pollut. Res.* **2016**, *23*, 3871–3883. [CrossRef] [PubMed]

20. Yu, J.H.; Ding, S.M.; Zhong, J.C.; Fan, C.X.; Chen, Q.W.; Yin, H.B.; Zhang, L.; Zhang, Y.L. Evaluation of simulated dredging to control internal phosphorus release from sediments: Focused on phosphorus transfer and resupply across the sediment-water interface. *Sci. Total Environ.* **2017**, *592*, 662–673. [CrossRef] [PubMed]

21. Annadotter, H.; Cronberg, G.; Aagren, R.; Lundstedt, B.; Nilsson, P.Å.; Ströbeck, S. Multiple techniques for lake restoration. *Hydrobiologia* **1999**, *395*, 77–85. [CrossRef]

22. Chen, C.; Zhong, J.C.; Fan, C.X.; Kong, M.; Yu, J.H. Simulation research on the release of internal nutrients affected by different dredging methods in lake. *Environm. Sci.* **2013**, *34*, 3872–3978. (In Chinese)

23. Liu, C.; Shao, S.G.; Shen, Q.S.; Fan, C.X.; Zhou, Q.L.; Yin, H.B.; Xu, F.L. Use of multi-objective dredging for remediation of contaminated sediments: A case study of a typical heavily polluted confluence area in China. *Environ. Sci. Pollut. Res.* **2015**, *22*, 17839–17849. [CrossRef] [PubMed]

24. Liu, C.; Zhong, J.C.; Wang, J.J.; Zhang, L.; Fan, C.X. Fifteen-year study of environmental dredging effect on variation of nitrogen and phosphorus exchange across the sediment-water interface of an urban lake. *Environ. Pollut.* **2016**, *219*, 639–648. [CrossRef] [PubMed]

25. Chen, C.; Zhong, J.C.; Yu, J.H.; Shen, Q.S.; Fan, C.X.; Kong, F.X. Optimum dredging time for inhibition and prevention of algae-induced black blooms in Lake Taihu, China. *Environ. Sci. Pollut. Res.* **2016**, *23*, 14636–14645. [CrossRef] [PubMed]

26. Graca, B.; Burska, D.; Matuszewska, K. The impact of dredging deep pits on organic matter decomposition in sediments. *Water Air Soil Pollut.* **2004**, *158*, 237–259. [CrossRef]

27. Luijn, F.V.; Boers, P.C.M.; Lijklema, L.; Sweerts, J.P.R.A. Nitrogen fluxes and processes in Sandy and Muddy sediments from a shallow eutrophic lake. *Water Res.* **1999**, *33*, 33–42. [CrossRef]

28. Michel, P.; Boutier, B.; Chiffolear, J.F. Net fluxes of Dissolved Arsenic, Cadmium, Copper, Zine, Nitrogen and phosphorus from the Gironde Estuary (France): Seasonal variations and trends. *Estuar. Coast Shelf Sci.* **2000**, *51*, 451–462. [CrossRef]

29. Brossard, D.M.; Lhoumear, C.; Jankowska, H.P. Seasonal variability of benthic ammonium release in the surface sediments of the Gulf of Gdańsk (southern Blatic Sea). *Oceanologia* **2001**, *43*, 113–136.

30. Malecki, L.M.; White, J.R.; Reddy, K.R. Nitrogen and Phosphorus Flux Rates from Sediment in the Lower St. Johns River Estuary. *J. Environ. Qual.* **2004**, *33*, 1545–1555. [CrossRef] [PubMed]

31. Jiang, X.; Jin, X.C.; Yao, Y.; Li, L.H.; Wu, F.C. Effects of biological activity, light, temperature and oxygen on phosphorus release processes at the sediment and water interface of Taihu Lake, China. *Water Res.* **2008**, *42*, 2251–2259. [CrossRef] [PubMed]

32. Madura, K.K.; Goldyn, R.; Dera, M. Spatial and seasonal changes of phosphorus internal loading in two lakes with different trophy. *Ecol. Eng.* **2015**, *74*, 187–195. [CrossRef]

33. Némery, J.; Gratiot, N.; Doan, P.T.K.; Duvert, C.; Vlvarado-Villanueva, R.; Duwig, C. Carbon, nitrogen, phosphorus, and sediment sources and retention in a small eutrophic tropical reservoir. *Aquat. Sci.* **2016**, *78*, 171–189. [CrossRef]

34. Chen, Y.W.; Fan, C.X.; Teubner, X.K.; Dokulil, M. Changes of nutrients and phytoplankton chlorophyll-a in a large shallow lake, Taihu, China: An 8-year investigation. *Hydrobiologia* **2003**, *506*, 273–279. [CrossRef]

35. Wang, X.J.; Liu, R.M. Spatial analysis and eutrophication assessment for chlorophyll a in Taihu Lake. *Environ. Monit. Assess.* **2005**, *101*, 167–174. [PubMed]

36. Hu, W.P.; Zhai, S.J.; Zhu, Z.C.; Han, H.J. Impacts of the Yangtze River water transfer on the restoration of Lake Taihu. *Ecol. Eng.* **2008**, *34*, 30–49. [CrossRef]

37. Zhong, J.C.; Fan, C.X.; Liu, G.F.; Zhang, L.; Shang, J.G.; Gu, X.Z. Seasonal variation of potential denitrificaiton rates of surface sediment from Meiliang Bay, Taihu, China. *J. Environ. Sci.* **2010**, *22*, 961–967. [CrossRef]

38. Fan, C.X.; Zhang, L.; Yang, L.Y.; Huang, W.Y.; Xu, P.Z. Simulation of internal loading of nitrogen and phosphorus in a lake. *Oceanol. Limnol. Sin.* **2002**, *33*, 370–378.

39. Hesslein, R.H. An in-situ sampler for close interval porewater studies. *Limnol. Oceanogr.* **1976**, *21*, 912–914. [CrossRef]

40. Jacobs, P.H. A new rechargeable dialysis pore water sampler for monitoring sub-aqueous in-situ sediment caps. *Water Res.* **2002**, *36*, 3121–3129. [CrossRef]

41. Urban, N.R.; Dinkel, C.; Wehrli, B. Solute transfer across the sediment surface of a eutrophic lake: I. Porewater profiles from dialysis samplers. *Aquat. Sci.* **1997**, *59*, 1–25. [CrossRef]

42. Ruban, V.; López-Sánchez, J.F.; Pardo, P.; Rauret, G.; Muntau, H.; Quevauviller, P. Selection and evaluation of sequential extraction procedures for the determination of phosphorus forms in lake sediment. *J. Environ. Monitor.* **1999**, *1*, 51–56. [CrossRef]

43. Adam, G.; Duncan, H. Development of a sensitive and rapid method for the measurement of total microbial activity using fluorescein diacetate (FDA) in a range of soils. *Soil Biol. Biochem.* **2001**, *33*, 943–951. [CrossRef]

44. Fisher, M.M.; Reddy, K.R. Phosphorus flux from wetland soils affected by long-term nutrient loading. *J. Environ. Qual.* **2001**, *30*, 261–271. [CrossRef] [PubMed]

45. Chinese EPA. *Methods for the Examination of Water and Wastewater*, 4th ed.; China Environmental Science Press: Beijing, China, 2002. (In Chinese)

46. Rasheed, M.; Badran, M.I.; Huettel, M. Particulate matter filtration and seasonal nutrient dynamics in permeable carbonate and silicate sands of the Gulf of Aqaba, Red Sea. *Coral Reefs* **2003**, *22*, 167–177. [CrossRef]

47. McCarthy, M.J.; Lavrentyev, P.J.; Yang, L.Y.; Zhang, L.; Chen, Y.W.; Qin, B.Q.; Gardner, W.S. Nitrogen dynamics and microbial food web structure during a summer cyanobacterial bloom in a subtropical, shallow, well-mixed, eutrophic lake (Lake Taihu, China). *Hydrobiologia* **2007**, *581*, 195–207. [CrossRef]

48. Wang, H.G.; Lu, J.W.; Wang, W.D.; Huang, P.S.; Yin, C.Q. Spatio-temporal distribution of nitrogen in the undulating littoral zone of Lake Taihu, China. *Hydrobiologia* **2007**, *581*, 97–108. [CrossRef]

49. Xu, H.; Paerl, H.W.; Qin, B.Q.; Zhu, G.W.; Gao, G. Nitrogen and phosphorus inputs control phytoplankton growth in eutrophic Lake Taihu, China. *Limmol. Oceanogr.* **2010**, *55*, 420–432. [CrossRef]

50. Mermillod-Blondin, F.; Nogaro, G.; Datry, T.; Malard, F.; Gibert, J. Do tubificid worms influence the fate of organic matter and pollutants in stormwater sediments? *Environ. Pollut.* **2005**, *134*, 57–69. [CrossRef] [PubMed]

51. Jing, L.D.; Liu, X.L.; Bai, S.; Wu, C.X.; Ao, H.Y.; Liu, J.T. Effects of sediment dredging on internal phosphorus: A comparative field study focused on iron and phosphorus forms in sediments. *Ecol. Eng.* **2015**, *82*, 267–271. [CrossRef]

52. Zhong, J.C.; You, B.S.; Fan, C.X.; Li, B.; Zhang, L.; Ding, S.M. Influence of sediment dredging on chemical forms and release of phosphorus. *Pedoshpere* **2008**, *18*, 34–44. [CrossRef]

53. Jing, L.D.; Wu, C.X.; Liu, J.T.; Wang, H.G.; Ao, H.Y. The effects of dredging on nitrogen balance in sediment–water microcosms and implications to dredging projects. *Ecol. Eng.* **2013**, *52*, 167–174. [CrossRef]

54. Liu, C.; Shao, S.G.; Shen, Q.S.; Fan, C.X.; Zhang, L.; Zhou, Q.L. Effects of riverine suspended particulate matter on the post-dredging increase in internal phosphorus loading across the sediment-water interface. *Environ. Pollut.* **2016**, *211*, 165–172. [CrossRef] [PubMed]

55. Zhong, J.C.; Liu, G.F.; Fan, C.X.; Li, B.; Zhang, L.; Ding, S.M. Environmental effect of sediment dredging in lake: II. The role of sediment dredging in reducing internal nitrogen release. *J. Lake Sci.* **2009**, *21*, 335–344. (In Chinese)

56. Yu, J.H.; Fan, C.X.; Zhong, J.C.; Zhang, Y.L.; Wang, C.H.; Zhang, L. Evaluation of in situ simulated dredging to reduce internal nitrogen flux across the sediment-water interface in Lake Taihu, China. *Environ. Pollut.* **2016**, *214*, 866–977. [CrossRef] [PubMed]

57. Liu, C.; Shen, Q.S.; Zhou, Q.; Fan, C.X.; Shao, S.G. Precontrol of algae-induced black blooms through sediment dredging at appropriate depth in a typical eutrophic shallow lake. *Ecol. Eng.* **2015**, *77*, 139–145. [CrossRef]

Article

Modified Local Soil (MLS) Technology for Harmful Algal Bloom Control, Sediment Remediation, and Ecological Restoration

Gang Pan [1,2,3,4,5,6,*], Xiaojun Miao [1,2], Lei Bi [1], Honggang Zhang [1], Lei Wang [1], Lijing Wang [1,5], Zhibin Wang [1,5], Jun Chen [1,2], Jafar Ali [1,2], Minmin Pan [1,6,7], Jing Zhang [1], Bin Yue [3,4,8] and Tao Lyu [3,4,*]

[1] Research Center for Eco-Environmental Sciences, Chinese Academy of Sciences, Beijing 100085, China; miaoxiaojun22@126.com (X.M.); leibi@rcees.ac.cn (L.B.); hgzhang@rcees.ac.cn (H.Z.); leiwang@rcees.ac.cn (L.W.); wanglijing@basic.cas.cn (L.W.); wangzhibin@basic.cas.cn (Z.W.); 13426003723@126.com (J.C.); jafaraliqau@gmail.com (J.A.); panminmin@hotmail.com (M.P.); jingzhang@rcees.ac.cn (J.Z.)

[2] University of Chinese Academy of Sciences, Beijing 100049, China

[3] School of Animal, Rural, and Environmental Sciences, Nottingham Trent University, Brackenhurt Campus, Nottinghamshire NG25 0QF, UK; bin.yue@ntu.ac.uk

[4] Centre of Integrated Water-Energy-Food studies (iWEF), Nottingham Trent University, Nottinghamshire NG25 0QF, UK

[5] Beijing Advanced Sciences and Innovation Center, Chinese Academy of Sciences, Beijing 101407, China

[6] Sino-Danish College of University of Chinese Academy of Sciences, Beijing 100049, China

[7] Department of Environmental Engineering, Technical University of Denmark, DK-2899 Lyngby, Denmark

[8] College of Geography and Environmental Engineering, Lanzhou City University, Lanzhou, Gansu 730070, China

[*] Correspondence: gang.pan@ntu.ac.uk (G.P.); tao.lyu@ntu.ac.uk (T.L.)

Received: 10 May 2019; Accepted: 25 May 2019; Published: 29 May 2019

Abstract: Harmful algal blooms (HABs), eutrophication, and internal pollutant sources from sediment, represent serious problems for public health, water quality, and ecological restoration worldwide. Previous studies have indicated that Modified Local Soil (MLS) technology is an efficient and cost-effective method to flocculate the HABs from water and settle them onto sediment. Additionally, MLS capping treatment can reduce the resuspension of algae flocs from the sediment, and convert the algal cells, along with any excessive nutrients in-situ into fertilisers for the restoration of submerged macrophytes in shallow water systems. Furthermore, the capping treatment using oxygen nanobubble-MLS materials can also mitigate sediment anoxia, causing a reduction in the release of internal pollutants, such as nutrients and greenhouse gases. This paper reviews and quantifies the main features of MLS by investigating the effect of MLS treatment in five pilot-scale whole-pond field experiments carried out in Lake Tai, South China, and in Cetian Reservoir in Datong city, North China. Data obtained from field monitoring showed that the algae-dominated waters transform into a macrophyte-dominated state within four months of MLS treatment in shallow water systems. The sediment-water nutrient fluxes were substantially reduced, whilst water quality (TN, TP, and transparency) and biodiversity were significantly improved in the treatment ponds, compared to the control ponds within a duration ranging from one day to three years. The sediment anoxia remediation effect by oxygen nanobubble-MLS treatment may further contribute to deep water hypoxia remediation and eutrophication control. Combined with the integrated management of external loads control, MLS technology can provide an environmentally friendly geo-engineering method to accelerate ecological restoration and control eutrophication.

Keywords: harmful algal blooms (HABs); eutrophication; internal loads control; oxygen nanobubble; shallow lakes; deep water hypoxia

---

1. Introduction

Harmful algal bloom (HAB) is one of the detrimental consequences of eutrophication in water bodies, which may pose serious threats to water quality, human and animal health, economic development, ecological balance, landscape aesthetics, and even social stability [1]. The cause of the HABs is complex, and can be mainly due to agriculture, industrial, and other anthropogenic activities, which cause an ever-increasing amount of nutrients to run off from inlands into natural waters [2]. Hence, the mitigation of the problem is complicated, and integrated management is often essential. External loads control and integrated basin management have been demonstrated to be important practices to prevent the continuous input of nutrients from land into water [3]. However, when external loads are under control, the internal loads from polluted sediment often limit the improvement of water quality [4], and the latter is becoming compulsory in many countries by local governance or environmental law. Consequentially, in some countries such as China, this has triggered a large environmental market which is initially driven by government investments and followed by the private sector. The natural restoration of lake ecology or water quality is often a lengthy process that is far beyond the scope of the targets set within one government's term. In this regard, geo-engineering materials, such as modified clays or soils, may accelerate the improvement of water and sediment quality in an ecologically-friendly manner [5,6]. This is because the particle-water interaction of suspended particles often represents an important and natural self-cleaning mechanism that scavenges pollutants from natural waters [7]. In return, the improvement in sediment environment and water quality paves the way for ecological and biodiversity restoration.

Over the last few decades, we have developed a series of modified local soil (MLS) materials which can be used to remove HABs and remediate polluted sediment [8]. The multidisciplinary and multifunctional principle of MLS technology is schematically illustrated in Figure 1. The MLS materials are made from clean local soil, or commercially available clay/solid particles [9–11], that are modified by a very small amount of natural products (usually less than 1% of the soil), such as chitosan [12,13], cationic starch [14], *Moringa oleifera* [15], and xanthan [16]. The solids can be local soils [12,13], local sands [9], clays [8], red soil, [17] or coal fly ash [10]. MLS is also used for capping and locking the algae flocs or nutrients in the sediment, so that the nutrient release from sediment to water can be reduced [18]. After the MLS flocculation and capping treatment, the decomposition of algae toxins remarkably increases in the sediment compared to bulk water treatment [19]. Subsequently, when the water clarity is achieved, the capped algal flocs and associated nutrients can be utilised in-situ as fertilisers for the growth of submerged macrophytes in shallow water systems [20,21], which is beneficial as macrophytes can support a healthier food chain than algal blooms. Therefore, MLS technology has made it possible to put algae along with the excessive nutrients into the food chain by flocculating and removing them from water into sediment, and converting them into submerged vegetation [20,21]. Recently, we have developed an interface oxygen nanobubble technology [22,23], which allows the loading of a substantial amount of oxygen into the microporous surfaces of clays. These clays can stably hold the nanobubbles for months, so that once the oxygen is delivered by settling the clay particles onto the sediment, an aerobic capping layer is formed on top of the anaerobic sediment. This oxidized capping layer can prevent the consumption of dissolved oxygen in the water column by the anaerobic substances from the sediment underneath the capping layer [24,25]. Furthermore, it has been demonstrated that emissions of the greenhouse gases CH_4 and CO_2 from eutrophic waters can be substantially reduced through the oxygen nanobubble-MLS treatment [26]. Thus, the oxygen nanobubble-MLS treatment may provide a principle for delivering oxygen down into deep waters or sediment without the need for mixing energy (e.g., pumps) to combat anoxia/hypoxia problems that are crucial for eutrophication control [27].

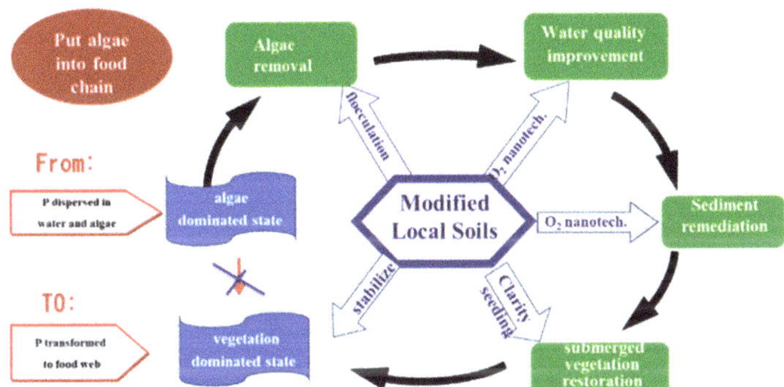

Figure 1. Multi-disciplinary principle of Modified Local Soil (MLS) technology.

The objective of this study is to examine the short-term and mid/long-term effects of MLS technology through a series of pilot whole-pond field experiments in the watersheds of South and North China, respectively. We anticipate that the replicated short term and long term mornitoring results of MLS engineering will provide further insights, and serve as a showcase for local governments, engineers, and researchers who require support for sustainable lake restoration strategies.

2. Materials and Methods

2.1. Materials

The soils/clays used in the field experiments were collected from the local banks of Lake Tai (South China) and the Cetian Reservoir (North China), respectively. Local soils were washed and screened with a self-designed flotation facility to remove floating substances, and to select suitable particle size fractions (~70 μm) of the soil [20]. Chitosan (solid) was obtained from Qingdao Haisheng Bioengineering Co., Ltd (deacetylation degree was 83.6%). Polyaluminum chloride (PAC) was supplied by Dagang Reagent Plant, Tianjin, China. The basicity (B = [OH]/[Al]) of PAC was 2.4 and its Al_2O_3 content was 30%.

2.2. Experimental Sites

The whole-pond experiments were conducted in two specifically constructed field experimental sites: 1) Lake Tai, Wuxi city, South China; and 2) Cetian Reservoir, Datong city, North China. Lake Tai is the third-largest freshwater lake in China, which has been suffering from annual harmful algal blooms (HABs) for the last several decades. Five enclosures with a total area of 5000 m^2 (4 x 400 m^2, 1 x 3400 m^2) were constructed in 2012 in Tanxi Bay for the proposed research (Figure 2). A further three open-water areas in Lake Tai, i.e. Meiliang Bay, Mashan Bay, and Shibai Bay, were also selected for field-scale studies. The Cetian research site consists of eight equally sized water ponds (800 m^2 each, with a maximum depth of 1.7 m), 12 large mesocosm systems (Φ2.5 m each with a height of 2.5 m), and an on-site indoor laboratory equipped with analytical instruments and monitoring facilities (Figure 3). The replicable ponds allowed for the monitoring of environmental responses at an ecologically meaningful scale for the study of Modified Local Soil (MLS) treatments. The two different field sites were chosen because the dominant phytoplankton and ecological status vary under different geological and climate conditions.

Figure 2. Experimental site in Tanxi bay of Lake Tai.

Figure 3. Experimental site in Cetian Reservoir.

2.3. Preparation of Modified Local Soil

After a jar test with in-situ water samples, chitosan alone and chitosan-PAC were selected to modify the soil for the experiment held in Lake Tai and Cetian Reservoir, respectively. For the chitosan-MLS preparation, the required amount of chitosan flakes were dissolved in 0.5% HAc and stirred until all the chitosan was dissolved. This solution was diluted with in-situ pond water to obtain a final concentration of 1 g/L before use. The required volume of chitosan solution was then added to the soil suspension (diluted using pond water) to make up a final concentration of 100 mg soil/L and 3 mg chitosan/L in the pond [12,13]. For the Cetian Reservoir pond experiment, chitosan-PAC MLS was used by adding dissolved PAC to chitosan-MLS to achieve a final concentration of 100 mg soil/L, 10 mg PAC/L, and 5 mg chitosan/L in the pond [9].

2.4. Treatment and Sample Analysis

This paper summarises five engineering projects conducted by us in Lake Tai (South China) and Cetian reservoir (North China), where MLS technology was used for HABs removal, sediment remediation, and ecology restoration (Table 1). A custom-made spraying facility was used to spray the suspension of

MLS materials on the surface of the water ponds. After the treatment, water and/or sediment samples were collected from each experiment site to evaluate: 1) Water quality, 2) sediment, and 3) ecology response. Following the flocculation, a 1.5 cm thick capping was applied in the Lake Tai experiment (projects 3 and 4) using local soil. For the Cetian experiment (project 5), a 1 cm capping was applied using oxygen nanobubble-modified zeolite, which was prepared following the method described by Zhang, et al. (2018) [24]. Briefly, zeolite was placed in a pressure-resistant, airtight container to create a vacuum (-0.08 to -0.1 MPa for 2 h), followed by oxygen nanobubble loading (0.12 to 0.15 MPa for 4 h), which was repeated three times to achieve super-saturation of O_2 in the zeolite micropores.

Table 1. The five engineering projects.

Project	Time	Site	MLS Material	Monitoring	Duration
1	2008	Lake Tai- Meiliang Bay	chitosan MLS	Water (visually)	1 day
2	2009	Lake Tai- Mashan Bay	chitosan MLS	Water (visually)	1 day
3	2012	Lake Tai- Tanxi Bay	chitosan MLS	Water	20 days
4	2010	Lake Tai- Shiba Bay	chitosan MLS	Sediment	1 year
5	2012	Cetian Reservoir	chitosan-PAC MLS	Water Ecology	70 days 3 years

Water samples were collected to analyse chlorophyll-*a*, turbidity, dissolved oxygen (DO), chemical oxygen demand (COD), transparency, and concentrations of nutrients (TP, TN, NH_4^+-N, NO_3^--N, and NO_2^--N) at different time intervals. Turbidity was analysed with a portable turbidity meter (HANNA, HI98713). The DO was measured using Yellow Springs Instruments (YSI, Proplus). The transparency was detected by a Secchi Disc. COD was determined through Hach test kits with a UV–vis Spectrophotometer (DR 5000, Hach, USA). TP was determined using the potassium persulfate digestion-Mo-Sb-Vc colorimetric method; TN using an alkaline potassium persulfate digestion–ultraviolet spectrometer; NH_4^+-N with Nessler's colorimetric; and NO_3^--N and NO_2^--N using the ultraviolet colorimetric method with and without cadmium column reduction, respectively [28]. Chlorophyll-*a* (Chl-*a*) was determined fluorometrically after filtering 1 L samples through Whatman GF/C filters and extracting the chlorophyll into cold methylated spirit (99% IMS: 95% ethanol, 4% methanol). The concentration of Chl-*a* was calibrated against direct microscope cell counts to monitor the concentration change of algae cells. Sediment samples were collected using plexiglass cylinders with an inner diameter of 8.4 cm and height of 50 cm. The concentration of PO_4^{3-}-P in the pore water of the sediment was determined using Peepers. The device was deployed for 2 days in the sediments for equilibration, to sample pore waters at a vertical resolution of 2 mm. The submerged vegetation samples were randomly collected from a 1 m^2 area of each pond. The fresh biomass of different plant species was measured using a weighing balance. Additionally, phytoplankton samples were fixed with Lugol's iodine solution (1.5% final conc.) and settled for 24 hours. Cell density was measured with a Sedgwich-Rafter counting chamber under a microscopic magnification of x400, and the phytoplankton species were identified according to Hu and Wei (2006) [29].

3. Results

3.1. Water Quality Improvement and Algal Bloom Control

3.1.1. Emergent HABs Removal

An engineering application was conducted under the witness of the Mayor and an expert panel of Wuxi City to clean up heavy algal blooms at two different areas of Lake Tai in 2008 (Figure 4a) and 2009 (Figure 4b), respectively. Chitosan-MLS (approximately 2000 kg) was sprayed for 30 minutes using a first generation MLS spraying boat over an enclosure of 50,000 m^2 in Meiliang Bay, Lake Tai in August 2008 (project 1 in Table 1). In 2009, a purpose-built automatic second generation MLS spraying boat was manufactured and used in open waters at Mashan Bay, Lake Tai for an emergency treatment of HABs on the request of the local government in preparing for an important local event (project 2 in Table 1). Visually, the HABs were rapidly eliminated, and the water clarity increased after a few hours of the MLS treatment.

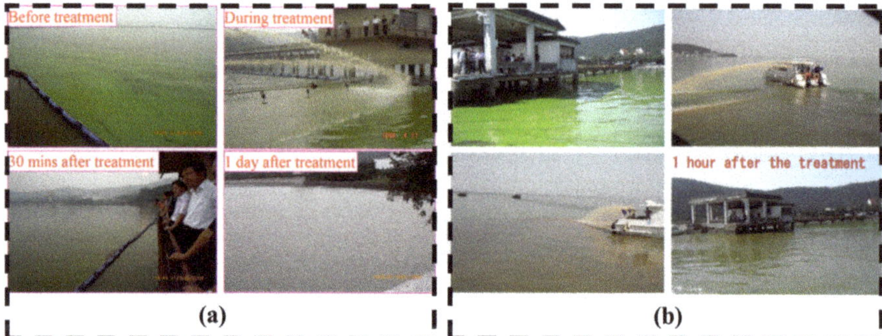

Figure 4. Application of MLS in (**a**) an enclosure with total area of 50,000 m^2 in Meiliang Bay in August 2008, and (**b**) in open water at Mashan Bay of Lake Tai in August 2009.

3.1.2. Short-Term Water Quality Improvement

In 2012, a controlled pilot field experiment was conducted in the water pond at the Tanxi bay (Figure 5) of Lake Tai (project 3 in Table 1). Approximately 16 kg of chitosan-MLS was sprayed into the water pond (400 m^2, 1.5 m depth). The Secchi depth of the pond water was less than 5 cm before the treatment. After the chitosan-MLS treatment, the algal blooms were removed from the pond within 2 hours, and the transparency of the pond water monitored by a Secchi Disc reached 1.5 m on the second day post treatment (Figure 5a).

Figure 5. Application of MLS treatment at the Tanxi Bay research site. (**a**) Visual change of the treatment pond in 1 day; and (**b**) control (left) and treatment (right) ponds after 20 days.

The short-term changes in water quality were monitored and compared between the treatment and control ponds (Figure 5b) for a period of 20 days. The Chl-*a* concentration in the treatment pond decreased from 85 µg/L to 13 µg/L, and was maintained below this level for 20 days after the treatment, while the Chl-*a* concentration in the control pond continually increased, reaching a concentration of 350 µg/L on day 20 (Figure 6). The turbidity was reduced from 95 NUT to 5.3 NUT in the treatment pond, while it was maintained above 100 NUT in the control pond during the same period (Figure 6). The COD and nutrient concentrations, including TN, TP, $NO_3^- -N$, $NO_2^- -N$, and $PO_4^{3-} -P$, reduced significantly in the treatment pond compared to the control pond (Figure 7). The significant increase of

Chl-a and TN concentrations at day 20 may be due to the fast growth of the algae in the control pond, which was shown to be largely removed in the treatment ponds.

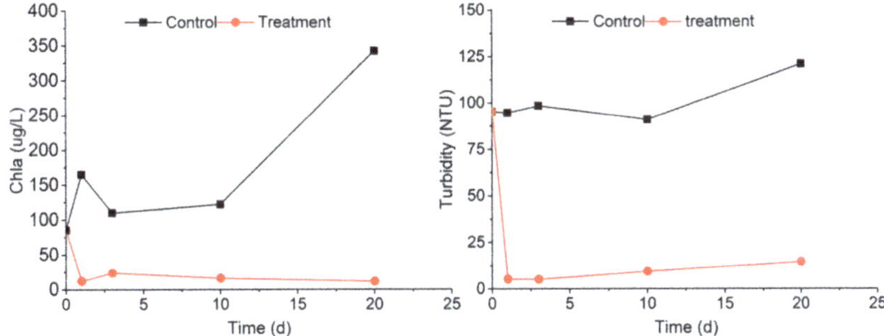

Figure 6. Chl-α concentration and Turbidity changes in the control and treatment ponds.

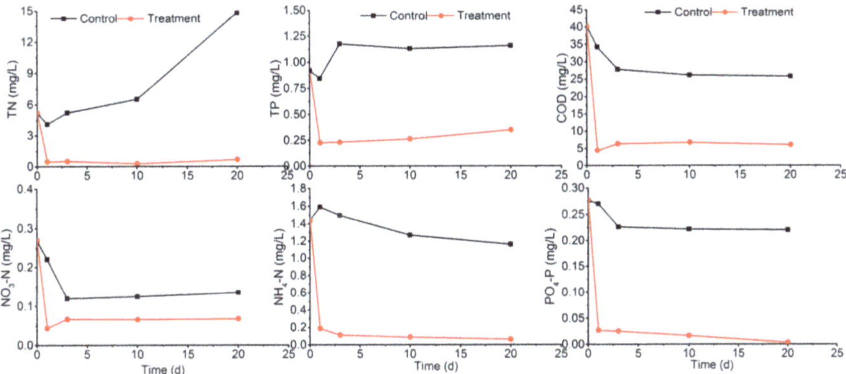

Figure 7. The concentrations of TN, TP, COD, NO_3^-–N, NO_2^-– N, and PO_4^{3-}– P in control and treatment ponds.

3.1.3. Mid-Term Water Quality Improvement

A mid-term whole-pond experiment was conducted in the Cetian Reservoir research site in July 2012 (Figure 8, project 5 in Table 1). Approximately 32 kg of chitosan-PAC MLS was sprayed into the treatment pond (800 m^2, 1.5 m depth). The water quality parameters in the treated and control ponds after 70 days are presented in Table 2. The water quality in the control ponds remained as a grade bad-V (a Chinese natural water quality standard) during the 70 days monitoring period. In the treatment ponds, the TN, TP, NH_4^+–N, and COD concentrations improved to a grade III (Table 2).

Table 2. The mid-term effect of MLS treatments on water quality after 70 days in the control and treatment ponds.

	pH	DO	TN	TP	COD	NH$_4$$^+$-N
		\multicolumn{5}{c}{mg/L}				
Treated Pond	9	10.8^I	0.8^{III}	0.08^{II}	14.5^{II}	0.4^{II}
Std.	0.4	1.4	0.3	0.03	2.1	0.2
Control Pond	8.9	5.9^{II}	3.1 ^{bV}	0.5 ^{bV}	58 ^{bV}	1.1^{IV}
Std.	0.5	0.9	1.8	0.3	8.5	0.3

Note: I, II, III, IV and b^V represent the water quality grades based on environmental quality standards for surface water of China.

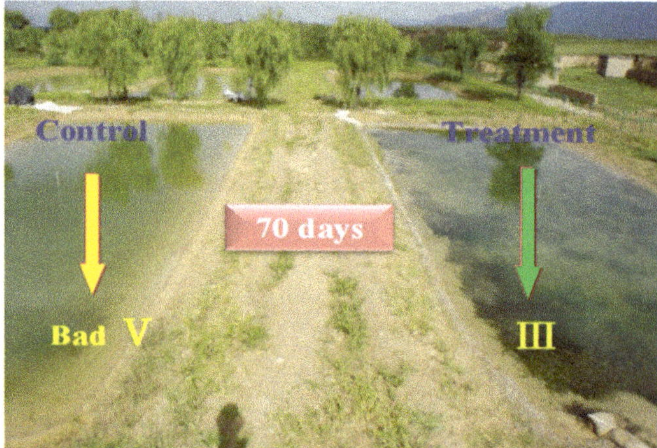

Figure 8. Control (**left**) and treatment (**right**) ponds at Cetian Reservoir.

3.2. Sediment Remediation

A one year monitoring experiment was conducted in open water at Shiba Bay of Lake Tai in 2010 (project 4 in Table 1). The chitosan-MLS capping material was applied to a 4 m^2 sediment in the open lake, so that the overlying water condition remained the same as the surrounding untreated algae water, but the 4 m^2 sediment differed from the surounding sediment throughout the year. After the MLS capping treatment, a 1.5 cm capping layer was formed (Figure 9). Sediment-water fluxes of nutrients in the capping area were monitored and compared with the surrounding untreated sediment. During summer and autumn, $PO_4^{3-}-P$ flux across the sediment-water interface was largely reduced in the treated area (4 m^2) compared to the untreated surrounding area (Figure 10).

Figure 9. Sediment sampling in the open water in Shiba Bay of Lake Tai after the MLS treatment.

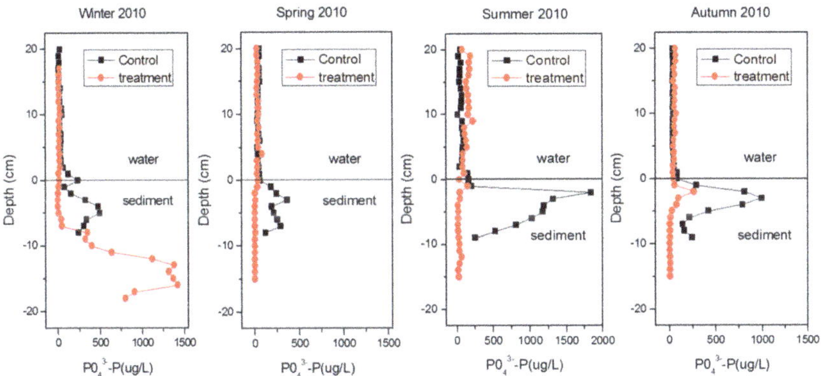

Figure 10. Changes of PO_4^{3-}–P profile throughout the year at the treated sediment area (**red**) and untreated sediment (**black**).

3.3. Long Term Ecological Restoration

The ecological responses of the MLS treatment was monitored at the Cetian research site (Figure 11, project 5 in Table 1). Compared to the control pond, the submerged vegetation had been successfully restored in the treatment ponds. After four months, following the chitosan-PAC MLS treatment, the total biomass of submerged vegetation increased from 150 to 1031 g/m^2 and the aquatic vegetation coverage rate exceeded 65% (Figure 12a). Whereas in the control pond, the total biomass of submerged vegetation was 75 g/m^2 and the aquatic vegetation coverage rate was less than 10%. The pond was successfully transformed from an algae-dominated state to a vegetation-dominated state. In order to maintain the balance of the aquatic ecological system, the submerged vegetation was harvested once annually by mechanical machines in the 2nd year and 3rd year, which helped to remove nutrients from the water, and maintain the ecological balance and water clarity. The dominant species of macrophytes in the treatment pond were *P. malaianus*, *P. pectinatus* and *P. crispus* (Figure 12b).

The composition and cell count of the phytoplankton in the control and treatment ponds after the MLS treatment are shown in Figure 13. *Chlorophytes* were the dominant phytoplankton in all the ponds before the treatment, which was kept the same in the control ponds from June to October. One month after the MLS treatment, about 75% of the algal cells were removed from the water column, and the percentage of the dominant Chlorophytes was reduced from 72% to 48%. During July and August, *Cyanophytes* became the dominant species, and *Chlorophytes*, along with *Bacillariophytes*, remained as the sub-dominant phytoplankton. During September and October, the percentage of *Bacillariophytes* increased due to the decreasing water temperature. Hence, throughout the monitoring period, the phytoplankton biodiversity increased, and the total algal cells decreased in the treatment pond.

Figure 11. Ecological responses monitoring experiment at Cetian research site after MLS treatment.

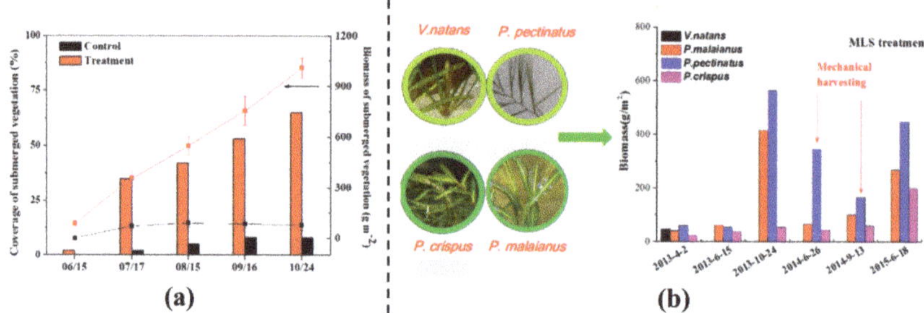

Figure 12. (**a**) The coverage and biomass of submerged vegetation in the control and treatment ponds after 4 months and (**b**) a 3-year monitoring result on submerged vegetation restoration in the treated pond.

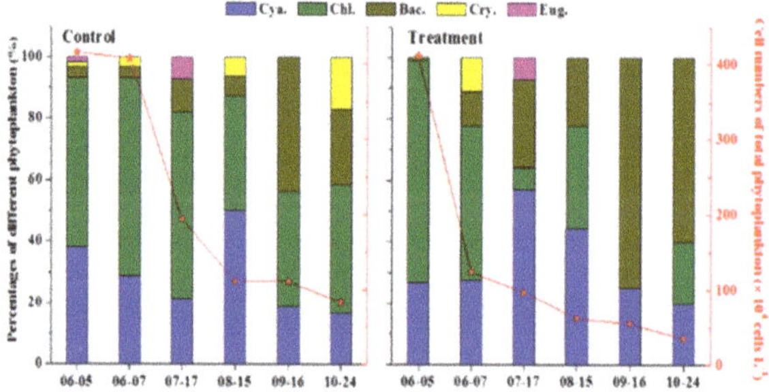

Figure 13. Changes of the dominant phytoplankton composition in the control and treatment ponds.

4. Disscussion

4.1. Water Quality

The principles used in MLS technology (Figure 1) for water quality improvement is flocculation for particulate pollutants (e.g., algae cells) and adsorption for dissolved pollutants. Suspended particles are ubiquitous in natural waters [7]; they play important roles in expelling pollutants through the aforementioned principles, i.e. particle-particle interactions (flocculation) or particle-water interactions (adsorption). We have found that many natural products, such as chitosan [12,13], cationic starch [14], *Moringa oleifera* extracts [15,30], and xanthan [16], can greatly enhance these interactions, and thus be used to modify solid particles for algae flocculation. These solid particles are important for increasing the number of collisions between particles and algae cells for flocculation, as well as providing a ballast for settling. MLS is an effective way to retain high concentrations of flocculants on solid surfaces whilst maintaining a relatively low concentration in solution. This can increase the effective collisions between particles so as to avoid a high concentration of flocculants needed in water. The solids used can be local soils [12,13], clay [8], sands [9], industrial by-products [10], and other cheap and safe solids [31]. The effect of flocculation is often affected by the type of algae and water composition (e.g., salinity levels [13,15]). Therefore, a jar test is always nessasary before any engineering work commences, in order to decide the optimized dosage and composition of the MLS materials. Understanding the flocculation mechanism is also important for finding the appropriate recipes. The formation of small floppy flocs using inorganic flocculants is often not enough to achieve an obvious engineering effect in the field due to natural disturbances (e.g., wind and current). The polymer modifiers with net and bridging properties are important to form large flocs, which can either sink the flocs to the bottom using clay/soil ballasts, or harvest the flocs via floating bubbles [1]. In Lake Tai, the dominant blue-green algae is the cyanobacteria such as *Microcystis aeruginosa*, which often exists in a colonised form. Chitosan-MLS using local soil is very effective in cleaning up colonised *Microcystis aeruginosa* in fresh waters (projects 1–3). In the Cetian reservoir (project 5), *Chlorophytes* were the dominant phytoplankton. For blooms in high salinity waters, a chitosan and PAC combined modification of local soil is needed for a more effective clean-up.

It is well documented that nutrient limitation is important for the management of eutrophication [2]. Using the same pond at the Tanxi research site (project 3 in Table 1), we found that nutrient limitation can be manipulated using MLS treatment. The nutrient limitation in the whole-pond experiment was shifted from nitrogen (N) and phosphorus (P) co-limitation to P limitation within six months after MLS treatment compared to the control pond [32].

4.2. Sediment Remediation

4.2.1. Shallow Waters

Internal sediment nutrient release is one of the main reasons for eutrophication in shallow lakes. When algae biomass flocculates from the water to the sediment, it reduces the immediate harmful effect of the HABs in water, but increases the sediment loading, which holds a high risk of returning the algae back to the water column via resuspension or diffusion. An MLS capping layer can not only prevent the resuspension of the algal flocs, but also reduce the nutrient release into the overlying water from the sediment [18]. Microorganism-modified MLS is also effective in decomposing algae toxins, because MLS flocculation will result in the intense accumulation of microbes and algae cells under the capping layer, as compared to that in the bulk water [19]. For shallow water sediment, the MLS capping aims to provide a 'window period' (usually for a few months in shallow waters) to reach appropriate water and sediment quality levels for the restoration of submerged vegetation [21].

4.2.2. Deep Waters

The flux of nutrients from sediment to water is very sensitive to redox reactions. Many elements such as P, N, C, S, and other metal ions tend to be released from sediment to water under anoxic conditions. Current remediation strategies for deep water hypoxia are mainly based on the delivery of oxygen using artificial energy (e.g., pumping). For deep large water bodies, the use of mixing energy (e.g., ventilation of surface water down to deep water) can be very laborious, expensive, and carry a risk of triggering the resuspension of anaerobic substances and a change of water stratification. The interfacial oxygen nanobubble technology can cheaply and safely deliver oxygen to deep water sediment due to the clay gravity, without the need for these mixing energies [24]. The oxygen-locking layer between the overlying water and sediment could play important roles in both the physical isolation and chemical fixation of pollutants in the sediment. The oxygen-locking layer could affect many geochemical and microbial processes, thereby opening a new possibility to remediate sediment, especially for deep waters, which are usually very difficult. For instance, oxygen nanobubble–MLS can be used to remediate the sediment to reduce methane emissions [26]. Another effective way to lock phosphorus in the sediment is to use Phoslock, which has so far been tested in about 200 water bodies worldwide [31]. We have recently developed new geo-engineering materials, which have a P removal capacity 5 to 8 times higher than the commercially available Phoslocks [33,34]. The principle of MLS for sediment remediation (Figure 1) is to use natural geo-engineering materials to reduce the resuspension of the sediment, to adsorb and lock phosphorus, and alter redox conditions at sediment-water interfaces to control the internal loads. We believe the oxygen nanobubbles-MLS will attract particular attention in the future, due to its great potential for deep-water sediment remediation, which is a serious and challenging issue worldwide.

4.3. Ecological Restoration

It is well recognised that TP reduction is one of the key factors in controlling the alternation between an algae-dominated state and a vegetation-dominated state [35]. The restoration of submerged macrophytes in polluted water can be very slow under poor water and sediment conditions. MLS can be used to decrease the TP quickly so as to accelerate the ecological restoration on a large engineering scale. Once a stabilised and optimised vegetation state is established in shallow waters, it will help to maintain the water quality on a long-term basis. Nutrients are needed for the growth of submerged vegetation, but high TP levels will only lead to the degradation of submerged macrophytes, as the algae holds the priority to use the nutrients to become a bloom. MLS flocculation and capping can help redistribute the nutrients in such a way that once nutrients are taken up by the algae cells, they can be flocculated and buried in the sediment. The buried nutrients under the MLS capping layer can be utilised by submerged macrophytes compared to that in water. This was demonstrated in the treatment pond at the Cetian research site (project 5 in Table 1).

The restored macrophytes of *P. malaianus*, *P. pectinatus* and *P. crispus* in project 5 have a strong nutrient assimilation ability. The concentrations of TN and TP in the treatment pond were much lower than those in the control pond, and the TP in the treatment pond remained lower than 0.1 mg/L (Table 2) during the monitoring period. Meanwhile, the total biomass of the submerged vegetation increased and the cell numbers of phytoplankton decreased continuously (Figures 11 and 12). Submerged vegetation was responsible for the water quality in the mid- and long-term, because they assimilated nutrients and inhibited algae reproduction in the ecosystem. For shallow water engineering, the restored submerged macrophytes triggered by MLS could be regularly harvested and managed, so that the nutrients can be continually removed from the water while sustaining the balance and health of the ecological system and sediment respectively. Once the algae is turned into vegetation, it is then much easier to harvest macrophytes using mechanical management than it would have been to harvest the algal blooms. Additionally, the levels of indigenous macrophyes may be kept naturally balanced by fish through the food chain.

In a previous engineering study, a 100,000 m^2 whole bay with heavy algal blooms located in Liangyangyuan Bay in Lake Tai was treated by MLS, and the biodiversity index steadily increased within the six months monitoring period after the treatment [20]. In practice, it is often necessary to selectively control fish, as they can seriously influence the treatment effect of MLS especially at the sediment remediation stage before the vegetation system has established in shallow waters.

4.4. Cost Analysis

MLS technology provides several functions, including HABs removal, water quality improvement, sediment remediation, O_2 supply, and ecological restoration (Fig. 1). The cost of implementation depends on the needed functions of individual engineering project. For HABs flocculation, various particle materials, such as kaolinite and local soil, may be used depending on their local price and availability. Short-term HABs removal using MLS is very cost-effective compared to other existing methods. The price of flocculating apparent HABs in 0.5 m surface water is between US\$ 30–60K per km^2; the price is mainly affected by the modifier used (Table 3). The modification recipe may vary depending on the water quality (e.g., salinity levels) and the dominant algal species. For example, the chitosan-MLS and chitosan-PAC MLS is appropriate for the eutrophic lakes dominated by *Microcystis aeruginosa* [8] and *Chlorophytes* in projects 3 and 5 (Table 1), respectively. As the treatment area increases, it is more feasible to make use of local soil rather than commercial clays with respect to effect and cost. Submerged macrophytes are key for long-term maintenance of water clarity in shallow waters, but they can be hardly restored in algae dominated systems. To achieve such long-term effects, capping is often necessary following flocculation, especially in shallow waters. Table 3 shows that capping using clean local soils or sands (US\$ 0.9 M/km^2) is much cheaper than using commercial clays such as granule zeolite (US\$ 1.8 M/km^2). For deep water systems, the effect of flocculation alone may last much longer than in shallow waters. However, to prevent the reoccurrence of HABs, the sediment nutrient release needs to be fixed, which is where interfacial oxygen nanobubbles may be particularly useful [24]. The cost of oxygen needed in oxygen nanobubble-modified zeolite is much lower than the cost of zeolite itself, which is about US\$ 30K/km^2 (Table 3). The operational costs (including the boat, fuel, and labour) to mechanically spray the MLS materials is about US\$ 1.5K/km^2. To sumarize, for a short-term algal clean-up using MLS and mid-term capping treatment using zeolite, the total cost is about US\$ 1.9 M/km^2. The cost heavily depends on the thickness of the capping materials needed, rather than the flocculation of the HABs.

Table 3. Summary of the cost per km^2 of field implementation of Modify Local Soil (MLS) technology.

Particle Materials		Flocculation by MLS[1]		Capping[4]	Extra O_2 Supply[5]	Operation
Types	Price ($/t)	Chitosan MLS[2] (k$/km²)	Chitosan-PAC MLS[3] (k$/km²)	Price (k$/km²)	O_2 nanobubble Modification (k$/km²)	Mechanic Spray (k$/km²)
Kaolinite	135	38	59	2024	-	1.5
Local soil	60	35	56	900	-	1.5
zeolite	120	-	-	1799	30	1.5

Note: [1] Flocculation of apparent harmful algal blooms (HABs) in 0.5 m surface water; [2] Based on Lake Tai projects (project 1–3 in Table 1); [3] Based on the Cetian Reservoir project (project 5 in Table 1); [4] Based on 1 cm depth of capping; [5] Based on project 5 of 1 cm capping of O_2 nanobubble-modified zeolite.

4.5. Ecological Safety

So far, natural clay or soils used as a ballast have not shown any negative side-effect on aquatic ecological systems. Continuous supply of high concentration suspended particles in natural rivers and lakes may cause siltation, but not ecological problems. Rather, suspended particles are important for the transport and transformation of elements in natural waters [7,36]. The amount of soil/clay used in MLS treatments is lowcompared to the natural conditions of many rivers (e.g., Yellow River). When clean soil/clay settles on heavily polluted sediment, it often represents an improvement of sediment quality. There have been concerns regarding the use of PAC, a chemical reagent used in drinking water treatment, in natural waters. Existing toxicity studies illustrate that PAC is among the most non-toxic flocculants for aquatic organisms [37,38]. However, the long-term ecological effect for the non-biodegradable compound remains a concern, and requires further studies. When chitosan is not used together with clay/soil, it may have antimicrobial properties against some bacteria [39] including cyanobacteria species [40]. Our recent study found that the combination of chitosan with natural soils could greatly reduce the toxicity to the aquatic organisms exerted by chitosan alone [38]. This is because MLS makes it possible to keep the concentration of chitosan low in the water, while remaning high on soil particles, which is essential to maintain a high flocculation efficiency whilst maintaining a low toxic effect. When capping is used following the flocculation treament, lysis of algae cells may take place, which is benefcial in avoiding a re-growth of the algae and the degradation of algae toxins [19]. We have also found that some industrial by-products with a relatively high content of aluminum or iron ions, such as coal fly ash, could also be used to modify soils [10], although their potential environmental risk requires further invsetigation.

For practical lake restoration engineering, conditions are often complex and variable. There are many interfering factors in the field (such as wind and bio-disturbance) and it is very important to study these influences and their counter measures before implementing an artificial geo-engineering treatment.

5. Conclusions

The short and mid/long-term effects of MLS treatments were investegated through 5 whole-pond or open-lake scale experiments. Results showed that, combined with an integrated management of external loads control, MLS can be used as an in-lake technology for lake restoration through its multiple functions of water quality improvement (projects 1–3), sediment remediation (project 4), and ecological restoration (project 5). These findings can serve as useful guidelines for researchers, engineers, and local governments for controlling eutrophication and accelaerating lake restoration in an eco-friendly and sustainable manner.

Author Contributions: Conceptualization, G.P.; Funding acquisition, G.P.; Investigation, G.P.; Methodology, L.B., H.Z., Lei Wang, Lijing Wang, Z.W., X.M., J.C., J.A., M.P., J.Z., and B.Y.; Project administration, G.P.; Writing—original draft, G.P., L.B., H.Z., Lei Wang, and T.L.; Writing—review & editing, G.P., Lijing Wang, B.Y. and T.L.

Acknowledgments: This research was supported by the National Key Research and Development Program of China (2017YFA0207204), the Key R&D program of MoST (2002AA601011), Chinese National Basic Research Program (2002CB412308; 2008CB418105; 2010CB933600; and the Strategic Priority Research Program of CAS (XDA09030203). We thank Bashaer Shariff for proof reading.

Conflicts of Interest: The authors declare the following competing financial interests: Several authors are co-inventors of patents related to the technology described in this paper for which they are entitled to receive royalties. GP and HZ are co-inventors of Chinese Patent No. 201810350928.0; GP, LW, ZW and LB are co-inventors of Chinese Patent Nos. 201210325832.1 and CN201110190392.9. GP is the first inventor of Chinese Patents Nos. ZL02155284.3; ZL200910080563.5; 201410141698.9; ZL2008100576735; 201210180985.1; ZL200510099736.X; CN201110194044.9; CN201010104197.5; CN201010104186.7; CN2010101041890; CN201010214522.3; CN201010288173.X. GP is the first inventor of U.S. Patent No. US 7,758,752 B2 and Australia Patent No. 2005336317. Above mentioned patents are originally issued to Research Center for Eco-Environmental Sciences, Chinese Academy of Sciences.

References

1. Pan, G.; Lyu, T.; Mortimer, R. Comment: Closing phosphorus cycle from natural waters: Re-capturing phosphorus through an integrated water-energy-food strategy. *J. Environ. Sci.* **2018**, 375–376. [CrossRef] [PubMed]
2. Conley, D.J.; Paerl, H.W.; Howarth, R.W.; Boesch, D.F.; Seitzinger, S.P.; Havens, K.E.; Lancelot, C.; Likens, G.E. Controlling eutrophication: Nitrogen and phosphorus. *Science* **2009**, *323*, 1014–1015. [CrossRef] [PubMed]
3. Khare, Y.; Naja, G.M.; Stainback, G.A.; Martinez, C.J.; Paudel, R.; van Lent, T. A phased assessment of restoration alternatives to achieve phosphorus water quality targets for Lake Okeechobee, Florida, USA. *Water* **2019**, *11*, 327. [CrossRef]
4. Lürling, M.; Mackay, E.; Reitzel, K.; Spears, B.M. Editorial–A critical perspective on geo-engineering for eutrophication management in lakes. *Water Res.* **2016**, *97*. [CrossRef]
5. Mackay, E.B.; Maberly, S.C.; Pan, G.; Reitzel, K.; Bruere, A.; Corker, N.; Douglas, G.; Egemose, S.; Hamilton, D.; Hatton-Ellis, T. Geoengineering in lakes: Welcome attraction or fatal distraction? *Inland Waters* **2014**, *4*, 349–356. [CrossRef]
6. Spears, B.M.; Maberly, S.C.; Pan, G.; Mackay, E.; Bruere, A.; Corker, N.; Douglas, G.; Egemose, S.; Hamilton, D.; Hatton-Ellis, T. Geo-engineering in lakes: A crisis of confidence? *Environ. Sci. Technol.* **2014**. [CrossRef]
7. Pan, G.; Krom, M.D.; Zhang, M.; Zhang, X.; Wang, L.; Dai, L.; Sheng, Y.; Mortimer, R.J. Impact of suspended inorganic particles on phosphorus cycling in the Yellow River (China). *Environ. Sci. Technol.* **2013**, *47*, 9685–9692. [CrossRef]

8. Pan, G.; Zhang, M.-M.; Chen, H.; Zou, H.; Yan, H. Removal of cyanobacterial blooms in Taihu Lake using local soils. I. Equilibrium and kinetic screening on the flocculation of Microcystis aeruginosa using commercially available clays and minerals. *Environ. Pollut.* **2006**, *141*, 195–200. [CrossRef]

9. Pan, G.; Chen, J.; Anderson, D.M. Modified local sands for the mitigation of harmful algal blooms. *Harmful Algae* **2011**, *10*, 381–387. [CrossRef]

10. Yuan, Y.; Zhang, H.; Pan, G. Flocculation of cyanobacterial cells using coal fly ash modified chitosan. *Water Res.* **2016**, *97*, 11–18. [CrossRef]

11. Biswas, B.; Warr, L.N.; Hildera, E.F.; Goswami, N.; Rahman, M.M.; Churchman, J.G.; Vasilev, K.; Pan, G.; Naidub, R. Biocompatible functionalisation of nanoclays for improved environmental remediation. *Chem. Soc. Rev.* **2019**. (under revision).

12. Zou, H.; Pan, G.; Chen, H.; Yuan, X. Removal of cyanobacterial blooms in Taihu Lake using local soils II. Effective removal of *Microcystis aeruginosa* using local soils and sediments modified by chitosan. *Environ. Pollut.* **2006**, *141*, 201–205. [CrossRef]

13. Pan, G.; Zou, H.; Chen, H.; Yuan, X. Removal of harmful cyanobacterial blooms in Taihu Lake using local soils III. Factors affecting the removal efficiency and an in situ field experiment using chitosan-modified local soils. *Environ. Pollut.* **2006**, *141*, 206–212. [CrossRef]

14. Shi, W.; Tan, W.; Wang, L.; Pan, G. Removal of *Microcystis aeruginosa* using cationic starch modified soils. *Water Res.* **2016**, *97*, 19–25. [CrossRef] [PubMed]

15. Li, L.; Pan, G. A universal method for flocculating harmful algal blooms in marine and fresh waters using modified sand. *Environ. Sci. Technol.* **2013**, *47*, 4555–4562. [CrossRef] [PubMed]

16. Chen, J.; Pan, G. Harmful algal blooms mitigation using clay/soil/sand modified with xanthan and calcium hydroxide. *J. Appl. Phycol.* **2012**, *24*, 1183–1189. [CrossRef]

17. Dai, L.; Pan, G. The effects of red soil in removing phosphorus from water column and reducing phosphorus release from sediment in Lake Taihu. *Water Sci. Technol.* **2014**, *69*, 1052–1058. [CrossRef]

18. Pan, G.; Dai, L.; Li, L.; He, L.; Li, H.; Bi, L.; Gulati, R.D. Reducing the recruitment of sedimented algae and nutrient release into the overlying water using modified soil/sand flocculation-capping in eutrophic lakes. *Environ. Sci. Technol.* **2012**, *46*, 5077–5084. [CrossRef] [PubMed]

19. Li, H.; Pan, G. Simultaneous removal of harmful algal blooms and microcystins using microorganism-and chitosan-modified local soil. *Environ. Sci. Technol.* **2015**, *49*, 6249–6256. [CrossRef] [PubMed]

20. Pan, G.; Yang, B.; Wang, D.; Chen, H.; Tian, B.-H.; Zhang, M.-L.; Yuan, X.-Z.; Chen, J. In-lake algal bloom removal and submerged vegetation restoration using modified local soils. *Ecol. Eng.* **2011**, *37*, 302–308. [CrossRef]

21. Zhang, H.; Shang, Y.; Lyu, T.; Chen, J.; Pan, G. Switching harmful algal blooms to submerged macrophytes in shallow waters using geo-engineering methods: Evidence from a 15N tracing study. *Environ. Sci. Technol.* **2018**, *52*, 11778–11785. [CrossRef]

22. Pan, G.; He, G.; Zhang, M.; Zhou, Q.; Tyliszczak, T.; Tai, R.; Guo, J.; Bi, L.; Wang, L.; Zhang, H. Nanobubbles at hydrophilic particle–Water interfaces. *Langmuir* **2016**, *32*, 11133–11137. [CrossRef] [PubMed]

23. Wang, L.; Miao, X.; Pan, G. Microwave-induced interfacial nanobubbles. *Langmuir* **2016**, *32*, 11147–11154. [CrossRef]

24. Zhang, H.; Lyu, T.; Bi, L.; Tempero, G.; Hamilton, D.P.; Pan, G. Combating hypoxia/anoxia at sediment-water interfaces: A preliminary study of oxygen nanobubble modified clay materials. *Sci. Total Environ.* **2018**, *637*, 550–560. [CrossRef]

25. Wang, L.; Miao, X.; Lyu, T.; Pan, G. Quantification of oxygen nanobubbles in particulate matters and potential applications in remediation of anaerobic environment. *Am. Chem. Sci. Omega* **2018**, *3*, 10624–10630. [CrossRef]

26. Shi, W.; Pan, G.; Chen, Q.; Song, L.; Zhu, L.; Ji, X. Hypoxia remediation and methane emission manipulation using surface oxygen nanobubbles. *Environ. Sci. Technol.* **2018**, *52*, 8712–8717. [CrossRef] [PubMed]

27. Conley, D.J.; Carstensen, J.; Vaquer-Sunyer, R.; Duarte, C.M. Ecosystem thresholds with hypoxia. *Hydrobiologia* **2009**, *629*, 21–29. [CrossRef]

28. APHA. *Standard Methods for the Examination of Water and Wastewater*; American Public Health Association (APHA): Washington, DC, USA, 2005.

29. Hu, H. *The Freshwater Algae of China: Systematics, Taxonomy and Ecology*; Science Press: Beijing, China, 2006.

30. Oladoja, N.A.; Pan, G. Modification of local soil/sand with Moringa oleifera extracts for effective removal of cyanobacterial blooms. *Sustain. Chem. Pharm.* **2015**, *2*, 37–43. [CrossRef]

31. Douglas, G.; Hamilton, D.; Robb, M.; Pan, G.; Spears, B.; Lurling, M. Guiding principles for the development and application of solid-phase phosphorus adsorbents for freshwater ecosystems. *Aquat. Ecol.* **2016**, *50*, 385–405. [CrossRef]

32. Wang, L.; Pan, G.; Shi, W.; Wang, Z.; Zhang, H. Manipulating nutrient limitation using modified local soils: A case study at Lake Taihu (China). *Water Res.* **2016**, *101*, 25–35. [CrossRef]

33. Yuan, X.-Z.; Pan, G.; Chen, H.; Tian, B.-H. Phosphorus fixation in lake sediments using LaCl3-modified clays. *Ecol. Eng.* **2009**, *35*, 1599–1602. [CrossRef]

34. Xu, R.; Zhang, M.; Mortimer, R.J.; Pan, G. Enhanced phosphorus locking by novel lanthanum/aluminum–hydroxide composite: Implications for eutrophication control. *Environ. Sci. Technol.* **2017**, *51*, 3418–3425. [CrossRef]

35. Scheffer, M.; Carpenter, S.; Foley, J.A.; Folke, C.; Walker, B. Catastrophic shifts in ecosystems. *Nature* **2001**, *413*, 591–596. [CrossRef]

36. Pan, G.; Krom, M.D.; Herut, B. Adsorption–desorption of phosphate on airborne dust and riverborne particulates in East Mediterranean seawater. *Environ. Sci. Technol.* **2002**, *36*, 3519–3524. [CrossRef]

37. Parkyn, S.M.; Hickey, C.W.; Clearwater, S.J. Measuring sub-lethal effects on freshwater crayfish (*Paranephrops planifrons*) behaviour and physiology: Laboratory and in situ exposure to modified zeolite. *Hydrobiologia* **2011**, *661*, 37–53. [CrossRef]

38. Wang, Z.; Zhang, H.; Pan, G. Ecotoxicological assessment of flocculant modified soil for lake restoration using an integrated biotic toxicity index. *Water Res.* **2016**, *97*, 133–141. [CrossRef] [PubMed]

39. Kong, M.; Chen, X.G.; Xing, K.; Park, H.J. Antimicrobial properties of chitosan and mode of action: A state of the art review. *Int. J. Food Microbiol.* **2010**, *144*, 51–63. [CrossRef]

40. Mucci, M.; Noyma, N.P.; de Magalhaes, L.; Miranda, M.; van Oosterhout, F.; Guedes, I.A.; Huszar, V.L.; Marinho, M.M.; Lürling, M. Chitosan as coagulant on cyanobacteria in lake restoration management may cause rapid cell lysis. *Water Res.* **2017**, *118*, 121–130. [CrossRef]

Article

Enhanced Phosphate Removal from Water by Honeycomb-Like Microporous Lanthanum-Chitosan Magnetic Spheres

Rong Cheng [1], Liang-Jie Shen [1], Ying-Ying Zhang [1], Dan-Yang Dai [1], Xiang Zheng [1], Long-Wen Liao [2], Lei Wang [3,*] and Lei Shi [1,*]

[1] School of Environment and Natural Resources, Renmin University of China, Beijing 100872, China; chengrong@ruc.edu.cn (R.C.); shenliangjie@ruc.edu.cn (L.-J.S.); zhangyingying@ruc.edu.cn (Y.-Y.Z.); daidanyang@ruc.edu.cn (D.-Y.D.); zhengxiang@ruc.edu.cn (X.Z.)
[2] Northwest Institute of Nuclear Technology, Xi'an 710024, China; liaolw16@tsinghua.org.cn
[3] Research Center for Eco-Environmental Sciences, Chinese Academy of Sciences, Beijing 100872, China
[*] Correspondence: leiwang@rcees.ac.cn (L.W.); shil@ruc.edu.cn (L.S.);
Tel.: +86-131-6103-1804 (L.W.); +86-138-1138-4632 (L.S.)

Received: 13 October 2018; Accepted: 12 November 2018; Published: 14 November 2018

Abstract: The removal of phosphate in water is crucial and effective for control of eutrophication, and adsorption is one of the most effective treatment processes. In this study, microporous lanthanum-chitosan magnetic spheres were successfully synthetized and used for the removal of phosphate in water. The characterization results show that the dispersion of lanthanum oxide is improved because of the porous properties of the magnetic spheres. Moreover, the contact area and active sites between lanthanum oxide and phosphate were increased due to the presence of many honeycomb channels inside the magnetic spheres. In addition, the maximum adsorption capacity of the Langmuir model was 27.78 mg P·g^{-1}; and the adsorption kinetics were in good agreement with the pseudo-second-order kinetic equation and intra-particle diffusion model. From the results of thermodynamic analysis, the phosphate adsorption process of lanthanum-chitosan magnetic spheres was spontaneous and exothermic in nature. In conditional tests, the optimal ratio of lanthanum/chitosan was 1.0 mmol/g. The adsorption capacity of as-prepared materials increased with the augmentation of the dosage of the adsorbent and the decline of pH value. The co-existing anions, Cl$^-$ and NO$_3^-$ had little effect on adsorption capacity to phosphate, while CO$_3^{2-}$ exhibited an obviously negative influence on the adsorption capacity of this adsorbent. In general, owing to their unique hierarchical porous structures, high-adsorption capacity and low cost, lanthanum-chitosan magnetic spheres are potentially applicable in eutrophic water treatment.

Keywords: porous structure; lanthanum; La-chitosan magnetic spheres; adsorption capacity; phosphate; adsorption isotherm

---

1. Introduction

The eutrophication of water bodies has become one of the major ecological environmental issues all over the world [1–3]. Especially, phosphorus, an essential element for the growth of aquatic organisms, is one of the limiting elements for the eutrophication of water bodies [4]. Phosphate is involved in the aqueous environment by means of both natural activities and the affairs of people, such as significant erosion, mining, agricultural fertilization, and industrial operation. The presence of excess phosphorus in water further results into the outbreak of algal blooms, death of aquatic organisms, and sharp deterioration of water quality. In severe cases, cyanobacterial blooms, caused by an excess phosphorus in water, generally poses many difficult challenges to the stability of aquatic ecosystems and the protection of domestic water quality of residents. Therefore, the removal of excess phosphate in water is of great significance for controlling and preventing the deterioration of water bodies.

At present, the methods for phosphorus removal mainly include chemical precipitation [5], biological phosphorus removal [6], membrane separation [7], adsorption [8–10], and ion exchange [11]. For example, chemical removal including precipitation with aluminum, iron, and calcium components is a common means for effective removal of phosphorus from water. However, higher sludge production is led by this method, and harmless sludge treatment needs to be further considered [12]. Biological phosphorus removal has attracted much attention due to its advantages of no chemical additions. However, it is worth pointing out that the phosphorus removal effect of biotechnology is greatly affected by the operating conditions [13]. In many of the above methods, adsorption is a very promising method for efficiently removing phosphate from water. It presents several advantages of simple operation, less sludge generation, and the ability to recover phosphorus element without secondary pollution. However, due to the limited adsorption capacity of natural adsorbents, it is necessary to select new and efficient adsorbents.

Recent studies have found that metal oxides have a good adsorption effect on phosphate in the water bodies, especially oxides of transition metal elements (zirconium, hafnium, and lanthanum) which effectively increases the adsorption capacity of phosphate, such as hydrous lanthanum oxide, zirconium oxide [14,15]. Especially, lanthanum has attracted the interest of many researchers due to its non-toxicity, chemically stable, and extremely strong affinity for phosphate. To date, several studies about the phosphate adsorption performances of La-based adsorbents have been reported in succession all over the world. For example, the phosphate removal efficiency of $LaCl_3$-modified kaolinite or pumice clays were almost 6–37% higher than those of pure clays [16]. The maximum adsorption of 25 mg P·g^{-1} and the phosphorus removal rate of 99% were reached by lanthanum-modified zeolite when the initial concentration of phosphate and pH is 30 mg·L^{-1} and 4.0, respectively. The maximum monolayer adsorption capacity of lanthanum-iron complex materials is 208.33 mg P·g^{-1} and it also has a high adsorption capacity for high concentration of phosphorous water. Although the above materials have good adsorption and stability, there are also several shortcomings, which also includes the difference between this lanthanum-chitosan magnetic spheres and above materials. For example, most of the adsorbents mentioned above are in powders state, which are difficult to be recovered and reused in water. The phosphate, adsorbed by the disposable adsorbents, will still be resolved into the water with time going on, resulting in eutrophication again. The problem has not been solved fundamentally.

Based on the above analysis, lanthanum-chitosan magnetic spheres were synthesized by an in situ chemical precipitation method. The combination of the advantages of lanthanum oxide and shaping characteristic of cheap and readily available magnetic chitosan was achieved to apply phosphate removal in water for the purpose of magnetic recovery of adsorbents and separation of phosphate from water. In addition, the honeycomb-like structures of as-prepared lanthanum-chitosan magnetic spheres in this study were formed to effectively improve the dispersity and utilization of lanthanum oxides. Moreover, very large specific surface area and active sites for phosphate adsorption are provided. Moreover, it is worthy of attention that the low temperature cryodesiccation technique, one of the means of maintaining the chemical composition and porous structure of materials, is applied to the preparation of the uniform honeycomb-like microporous spheres, different from the solid sphere in order to significantly increase the specific surface area of the composite spheres. And as-prepared spheres were characterized with different techniques (X-ray diffraction (XRD), Scanning electron microscopy (SEM), Fourier transform infrared spectroscopy (FTIR), and Vibrating sample magnetometer (VSM)) and the adsorption kinetics, isotherm, and thermodynamic analysis of phosphate by as-prepared adsorbents were further studied. Meanwhile, the influences of different experimental parameters (lanthanum/chitosan ratio, dosage of adsorbents, pH of solution, and coexisting anions) were also investigated to provide the reference for the practical application.

2. Materials and Methods

2.1. Chemicals

Chitosan (analytical reagent, AR) was produced by Shanghai Aladdin Biochemical Technology Co., Ltd., Shanghai, China. Lanthanum (III) nitrate hexahydrate (guaranteed reagent, GR), was produced by Xilong Scientific Co., Ltd., Shanghai, China. Ferrous sulfate (GR), iron (III) chloride hexahydrate (GR), potassium phosphate monobasic (AR), hydrochloric acid (AR), acetic acid (AR), and sodium hydroxide (GR) were produced by Sinopharm Chemical Reagent Beijing Co., Ltd., Beijing, China.

2.2. Preparation of Lanthanum-Chitosan Magnetic Spheres

In general, chitosan magnetic spheres were prepared by the chemical precipitation method using alkali liquor as the firming agent. Firstly, approximately 2.0 g of chitosan was dissolved in 30 mL deionized water and 1.50 mL glacial acetic acid was subsequently added to create sticky liquid (solution A). Then 2.70 g $FeCl_3 \cdot 6H_2O$ and 1.65 g $FeSO_4 \cdot 7H_2O$ were dissolved in 10 mL ultrapure water and mixed (solution B). A several amount of La $(NO_3)_3 \cdot 6H_2O$ (0, 1.0 mmol, 2.0 mmol, 3.0 mmol, 6.0 mmol, and 10.0 mmol, respectively) was dissolved into 10 mL deionized water (solution C). Then solution B and solution C were added to solution A and then stirred for 30 min. Thirdly, with a constant current syringe pump, the mixed solution was dropped into a beaker containing 300 mL of 15 wt% NaOH at a rate of 2.0 mL min^{-1} using a 20 mL syringe. The prepared spheres were shaken for 1.0 h, and then were frozen for 12 h by freeze drier. Finally, the material was placed in a constant-temperature drying oven for 2.0 h at 80 °C and lanthanum-chitosan (La-chitosan) magnetic spheres were prepared. In addition, chitosan magnetic spheres were prepared by the same process without adding La $(NO_3)_3 \cdot 6H_2O$.

2.3. Characterization of La-Chitosan Magnetic Spheres

The surface morphology of as-prepared samples was concretely analyzed by scanning electron microscopy (SEM) (Nova 400 Nano, FEI Company, Hillsboro, OR, USA). The phase composition of the samples was monitored by X-ray diffraction (XRD) (D/MAX-AX, Rigaku Corporation, Tokyo, Japan). The functional groups of the samples were parsed in detail by Fourier transform infrared spectroscopy (FTIR) (IRTracer-100, Hitachi, Ltd., Tokyo, Japan). The magnetization intensity of the samples was analyzed by vibrating sample magnetometer (VSM) (Quantum Design MPMS XL7, Quantum Design, Inc., San Diego, CA, USA).

2.4. Phosphate Adsorption Experiments

2.4.1. Adsorption Kinetics

Phosphate adsorption experiments of this study were conducted by evenly mingling 0.5 g·L^{-1} of the La-chitosan magnetic spheres with 100 mL of a K_2HPO_4 solution at 100 rpm and 25 °C. In addition, similar conditions have been also taken in kinetic experiments. The residual phosphate concentration of solution was determined accurately by a UV–VIS spectrophotometer (DR6000, HACH, Loveland, CO, USA) by the aid of the ammonium molybdate spectrophotometric method after filtration [17]. Then the q_e (mg P·g^{-1}) of phosphate were calculated as following Equation (1):

$$\text{Adsorption capacity}: \quad q_e = \frac{(C_0 - C_e)V}{m} \tag{1}$$

where C_0 represents the initial concentration of phosphate solution (mg·L^{-1}), C_e is the actual phosphate concentration of solution (mg·L^{-1}) at adsorption equilibrium, V indicates the total volume of reaction liquid (L), and m equals the dosage of as-prepared materials (g).

For the research on adsorption kinetics of as-prepared materials, the experiments were carried out with 20 mg·L^{-1} initial concentration of phosphate and 0.5 g·L^{-1} of La-chitosan magnetic spheres.

The adsorption data were further fitted through a pseudo-first order kinetic model, pseudo-second order kinetic model and intra-particle diffusion model to probe into the kinetic mechanism, according to Equations (2)–(4), respectively [18]:

$$\text{The pseudo-first order kinetic model}: \ log(q_e - q_t) = logq_e - \frac{k_1 t}{2.303} \tag{2}$$

$$\text{The pseudo-second order kinetic model}: \ \frac{t}{q_t} = \frac{1}{k_2 q_e^2} + \frac{t}{q_e} \tag{3}$$

$$\text{Intraparticle diffusion model}: \ q_t = k_p t^{1/2} + C \tag{4}$$

where t is the adsorption time (min), q_e indicates the adsorption capacity at equilibrium (mg P·g^{-1}), q_t presents the adsorption capacity at t moment (mg P·g^{-1}), k_1 is an adsorption rate constant of the pseudo-first order kinetic (min^{-1}), k_2 is an adsorption rate constant related to the pseudo-second order kinetic (g·mg·min^{-1}), k_p is a rate constant related to the intra-particle diffusion (mg·g^{-1}·min$^{-1/2}$), and C presents a constant related to the boundary layer thickness.

2.4.2. Adsorption Isotherm

For the study on adsorption isotherms, experiments were carried out with 100 mL of phosphate solutions of different initial concentrations (5, 10, 15, 20, 30, 40 mg·L^{-1}). The dosage of adsorbent was 0.05 g, and the pH was adjusted to 7.0. After adsorption, the adsorption data of La-chitosan magnetic spheres were fitted by Langmuir Equation (5) and Freundlich Equation (6) [19,20]:

$$\text{Langmuir equation}: \ q_e = q_m K_L \frac{C_e}{1 + C_e K_L} \tag{5}$$

$$\text{Freundlich equation}: \ q_e = K_F C_e^{1/n} \tag{6}$$

where q_e is the adsorption capacity at equilibrium (mg·g^{-1}), q_m is equal to the adsorption capacity at saturation (mg·g^{-1}), C_e indicates the concentration of phosphate in equilibrium state (mg·L^{-1}), K_L is a constant related to thermodynamics (L·mg^{-1}), K_F is a constant related to adsorption strength, n is a constant related to adsorptivity.

In addition, the characteristics of the Langmuir isotherm model can be visually represented by the equilibrium parameter (R_L), as shown in Equation (7):

$$R_L = \frac{1}{1 + K_L C_0} \tag{7}$$

where C_0 is the initial concentration of phosphate in solution (mg P·L^{-1}); K_L presents the Langmuir's adsorption constant (L·mg^{-1}).

2.4.3. Thermodynamic Analysis

Change in the values of entropy (ΔS°), Gibbs free energy (ΔG°) and enthalpy (ΔH°), were usually calculated to evaluate the thermodynamic direction and the behavior of adsorption process. The values of ΔH° and ΔS° can be calculated by means of Van't Hoff equation. Therefore, ΔG° can be obtained by the following Equations (8)–(10) [21]:

$$\Delta G^\circ = -RTln_{K_d} = -RTlnb \tag{8}$$

$$lnb = \frac{\Delta S^\circ}{R} - \frac{\Delta H^\circ}{RT} \tag{9}$$

$$b = q_e - C_e \tag{10}$$

where R presents the gas constant (8.314 J/(mol·K)); T indicates the temperature value (K); and K_d is the thermodynamic equilibrium constant of adsorption process, b presents the Langmuir equilibrium constant (L/mol).

2.4.4. Conditional Factor Experiments

The effect of lanthanum/chitosan (La/CS) ratio on the adsorption of phosphate was examined in a 20 mg·L^{-1} K_2HPO_4 solution. La-chitosan ratios of 0.5, 1.0, 1.5, 3.0, and 5.0 mmol/g of La-chitosan magnetic spheres were added to the solution, respectively. The effect of dosage of La-chitosan magnetic spheres on the adsorption was examined in a 20 mg·L^{-1} K_2HPO_4 solution. A certain amount of La-chitosan magnetic spheres was added to the solution so that the concentrations of the La-chitosan magnetic spheres were 0.5, 1.0, 1.5 g·L^{-1}, respectively. The effect of pH on the sorption of phosphate by the La-chitosan magnetic spheres was examined in a 20 mg·L^{-1} K_2HPO_4 solution at pH (3.0, 5.0, 7.0, 9.0), which was adjusted by using 0.1 M hydrochloric acid or 0.1 M sodium hydroxide solution. The effect of coexisting anions on the adsorption was examined by adding 0.05 g of La-chitosan magnetic spheres into the solution containing 0.1 M co-existing anions, which were prepared by dissolving sodium salts of Cl^-, NO_3^-, and CO_3^{2-} into 20 mg·L^{-1} K_2HPO_4 solution. Except for studying the effect of pH value on adsorption capacity, the other adsorption studies were performed at a pH of 7.0.

3. Results and Discussion

3.1. Characterization

3.1.1. SEM Analysis

Figure 1 presents the SEM images and energy dispersive spectrum (EDS) diagrams of the chitosan magnetic spheres and La-chitosan magnetic spheres. The samples both exhibit a typical uniform and regular porous structure with 15 μm of pore size (Figure 1a,d). More vividly, the structure of as-prepared spheres is similar to honeycomb-like shape, which is similar to what is reported in the literature [22]. In addition, the surface of La-chitosan magnetic spheres seems rougher and presents an arc sheet shape (Figure 1b,e). And it's obvious and intuitive that the lamellar surface of La-chitosan magnetic spheres is uniformly covered with a mesh of rod-like particles (Figure 1b,e), which is more favorable for adsorption of phosphate. As can be seen from the EDS diagrams in Figure 1c,f, it's clear that lanthanum elements could be clearly observed and contained inside the La-chitosan magnetic spheres compared to the chitosan magnetic spheres.

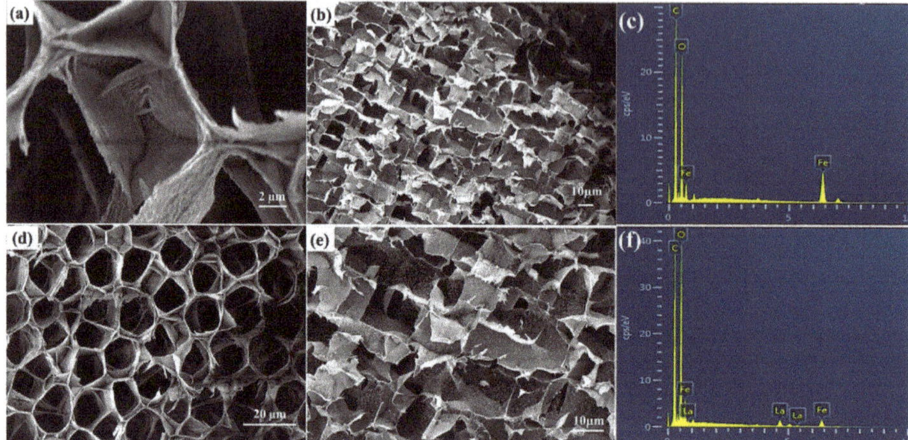

Figure 1. SEM images and EDS diagrams of the as-prepared chitosan magnetic spheres (**a–c**) and La-chitosan magnetic spheres (**d–f**).

3.1.2. VSM Analysis

The magnetization curves of La-chitosan magnetic spheres from the analysis of a vibrating sample magnetometer (VSM) at room temperature is shown in Figure 2. It can be seen that the saturation magnetization intensity of La-chitosan magnetic spheres is 7.90 emu/g. Meanwhile, as shown in the inset photo of Figure 2, with a magnet placed at the bottom of the bottle, the adsorbents that originally floated on the surface of the water would quickly sink to the bottom within 2 s, which intuitively shows the good magnetic properties and completely achieves solid-liquid separation for recovery and reuse from the treated solution.

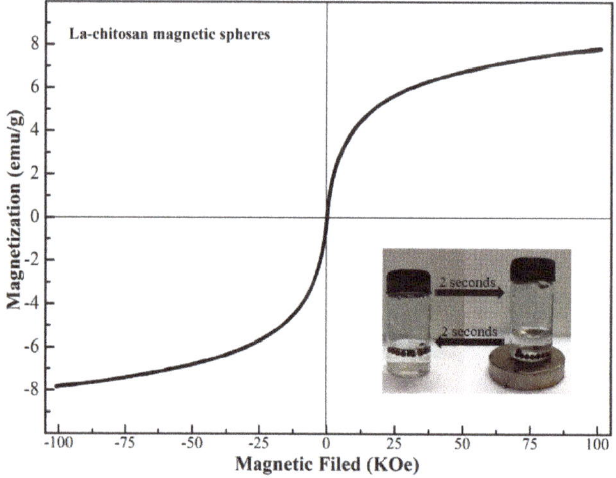

Figure 2. The magnetization curves of La-chitosan magnetic spheres (inserted with separation-redispersion process).

3.1.3. X-ray Diffraction Analysis

XRD results of the chitosan magnetic spheres and La-chitosan magnetic spheres are shown in Figure 3. It could be seen that there were three obvious characteristic peaks at 20°, 33.3°, and 43°, corresponding to the (110), (220), and (400) crystal plane positions of FeOOH of which hydroxyl groups are coordination groups and have better adsorption effects on phosphate ions. In addition, two obvious characteristic peaks at 35°, and 62.5° were presented, which certifies the existence of Fe_3O_4. Based on the above, the main forms of the metal compounds in chitosan magnetic spheres are FeOOH and Fe_3O_4. From the XRD pattern in red curve, it's found that the distinct characteristic peaks at 28°, 43° and 58° corresponded with (222), (431) and (541) crystal plane positions of lanthanum oxide. Thus, the Fe and La elements in the La-magnetic chitosan spheres were mainly present in the form of Fe_3O_4, FeOOH, and La_2O_3, which are consistent with the weak magnetism of the material (Figure 2). Moreover, the presence of La_2O_3 (marked blue peaks in Figure 3) is also a key factor that increased the adsorption capacity of phosphate on the adsorbent.

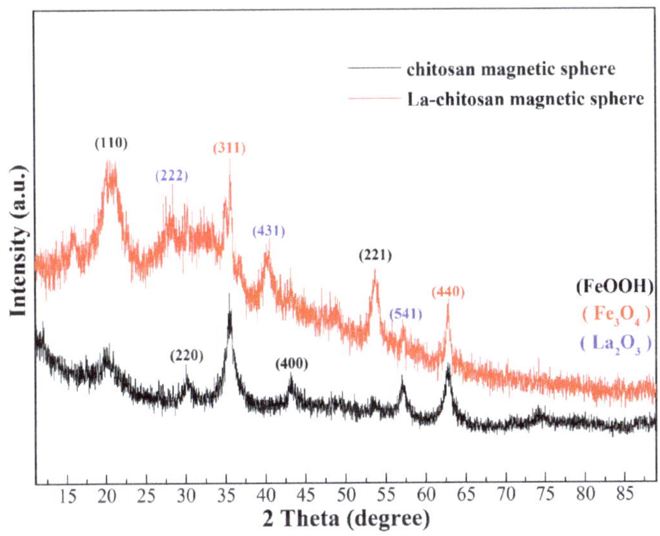

Figure 3. XRD spectra of chitosan magnetic spheres and La-chitosan magnetic spheres.

3.1.4. FTIR Analysis

Figure 4 shows the FTIR spectra of chitosan magnetic spheres and La-chitosan magnetic spheres before and after the reaction. The peak at 3435 cm^{-1} of the chitosan magnetic spheres represents the stretching vibration peak of –NH/–OH. The characteristic absorption peak of –CH and –CH$_2$ derived from chitosan is located at 2915 cm^{-1}. The peak at 570 cm^{-1} shows the Fe–O vibration peak of Fe_3O_4, which is consistent with the results of XRD spectrogram (Figure 3). Compared with the chitosan magnetic spheres, the bending vibration strength of the N–H is much larger before the reaction, which may be due to the bending of the N–H after the La^{3+} coordinates with the nitrogen atom of the chitosan to increase steric hindrance. C–O stretching vibration peak moves from 1029 cm^{-1} to 1031 cm^{-1}, and the intensity increases, probably because La^{3+} coordinates with hydroxyl oxygen atoms of chitosan afterwards, the electron density of the oxygen atoms was reduced, and C–O bond was weakened. The vibration peak of Fe–O is shifted to 626 cm^{-1}. Combined with XRD analysis, it is presumably due to the formation of FeOOH. Compared with the results before the reaction, the peak at 1432 cm^{-1} of La-chitosan magnetic spheres almost disappeared and moved to 1381 cm^{-1} after the reaction, indicating that hydroxyl was involved in the adsorption process of phosphates.

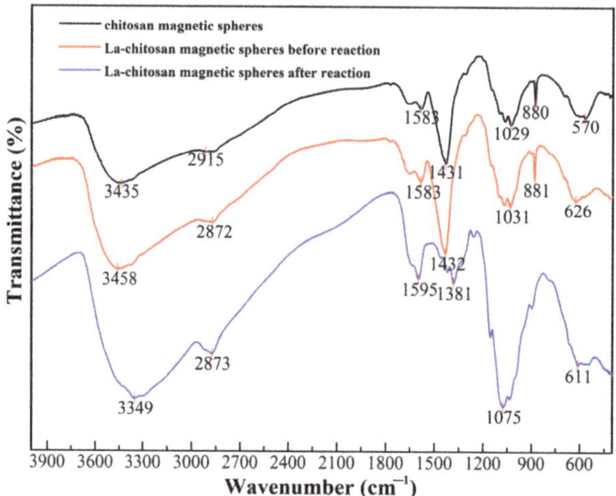

Figure 4. FTIR spectra of chitosan magnetic spheres and La-chitosan magnetic spheres before and after reaction.

3.2. Adsorption Experiments

3.2.1. Adsorption Kinetics

As shown in Figure 5a, the adsorption capacity was rapidly increased during the initial 5 h, reaching more than 80% of the saturated adsorption amount. However, as time further increased, the increase of adsorption slowed down due to the augment of the available adsorption sites on the surface of the adsorbent. After 5 h, the adsorption went into a more gradual phase, achieving adsorption equilibrium within 10 h.

For the further investigation of the phosphate adsorption process by La-chitosan magnetic spheres, both of the pseudo-first-order and pseudo-second-order models are used to the processing of kinetic data recorded in Figure 5a, as shown in Figure 5b,c. Meanwhile, the corresponding parameters and correlation coefficients obtained by fitting means are clearly presented in Table 1. Compared with the pseudo-first-order model (coefficient of determination (R^2) = 0.97), the pseudo-second-order model (R^2 = 0.99) can better depict and explicate the adsorption process, indicating that the chemisorption or chemical bonding between active sites of La-chitosan magnetic spheres and phosphate might play a leading role in the adsorption process.

As is shown in Figure 5d, the linear trend of intra-particle diffusion model clearly presents a three-stage type, which indicates that the adsorption process of as-prepared adsorbents consists of several stages [23]. The first region (the black line in Figure 5d) is mainly facilitated by the external surface or instantaneous adsorption. The main driving force in the first region is the concentration differences of phosphate. The gradual adsorption stage is presented by the second linear portion as shown in the red line in Figure 5d. Meanwhile, it is seen that the gradual adsorption stage is the rate-limiting step. There is no doubt that the third line segment (blue line in Figure 5d) is the final equilibrium stage. The deceleration of above stage is put down to the low content of residual phosphate in solution.

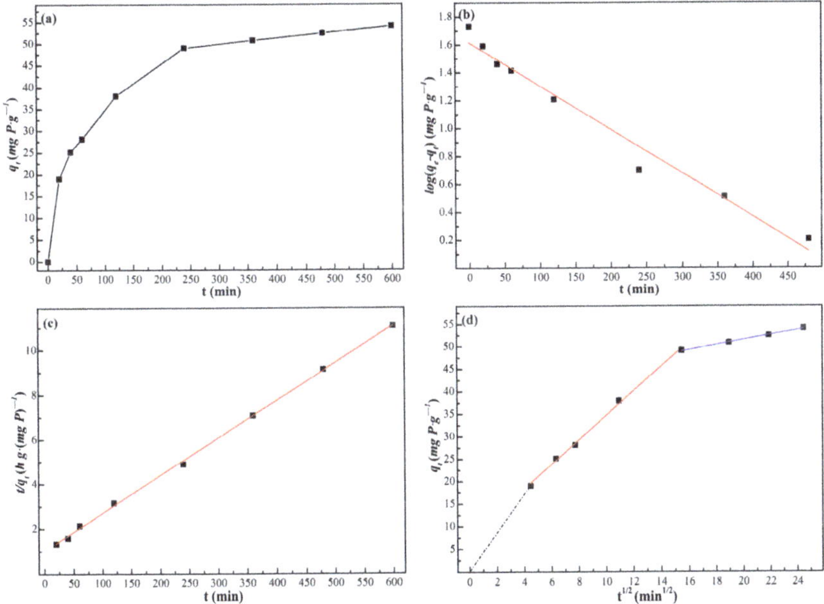

Figure 5. (a) Effect of contact time on the adsorption capacity of La-chitosan magnetic spheres, (b) pseudo-first-order model, (c) pseudo-second-order model, and (d) intra-particle diffusion model.

Table 1. Adsorption kinetic parameters of phosphorus onto La-chitosan magnetic spheres.

C_0 (mg·L^{-1})	$q_{e,exp}$ (mg g^{-1})	Pseudo-First-Order			Pseudo-Second-Order			Intra-Particle Diffusion		
		$q_{e,cal}$ (mg g^{-1})	k_1 (min^{-1})	R^2	$q_{e,cal}$ (mg g^{-1})	k_2 (g mg^{-1} min^{-1})	R^2	k_{p1} (g mg^{-1} min^{-1})	k_{p2} (g mg^{-1} min^{-1})	R^2
20	54.08	41.69	0.01	0.97	59.52	0.02	0.99	2.72	0.56	0.97

3.2.2. Adsorption Isotherm

As shown in Figure 6, Langmuir (a) and Freundlich (b) isotherm models were used to evaluate the adsorption isotherms of phosphate by La-chitosan magnetic spheres. Meanwhile, the estimated model parameters with the coefficient of determination (R^2) for the different models are shown in Table 2. The coefficient of determination, R^2, given in the Table 2, shows that the both Freundlich and Langmuir isotherm models can be used to explain the adsorption isotherms of La-chitosan magnetic spheres well. In the cases of R^2 values, the applicability of the above models, based on current experimental data, follows the order: Langmuir > Freundlich. Clearly, The R^2 values obtained for the Freundlich and Langmuir models were both above 0.98, indicating that both monolayer adsorption and multilayer adsorption occur in the system as shown in Figure 6. Results similar to those in this paper have been reported for the adsorption of phosphate in other literature [24]. According to the Langmuir model, the maximum phosphorus adsorption capacity reached 27.78 mg P·g^{-1}. Moreover, R_L values (0.0379) fall within the range of 0–1.0, implying that the phosphate adsorption onto La-chitosan magnetic spheres is favorable.

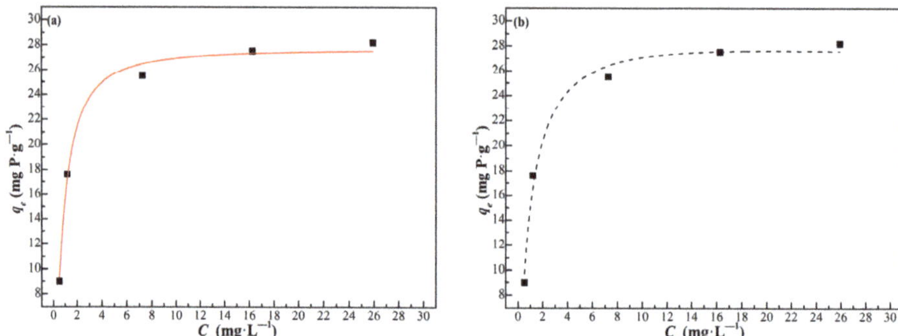

Figure 6. Langmuir (**a**) and Freundlich (**b**) models for the isotherm adsorption experiment.

Table 2. Adsorption isotherm parameter fitting result of Langmuir and Freundlich model.

Adsorbents	Langmuir Model				Freundlich Model		
	q_{max} (mg P·g^{-1})	K_L (L·mg^{-1})	R_L	R^2	K_F	$1/n$	R^2
La-chitosan magnetic spheres	27.78	1.27	0.0379	0.99	15.16	0.54	0.98

3.2.3. Thermodynamic Analysis

As we all known, the spontaneous nature of the processes is usually judged by three valuable thermodynamic parameters ($\Delta G°$, $\Delta S°$, and $\Delta H°$). Generally, values of $\Delta G°$ in the range of -20 to 0 kJ/mol presents a physical adsorption process; while $\Delta G°$ values in range of -80 to -400 kJ/mol presents a process of chemical adsorption [25]. Figure 7 shows the linear diagram of $\ln(q_e/C_e)$ (vertical ordinate) vs. $1/T$ (horizontal ordinate). The calculated corresponding parameter values are listed in Table 3. As shown in Table 3, the $\Delta G°$ values are less than zero at all temperatures, which means that the adsorption of phosphate onto La-chitosan magnetic spheres is favorable and spontaneous behavior. Moreover, the value of $\Delta G°$ is increased gradually from -32.04 kJ/mol to -26.88 kJ/mol with the alteration of temperature from 25 °C to 45 °C, illustrating that the spontaneity of this adsorption process is declined at elevated temperature. From Table 3, it is obvious that values of $\Delta G°$ are neither in the ranges of $\Delta G°$ in physical adsorption nor chemical adsorption process. Relatively speaking, the values of $\Delta G°$ are closer to the process of physical adsorption. The possible explanation is that with the exception of physical adsorption, many other mechanisms may work together, such as electrostatic interaction and ligand exchange [26]. In Table 3, the negative value of $\Delta H°$ (-77.73 KJ/mol) confirms the exothermic nature of phosphate adsorption [27]. In addition, The negative value of $\Delta S°$ (-0.1615 KJ/(mol·K)) definitely illuminates the decline in randomness during the adsorption process.

Table 3. Thermodynamic parameters for phosphate adsorption on La-chitosan magnetic spheres at 15 °C, 25 °C, 35 °C, and 45 °C, respectively.

Temperature (°C)	b (L/mol)	$\Delta G°$ (KJ/mol)	$\Delta S°$ (KJ/(mol·K))	$\Delta H°$ (KJ/mol)
15	646,695	-32.04		
25	90,932	-8.29		
35	55,181	-27.96	-0.1615	-77.73
45	26,061	-26.88		

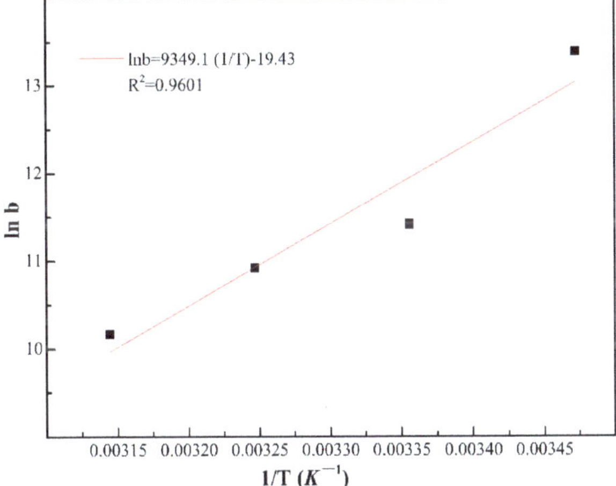

Figure 7. Van't Hoff diagram for the adsorption of phosphate with La-chitosan magnetic spheres.

Table 4 shows the concrete comparison of phosphate adsorption capacity of different reported adsorbents, which showed that La-chitosan magnetic spheres had higher adsorption capacity for phosphate compared to other adsorbents reported. Moreover, the adsorption capacity of La-chitosan magnetic spheres is obviously higher than that of commercial Phoslock®.

Table 4. The comparison of phosphate adsorption capacity of different adsorbents.

Type of Adsorbents	q_m (mg P·g^{-1})	Isotherm Model	Reference
La-chitosan magnetic spheres	27.78	Langmuir	This work
Fe_3O_4@SiO_2 with La_2O_3	27.8	Langmuir	[14]
Phoslock® (lanthanum-modified bentonite)	9.5–10.5	Langmuir	[25]
NT-25La	14.0	Langmuir	[28]
Activated aluminium oxide	13.8	Langmuir	[29]
Fe (III)-modified bentonite	11.2	Langmuir	[30]
Al (III)-modified bentonite	12.7	Langmuir	[31]

3.2.4. Effect of La/CS Ratios

Figure 8 shows the adsorption efficiency of the La-chitosan magnetic spheres with different La/CS ratios. It is obvious that the adsorption capacity of chitosan magnetic spheres reached only 10 mg P·g^{-1}, while the adsorption capacity of La-chitosan magnetic spheres generally increased with the change of La/CS ratio. In addition, it can be seen that the optimal La/CS ratio is 1.0 mmol/g, and the adsorption capacity reaches 27.49 mg P·g^{-1}. We can find that the fluctuation of adsorption capacity was occurred during the first 1.0–3.0 h. The reason for this phenomenon may be due to the freeze-drying process in the preparation of materials. When the adsorbent is just in contact with the phosphorus solution at the beginning, it adsorbed a large amount of phosphorus solution, resulting in a sudden increase in the amount of adsorption. When the adsorbent is moist enough, part of the solution will overflow, resulting in a slight decrease in the amount of adsorption. It is worth pointing out that the irregular changes of adsorption capacity of phosphate in water were showed in the first 1.0 h due to the incipient wetting process of freeze-dried materials. Thus, with the completion of wetting process, the amount of phosphate adsorbed in water gradually increased, as shown in Figure 8.

When the material is added to the water, the following steps may occur: the particle wetting process (first 1.0–3.0 h), rapid adsorption process (3–10 h), and adsorption saturation process (10–20 h). As a whole, with the augment of La/chitosan ratio, the corresponding adsorption capacity gradually increased. The ultimate adsorption capacity (27.49 mg P·g^{-1}) was gained when the La/chitosan ratio is equal to 1.0 mmol/g. When the dosage is less than the maximum value, the adsorption capacity will increase with the increase of the dosage. And when the dosage is greater than the maximum, the adsorption capacity is still equal to the maximum value. This is because the pore size, specific surface area, and shape of the adsorbents are affected when too much lanthanum is added. Meanwhile, the optimal ratio was selected in the subsequent experiments.

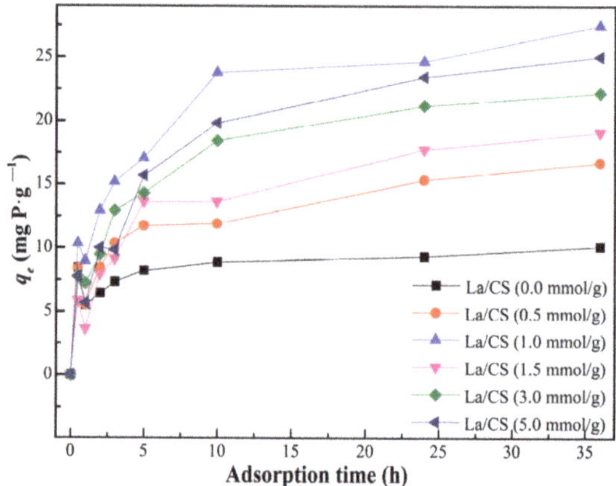

Figure 8. Effect of La/CS ratios on phosphate adsorption capacity of La-chitosan magnetic spheres.

3.2.5. The Effect of Dosage

Figure 9 shows the adsorption efficiency of different dosages of La-chitosan magnetic spheres. We can clearly see that when lanthanum was doped into magnetic spheres, the adsorption capacity increased by 23.6% compared with the chitosan magnetic spheres. Meanwhile, the adsorption capacity increased significantly to 33.1 mg P·g^{-1}, 40.9 mg P·g^{-1}, 51.3 mg P·g^{-1}, and 54.7 mg P·g^{-1}, respectively, with the dosage of adsorbents generally increased. Along with the enhancement of adsorbent dosage, the adsorption activity points increased, leading to more phosphate being adsorbed. It is worth mentioning that the addition in as-prepared spheres may increase the cost of preparation, while the adsorption capacity of La-chitosan magnetic spheres increases by 21.6%. The cost of composite materials is further reduced indirectly because of the recoverability and reusability of magnetic spheres.

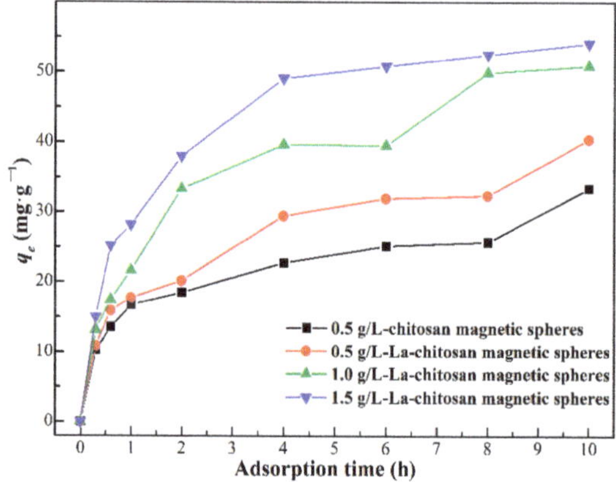

Figure 9. Effect of dosage of La-chitosan magnetic spheres on the phosphate adsorption capacity.

3.2.6. The Effect of pH

In general, pH of the aqueous solution affects the existing state of the substance in the solution and the surface charge is usually considered as a vital and meaningful variable that interferes with the adsorption of ions at water–adsorbent interfaces in the adsorption process. The adsorption of phosphate on La-chitosan magnetic spheres were studied at various pH values ranging from 3.0 to 9.0, as shown in Figure 10. When the pH value is 3.0, the adsorbent capacity of La-chitosan magnetic spheres reaches the maximum equilibrium adsorption and is significantly higher than that under the other pH conditions. When the pH value increased from 3.0 to 9.0, the adsorption capacity gradually decreased. On the one hand, the surface of the adsorbent will carry more negative charges at higher pH value, making it reject negatively charged ions in the solution. In the alkaline solution, the competitive relation adsorption active center of adsorbents between the phosphate ions and OH^- ions was significantly enhanced and resulted in a significant decrease in the capacity of phosphate. On the other hand, when the pH of the solution is greater than the point of zero charge, the common ion repulsion or electrostatic repulsion force is dominant. In addition, phosphate acid exists as four different chemical forms (H_3PO_4, $H_2PO_4^-$, HPO_4^{2-}, and PO_4^{3-}) at different pH ranges ($H_3PO_4 \leftrightarrow H_2PO_4^- + H^+$ (pK_{a1}); $H_2PO_4^- \leftrightarrow HPO_4^{2-} + H^+$ (pK_{a2}); $HPO_4^{2-} \leftrightarrow PO_4^{3-} + H^+$ (pK_{a3})). The dissociation constants for above three reactions are $pK_{a1} = 2.15$, $pK_{a2} = 7.20$ and $pK_{a3} = 12.33$, respectively [31]. When the pH of solution is between 2.15 and 7.20, the phosphate acid exists mainly in the form of $H_2PO_4^-$, which is the greatest affinity for the La-chitosan magnetic spheres to form La $(OH)_2^+$. As a result, the adsorbent has a large adsorption capacity, which is basically consistent with the experimental results (Figure 10).

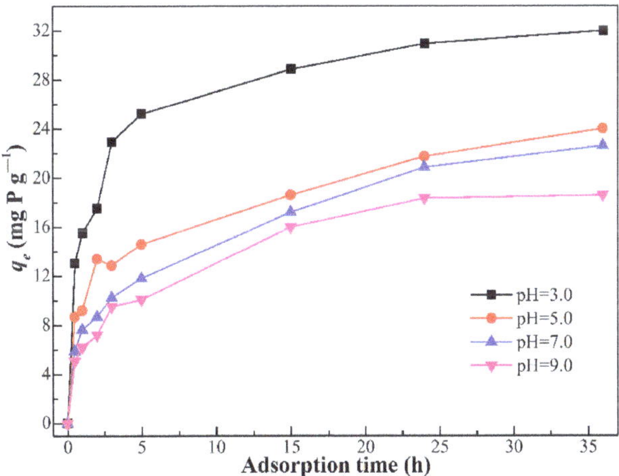

Figure 10. Effect of pH on the adsorption capacity of the La-chitosan magnetic spheres.

3.2.7. The Effect of Coexisting Anions

Several anions such as CO_3^{2-}, Cl^-, and NO_3^- are common in natural freshwater, and could participate in the phosphate adsorption process by means of competition of adsorption sites of adsorbents. Figure 11 obviously shows the effect of coexisting ions including CO_3^{2-}, Cl^-, and NO_3^- (0.1 mol·L^{-1}) on the adsorption of La-chitosan magnetic spheres. It is clearly found that Cl^- (13.51 mg P·g^{-1}), and NO_3^- (13.88 mg P·g^{-1}) has the slight/negligible influence on adsorption capacity of La-chitosan magnetic sphere, compared with blank group (13.69 mg P·g^{-1}), which means that there is almost no competitive adsorption between the above-mentioned ions (0.1 mol·L^{-1} Cl^- and NO_3^-) and the phosphate ions on La-chitosan magnetic spheres. However, CO_3^{2-} (5.92 mg P·g^{-1}) has wielded the most significant influence, which could be attributed to the smaller solubility product constant

(Ksp) (3.98×10^{-34}) of $La_2(CO_3)_3$ compared with the Ksp of $LaPO_4$ (3.70×10^{-23}). The competitive relationship for the adsorptive sites between co-existing CO_3^{2-} and phosphate in solution is formed, then resulting into a sharp decrease in the phosphate adsorption capacity [31].

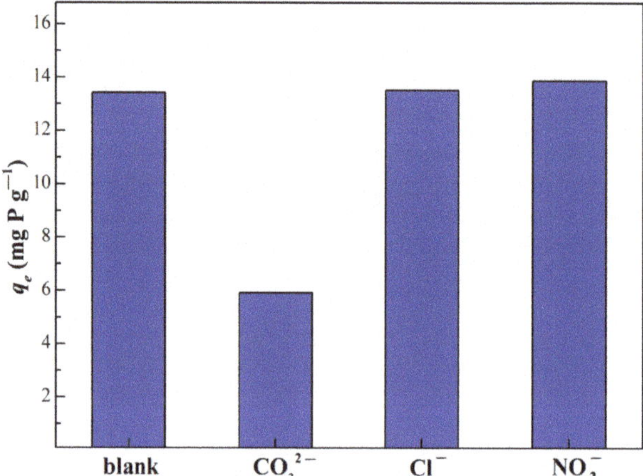

Figure 11. Effect of coexisting anions on the phosphate adsorption capacity of La-chitosan magnetic spheres.

4. Conclusions

In this paper, La-chitosan magnetic spheres were prepared and their adsorption performance for phosphates was studied. The prepared La-chitosan magnetic spheres adsorbents have many lamellar and regular porous structures with pore size of 15 μm. The main components of the adsorbents are FeOOH and La_2O_3, and the adsorption capacity of the La-chitosan magnetic spheres is obviously higher than that without lanthanum. La-chitosan magnetic spheres show a high adsorption rate for phosphate, and the adsorption process is in accordance with the pseudo-second order kinetic model and belongs to chemical adsorption, monolayer adsorption and electrostatic attraction, and ligand exchange from the values of $\Delta G°$ and the effect of pH on the adsorption capacity. In the lanthanum-doping process, both –NH$_2$ and –OH participate in the coordination. The ratio of La-chitosan magnetic spheres has an effect on the adsorption of La-chitosan magnetic spheres. The results of conditional experiments show that the optimal La/CS ratio is 1.0 mmol/g, of which the adsorption capacity is 27.49 mg P·g^{-1}; the positive correlation between the dosage of adsorbents and the capacity of the adsorbents; the initial pH of the solution has a significant effect on the adsorption capacity of La-chitosan magnetic spheres adsorbents. With the increasing of pH value, the adsorption capacity will be decreased significantly. At pH of 3.0, the maximum adsorption capacity is achieved, which is 33.04 mg P·g^{-1}. When the concentration of coexisting ions is 0.1 mol·L^{-1}, CO_3^{2-} will markedly reduce the phosphate adsorption capacity of as-prepared adsorbents, while Cl^- and NO_3^- have little effect. Owing to their unique hierarchical porous structures, high-adsorption capacity, La-chitosan magnetic spheres are potentially applicable in water treatment.

Author Contributions: Conceptualization: L.W. and L.-J.S.; methodology: L.W., L.S.; validation: L.-J.S., Y.-Y.Z., and L.W.; formal analysis: L.-J.S. and R.C.; investigation, literature search, study design: Y.-Y.Z.; data curation: L.-J.S.; data collection and data analysis: D.-Y.D.; figure design: R.C., L.-W.L.; writing-original draft preparation: L.-J.S. and L.-W.L.; writing-review and editing: R.C., X.Z; visualization and data interpretation: L.W.; supervision: L.S.

Funding: This research was funded by the Special Funds of the Construction of World-class Universities (Disciplines) and Guidance of Characteristic Developments for the Central Universities (Renmin University of China, 2018).

Acknowledgments: Thank L.W. for the careful revision of this article. L.-J.S. especially wants to thank the great support from R.C. during studying.

Conflicts of Interest: The authors declare no conflict of interest.

References

1. Mahdavi, S.; Akhzari, D. The removal of phosphate from aqueous solutions using two nano-structures: Copper oxide and carbon tubes. *Clean Technol. Environ. Policy* **2016**, *18*, 817–827. [CrossRef]
2. Zhang, M.; Gao, B.; Yao, Y.; Xue, Y.; Inyang, M. Synthesis of porous MgO-biochar nanocomposites for removal of phosphate and nitrate from aqueous solutions. *Chem. Eng. J.* **2012**, *210*, 26–32. [CrossRef]
3. Tu, Y.; You, C. Phosphorus adsorption onto green synthesized nano-bimetal ferrites: Equilibrium, kinetic and thermodynamic investigation. *Chem. Eng. J.* **2014**, *251*, 285–292. [CrossRef]
4. Yang, J.; Zeng, Q.; Peng, L.; Lei, M.; Song, H.; Tie, B.; Gu, J. La-EDTA coated Fe_3O_4 nanomaterial: Preparation and application in removal of phosphate from water. *J. Environ. Sci.* **2013**, *25*, 413–418. [CrossRef]
5. Lürling, M.; Van, O.F. Controlling eutrophication by combined bloom precipitation and sediment phosphorus inactivation. *Water Res.* **2013**, *47*, 6527–6537. [CrossRef] [PubMed]
6. Jabari, P.; Munz, G.; Yuan, Q.; Oleszkiewicz, J.A. Free nitrous acid inhibition of biological phosphorus removal in integrated fixed-film activated sludge (IFAS) system. *Chem. Eng. J.* **2016**, *287*, 38–46. [CrossRef]
7. Zheng, X.; Pan, J.; Zhang, F.; Liu, E.; Shi, W.; Yan, Y. Fabrication of free-standing bio-template mesoporous hybrid film for high and selective phosphate removal. *Chem. Eng. J.* **2016**, *284*, 879–887. [CrossRef]
8. Dithmer, L.; Nielsen, U.G.; Lundberg, D.; Reitzel, K. Influence of dissolved organic carbon on the efficiency of P sequestration by a lanthanum modified clay. *Water Res.* **2016**, *97*, 39–46. [CrossRef] [PubMed]
9. Yao, Y.; Gao, B.; Inyang, M.; Zimmerman, A.R.; Cao, X.; Pullammanappallil, P.; Yang, L. Removal of phosphate from aqueous solution by biochar derived from anaerobically digested sugar beet tailings. *J. Hazard. Mater.* **2011**, *190*, 501–507. [CrossRef] [PubMed]
10. Zhang, X.; Sun, F.; He, J.; Xu, H.; Cui, F. Robust phosphate capture over inorganic adsorbents derived from lanthanum metal organic frameworks. *Chem. Eng. J.* **2017**, *326*, 1086–1094. [CrossRef]
11. Chen, L.; Zhao, X.; Pan, B.; Zhang, W.; Hua, M.; Lv, L.; Zhang, W. Preferable removal of phosphate from water using hydrous zirconium oxide-based nanocomposite of high stability. *J. Hazard. Mater.* **2015**, *284*, 35–42. [CrossRef] [PubMed]
12. Zhang, L.; Gao, Y.; Zhou, Q.; Kan, J.; Wang, Y. High-Performance Removal of Phosphate from Water by Graphene Nanosheets Supported Lanthanum Hydroxide Nanoparticles. *Water Air Soil Pollut.* **2014**, *225*. [CrossRef]
13. Singh, M.; Srivastava, R.K. Sequencing batch reactor technology for biological wastewater treatment: A review. *Asia-Pac. J. Chem. Eng.* **2011**, *6*, 3–13. [CrossRef]
14. Lai, L.; Xie, Q.; Chi, L.; Gu, W.; Wu, D. Adsorption of phosphate from water by easily separable Fe_3O_4@SiO_2 core/shell magnetic nanoparticles functionalized with hydrous lanthanum oxide. *J. Colloid Interface Sci.* **2016**, *465*, 76–82. [CrossRef] [PubMed]
15. Su, Y.; Cui, H.; Li, Q.; Gao, S.; Shang, J.K. Strong adsorption of phosphate by amorphous zirconium oxide nanoparticles. *Water Res.* **2013**, *47*, 5018–5026. [CrossRef] [PubMed]
16. Yuan, X.; Pan, G.; Chen, H.; Tian, B. Phosphorus fixation in lake sediments using $LaCl_3$-modified clays. *Ecol. Eng.* **2009**, *35*, 1599–1602. [CrossRef]
17. Tu, C.; Wang, S.; Qiu, W.; Xie, R.; Hu, B. Phosphorus Removal from Aqueous Solution by Adsorption onto La-modified Clinoptilolite. *MATEC Web Conf.* **2016**, *67*, 07013. [CrossRef]
18. Zong, E.; Liu, X.; Wang, J.; Yang, S.; Jiang, J.; Fu, S. Facile preparation and characterization of lanthanum-loaded carboxylated multi-walled carbon nanotubes and their application for the adsorption of phosphate ions. *J. Mater. Sci.* **2017**, *52*, 7294–7310. [CrossRef]
19. Langmuir, I. The adsorption of gases on plane surfaces of glass, mica and platinum. *J. Am. Chem. Soc.* **1918**, *40*, 1361–1403. [CrossRef]
20. Freundlich, H. Über die adsorption in lösungen (adsorption in solution). *Z. Phys. Chem.* **1906**, *57*, 384–470.
21. Butt, H.J.; Graf, K.; Kappl, M. *Physical and Chemistry of Interfaces*; Wiley-VCH Verlag GmbH & Co. KGaA: Weinheim, Germany, 2004.

22. Ma, H.; Pu, S.; Ma, J.; Yan, C.; Zinchenko, A.; Pei, X.; Chu, W. Formation of multi-layered chitosan honeycomb spheres via breath-figure-like approach in combination with co-precipitation processing. *Mater. Lett.* **2018**, *211*, 91–95. [CrossRef]

23. Zhou, J.; Yang, S.; Yu, J.; Shu, Z. Novel hollow microspheres of hierarchical zinc-aluminum layered double hydroxides and their enhanced adsorption capacity for phosphate in water. *J. Hazard. Mater.* **2011**, *192*, 1114–1121. [CrossRef] [PubMed]

24. Yan, L.; Xu, Y.; Yu, H.; Xin, X.; Wei, Q.; Du, B. Adsorption of phosphate from aqueous solution by hydroxy-aluminum, hydroxy-iron and hydroxy-iron–aluminum pillared bentonites. *J. Hazard. Mater.* **2010**, *179*, 244–250. [CrossRef] [PubMed]

25. Haghseresht, F.; Wang, S.; Do, D.D. A novel lanthanum-modified bentonite, Phoslock, for phosphate removal from wastewaters. *Appl. Clay Sci.* **2009**, *46*, 369–375. [CrossRef]

26. Zhou, Q.; Wang, X.; Liu, J.; Zhang, L. Phosphorus removal from wastewater using nano-particulates of hydrated ferric oxide doped activated carbon fiber prepared by Sol–Gel method. *Chem. Eng. J.* **2012**, *200–202*, 619–626. [CrossRef]

27. Huang, W.; Li, D.; Liu, Z.; Tao, Q.; Zhu, Y.; Yang, J.; Zhang, Y. Kinetics, isotherm, thermodynamic, and adsorption mechanism studies of La (OH)$_3$-modified exfoliated vermiculites as highly efficient phosphate adsorbents. *Chem. Eng. J.* **2014**, *236*, 191–201. [CrossRef]

28. Kuroki, V.; Bosco, G.E.; Fadini, P.S.; Mozeto, A.A.; Cestari, A.R. Use of a La (III)-modified bentonite for effective phosphate removal from aqueous media. *J. Hazard. Mater.* **2014**, *274*, 124–131. [CrossRef] [PubMed]

29. Genz, A.; Kornmüller, A.; Jekel, M. Advanced phosphorus removal from membrane filtrates by adsorption on activated aluminium oxide and granulated ferric hydroxide. *Water Res.* **2004**, *38*, 3523–3530. [CrossRef] [PubMed]

30. Zamparas, M.; Gianni, A.; Stathi, P.; Deligiannakis, Y.; Zacharias, I. Removal of phosphate from natural waters using innovative modified bentonites. *Appl. Clay Sci.* **2012**, *62–63*, 101–106. [CrossRef]

31. Huang, W.; Li, D.; Yang, J.; Liu, Z.; Zhu, Y.; Tao, Q.; Xu, K.; Li, J.; Zhang, Y. One-pot synthesis of Fe (III)-coordinated diamino-functionalized mesoporous silica: Effect of functionalization degrees on structures and phosphate adsorption. *Microporous Mesoporous Mater.* **2013**, *170*, 200–210. [CrossRef]

Article

Synergistic Recapturing of External and Internal Phosphorus for In Situ Eutrophication Mitigation

Minmin Pan [1,2,3], Tao Lyu [4,5,*], Meiyi Zhang [1], Honggang Zhang [1], Lei Bi [1], Lei Wang [1], Jun Chen [1], Chongchao Yao [1], Jafar Ali [1], Samantha Best [4,5], Nicholas Ray [4,5] and Gang Pan [1,2,4,5,*]

[1]   Research Center for Eco-Environmental Sciences, Chinese Academy of Sciences, Beijing 100085, China; panminmin@hotmail.com (M.P.); myzhang@rcees.ac.cn (M.Z.); hgzhang@rcees.ac.cn (H.Z.); leibi@rcees.ac.cn (L.B.); leiwang@rcees.ac.cn (L.W.); chenjunzhl@126.com (J.C.); ancico@126.com (C.Y.); jafaraliqau@gmail.com (J.A.)
[2]   Sino-Danish College of University of Chinese Academy of Sciences, Beijing 100049, China
[3]   Department of Environmental Engineering, Technical University of Denmark, DK-2899 Lyngby, Denmark
[4]   School of Animal, Rural, and Environmental Sciences, Nottingham Trent University, Brackenhurst Campus, Nottingham NG25 0QF, UK; samantha.best02@ntu.ac.uk (S.B.); nicholas.ray@ntu.ac.uk (N.R.)
[5]   Centre of Integrated Water-Energy-Food Studies (iWEF), Nottingham Trent University, Nottinghamshire, Nottingham NG25 0QF, UK
*   Correspondence: tao.lyu@ntu.ac.uk (T.L.); gang.pan@ntu.ac.uk or gpan@rcees.ac.cn (G.P.); Tel.: +44-115-848-5296 (T.L.); +44-0115-848-5348 (G.P.)

Received: 30 November 2019; Accepted: 16 December 2019; Published: 18 December 2019

Abstract: In eutrophication management, many phosphorus (P) adsorbents have been developed to capture P at the laboratory scale. Existing P removal practice in freshwaters is limited due to the lack of assessment of the possibility and feasibility of controlling P level towards a very low level (such as 10 μg/L) in order to prevent the harmful algal blooms. In this study, a combined external and internal P control approach was evaluated in a simulated pilot-scale river–lake system. In total, 0.8 m^3 of simulated river water was continuously supplied to be initially treated by a P adsorption column filled with a granulated lanthanum/aluminium hydroxide composite (LAH) P adsorbent. At the outlet of the column (i.e., inlet of the receiving tanks), the P concentration decreased from 230 to 20 μg/L at a flow rate of 57 L/day with a hydraulic loading rate of 45 m/day. In the receiving tanks (simulated lake), 90 g of the same adsorbent material was added into 1 m^3 water for further in situ treatment, which reduced and maintained the P concentration at 10 μg/L for 5 days. The synergy of external and internal P recapture was demonstrated to be an effective strategy for maintaining the P concentration below 10 μg/L under low levels of P water input. The P removal was not significantly affected by temperature (5–30 °C), and the treatment did not substantially alter the water pH. Along with the superior P adsorption capacity, less usage of LAH could lead to reduced cost for potation eutrophication control compared with other widely used P adsorbents.

Keywords: eutrophication control; phosphorus recapturing; lake restoration; phosphorus adsorbent

---

1. Introduction

Phosphorus (P), largely derived from phosphate rock, is an essential nutrient for crop growth and hence for global food supply [1]. However, global P reserves are dramatically depleting due to the increasing demand for P-based agricultural fertilizers [2]. After the application of fertilizer to agricultural soils, substantial amounts of P can run away from agricultural land and enter natural waters, causing eutrophication, where harmful algal blooms (HABs), death of fish, degradation of aquatic macrophytes, and water quality deterioration are likely to occur [3]. Therefore, P removal and recovery from surface waters is one of the most important, yet difficult objectives for eutrophication control [4].

There is no consensus yet on the threshold of P concentration for preventing eutrophication in natural waters, even though many studies have concluded the concentration of 100 µg P/L to be too high [5,6]. Various methods, such as constructed wetland systems, have been applied in attempts to remove P and safeguard rivers [7]. P concentrations can be reduced from the 4.5–19.7 mg/L level to around 23 µg/L before entering surface waters by means of such wetlands, [8]. However, recent studies show that HABs could happen even at P levels as low as of 30 µg/L [9]. The strictest regulation of P concentration for lakes and reservoirs has been set down by the US Environmental Protection Agency as 10 µg P/L [10]. Thus, the development of an effective approach to achieve ultra-low concentrations of P is needed.

P removal techniques can be generally classified into physical, biological, and chemical methods. A combination of these methods has been reported to achieve less than 100 µg P/L [11], and sometimes even lower than 10 µg P/L [12]. Among these methods, P adsorption has been recognized as a promising technology for reaching such low P concentrations [13]. Many P adsorbents derived from natural minerals and biochars are not desirable because they are usually associated with high equilibrium P concentration at zero net P sorption (EPC$_0$) due to negatively bound P [14,15]. Industrial byproducts used as P adsorbents are cheap; however, their application often alters the pH of the water and produce adverse effects [16]. Thus, synthetic P adsorbents such as Phoslock® [17] have received growing attention, as their characteristics can be manipulated in order to maximize the P adsorption ability (Table 1). Even though Phoslock® has been used for P removal in lakes, a relatively high solid dosage of 200:1 Phoslock:P weight ratio is necessary to reach the desired levels of less than 10 µg P/L [18]. Nevertheless, there is a demand for a highly efficient P adsorbent that can control P levels of natural waters at around 10 µg/L.

Table 1. The material-based phosphorous adsorbent categories and their adsorption capacities.

Category	Name	Adsorption Capacity (mg P/g)
Natural materials	Soil or sands	4.2–5.8 [19]
	Calcite	4.1 [20]
	Limestone	0.3 [21]
Industrial byproducts	Slags	2.0–2.3 [22]
	Fly ash	6.6 [23]
	Red mud	7 [23]
Modified/synthetic products	Phoslock®	7.2–75 [24]
	AlgalBLOCK®	50 [25]
	BaraClear®	25–55 [24]

We previously developed an effective P adsorbent, lanthanum/aluminium hydroxide composite (LAH), which has a P adsorption capability 5–8 times higher than Phoslock® [26]. However, like other new P adsorbent demonstration experiments [27], the P adsorption capability has been tested in small-scale experiments using water samples with high P concentrations of 30 mg/L or above [11,14]. Equilibrium adsorption studies are often conducted under conditions with relatively long contact time (hours or days) [28]. Runoff and river water represent a great challenge for P removal due to their high volume and flow rate, which makes some bench-scale results hardly applicable under up-scaled field conditions.

In this study, the P recapture capability of the P adsorbent LAH, previously demonstrated to be superior in lab-scale batch experiments, was further evaluated in a simulated pilot-scale river–lake system. The inflow of the system was designed as a more realistic situation, with an inflow rate of 57 L/day and P concentration of 230 µg/L. In order to achieve the strict target of 10 µg P/L in natural waters, the concept of a combined approach of external P recapture by a P adsorbent column and further internal P removal inside the simulated lake (receiving tank) by directly applied adsorbent was investigated. Moreover, the feasibility of the LAH adsorbent for further implementation is discussed against other widely used P adsorbents. The study aimed to explore an effective P adsorbent and a promising strategy for eutrophication management.

2. Materials and Methods

2.1. Experimental Setup

Based on the concept of a combined approach of external–internal P adsorption, the simulated river–lake system (Figure 1) was set up with an inflow rate of 57 L/day for 14 days during the experiment. First, 0.8 m^3 of simulated river water was flowed through the adsorption column (Ø 4 cm and height 60 cm) in order to test the external P-capturing treatment. The columns were filled with LAH–zeolite-coated adsorbent. The water/adsorbent contact time in the adsorption columns was 16 min and the hydraulic loading rate was 45 m/day [29]. To simulate river water, the P concentration of the influent was set at the level of 230 μg/L [9]. The effluent water passed through the column and then flowed into the receiving tank. The receiving tanks (1 m^3) were made of PVC and glass (panel only) with the dimension of 1 m × 1 m × 1 m in length × height × width. Additional P adsorbent was placed in the receiving tanks for in-lake P capturing. The treatment systems were designed in duplicate, and another system without any LAH adsorbent was run as the control. The size of the tank for the control system was 0.375 m^3 (0.5 m × 0.5 m × 1.5 m) due to the unavailability of the same sized tank as was used with systems A and B.

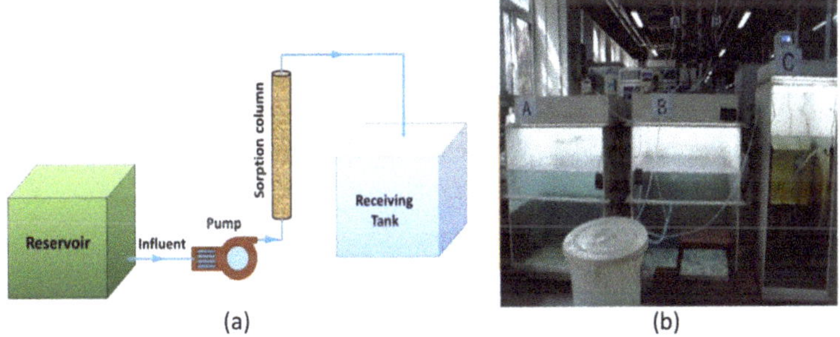

Figure 1. The conceptual arrangement (**a**) and picture (**b**) of the pilot scale experimental setup.

2.2. P Adsorbents and Inflow Water Preparation

The LAH adsorbent was prepared by precipitating lanthanum (La) with aluminium (Al) to obtain a La/Al hydroxide composite (5.3% La) following the procedure described by Xu [26]. In order to reduce the adsorbent loss through water flush in the column, the LAH adsorbent powders were coated onto zeolite particles (Ø 0.5–1 cm). In the adsorption column, 280 g of this modified adsorbent material (LAH mass percentage of 10.7%) was employed. At the beginning of the experiment, 836 g of LAH-modified zeolite material was placed in the bottom of receiving tank (a total of 90 g of LAH adsorbent in 1 m^3 water).

The simulated river water was prepared using tap water with a total phosphorus (TP) concentration of around 230 μg/L. To achieve this concentration, KH_2PO_4 and $C_6H_6Na_{12}O_{24}P_6$ (sodium phytate) were used, resulting in a composition of 87% inorganic and 13% organic P. Additionally, NH_4Cl, $NaNO_3$, and $NaCl$ were added and the ammonium, total nitrogen, and chloride concentrations were 0.65, 12, and 100 mg/L, respectively. The experiment was conducted at room temperature (around 28 °C). Receiving tanks for the duplicated treatment groups were filled with 0.75 m^3 tap water at the beginning of the experiment, while in the control system, the receiving tank was left empty as no further P adsorption treatment was applied.

2.3. Sampling and Analysis

Water samples from inflow water, adsorption column outflow, and receiving tanks in the duplicated treatment groups were collected every day. Under the same daily sampling campaign, only inflow and receiving tank water samples were taken for the control system. All water samples were collected in triplicate using 100 mL sterilized glass bottles and stored at 4 °C before being analyzed within 48 h. The pH and electrical conductivity (EC) were measured using portable meters (Multi-Parameter Meter HQ40d, and Sension + EC5, HACH, Loveland, CO, USA). After sample digestion with potassium persulfate, the TP concentration was determined by the Mo–Sb anti-spectrophotometer method [30]. Meanwhile, the TN concentration was measured by ultraviolet spectrophotometry following a digestion by potassium persulfate [31].

2.4. Adsorption Equilibrium Test

To understand the effect of temperature on P adsorption equilibrium and the kinetics of the LAH adsorbent, adsorption isotherm experiments were conducted in 50 mL centrifuge tubes with 1 g/L LAH suspensions at an ionic strength of 0.01 M NaCl. A series of initial phosphate concentrations of 100, 200, 300, 1000, 3000, 6000, and 10,000 µg/L were added and made up to a total volume of 30 mL in each tube. The centrifuge tubes were continually agitated for 24 h in a thermostatic shaker at 150 rpm at different temperatures (5, 10, 20, and 30 °C). After completion, the suspensions were collected and the P concentrations were determined. Adsorption data were fitted by the Freundlich isotherm model using the following equation [32]

$$q_e = K_F C_e^{1/n} \tag{1}$$

where q_e is the amount of phosphate adsorbed on the solid phase (mg/g), C_e (mg/L) is the equilibrium phosphate concentration, K_F is the constant of the adsorption capacity, and $1/n$ is the constant of the intensity of adsorption.

3. Results & Discussion

3.1. External and Internal P Recapture

In this study, the P concentrations of the outflow of the adsorption columns and in the receiving tanks decreased rapidly in the treatment systems (Figure 2a). However, the control system had no change in P concentration between the inflow and the effluent in the receiving tank. The P concentration of the adsorption column effluent decreased rapidly from 230 µg/L to around 20 µg/L in the first 2 days, and remained at this reduced level until Day 5. P concentrations then gradually increased until Day 14, reaching a concentration of approximately 100 µg/L. Based on the mass balance calculation for P, the adsorption was 56.1 mg P/g LAH adsorbent at Day 14. Even though the previously demonstrated equilibrium P-adsorption capacity of LAH was 76.3 mg P/g LAH in a batch experiment [26], the present study showed that the P-adsorption ability and rate was reduced around 25%. The result agrees with the previous study that the column adsorption capacity for P could be negatively affected by high flow rate and low inflow P concentration [33]. The current flow rate (57 L/day) was generally much higher than that in the reported column-mode studies [34]. Under such conditions, the adsorption capacity (56.1 mg P/g) still showed comparable value to other synthesized P adsorbents under the batch experiments.

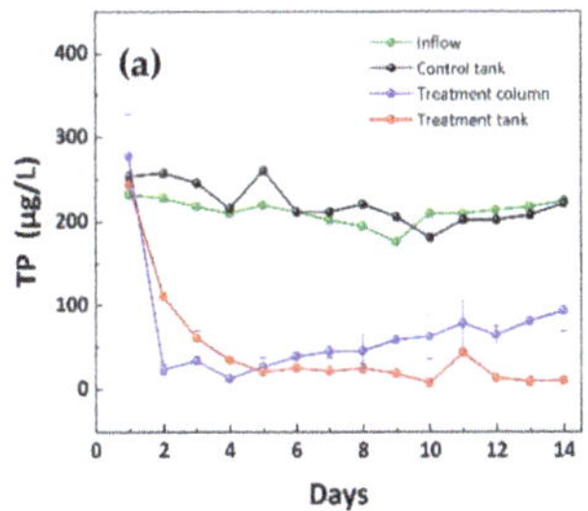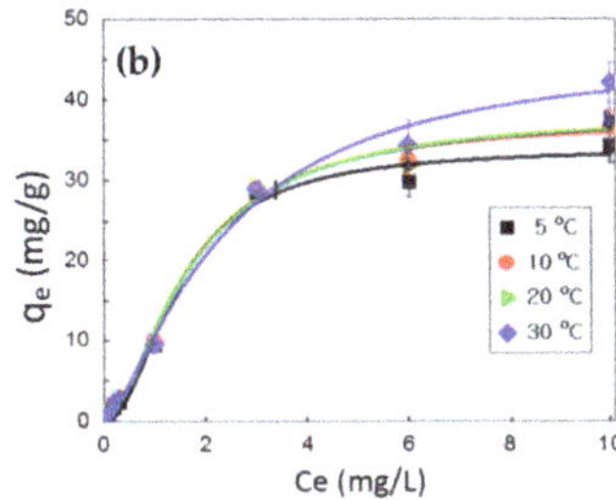

Figure 2. The dynamics of total phosphorus (TP) (**a**) in the inflow and outflow from the adsorption columns and tanks. (**b**) Freundlich adsorption isotherm fitting plots of phosphorus adsorption by LAH adsorbent.

The reduced P-adsorption ability may also have been due to the possibility of the LAH powder loss from the coated zeolite during water flushing. The majority of the synthetic P adsorbent consists of small particles; thus, the techniques used in a granulating process or one that is used to coat them onto a substrate are important for their effective retention in the adsorption columns [35]. In order for the field implementation of the LAH P adsorbent, it is important to optimize the granulation technique in the production of LAH adsorbent particles.

Geoengineered P adsorbent materials can be applied in situ in lakes for internal P recapture [36]; however, the target of 10 µg P/L in the water could hardly be achieved by a single operation. Previous studies have reported that in-lake treatment by the application of a P adsorbent, such as modified zeolite, directly into the water, could only decrease the P concentration from 769 µg/L to 310 µg/L [37]. Phoslock®, a synthesized P adsorbent, has been recommended at a dosage at 200 g adsorbent per m³ lake waters with an initial 100 µg P/L in order to reach the final P concentration of 10 µg P/L [18]. In the present study, the simulated river water was first treated by the P-adsorption column and then subjected to further P recapturing in the receiving tank with 90 g LAH adsorbent per m³ water. The P concentration in the tank gradually decreased to 20 µg/L on Day 5, and generally remained below 10 µg/L after Day 10 (Figure 2a). The results indicated that the highly effective LAH adsorbent could be used to successfully recapture ultra-low P.

3.2. P Equilibrium Adsorption Capacity and Potential Interaction with Water Quality

Natural water body treatment of P usually requires long-term and stable performance under multi-environmental changes such as pH and temperature. A previous study investigated the effect of pH on LAH performance and indicated adsorption stability against pH changes between 4 and 10 [26]. In this study, the P adsorption isotherm results (Figure 2b) fitted well with the Freundlich model (R^2 range of 0.91 to 0.96, Table 2). The equilibrium achieved a stable P adsorption capacity from 34 to 40 mg P/g, which supported that LAH adsorbent maintained a relatively stable P adsorption performance within the temperature range of 5–30 °C. This may be due to the wide temperature stability of the La and Al hydroxide in LAH, which could lead to a relatively stable positive charge of LAH surface to attract the negatively charged PO_4^{3-} [26]. Comparatively, the P adsorption capacity of Phoslock® can be significantly inhibited by low environmental temperatures [38]. The stable performance of the LAH adsorbent indicated long-term P adsorption effectiveness and a wider application than P adsorbents that can be easily affected by temperature fluctuation.

Figure 3a shows the similar pH values ($p > 0.05$) between the tank outflow (range of 7.6–8.0) and the inflow (pH = 7.5), which support the idea that the LAH adsorbent exhibited limited aquatic environmental disturbance. Besides P, the accumulation of nutrient N in water is also a significant factor that could cause water eutrophication [39]. In this study, TN concentrations generally remained at the same level (average of 12 mg/L) throughout the experiment (Figure 3b). This result agrees with the previous study on LAH, where the adsorbent had slight or no significant effect on the concentration of N [26]. When considering future engineering applications, improvements in coating materials could be made so that N may be removed simultaneously. For example, materials such as resins could be considered as a coating material, instead of zeolite, which may provide an additional function for N removal [40].

Table 2. Freundlich isotherm nonlinear curve fit parameters.

Temperatures	$q_e = kC_e{}^{1/n}$	Standard Error	R^2
5 °C	k = 16.8 1/n = 0.41	2.39 0.49	0.96
10 °C	k = 22.4 1/n = 0.29	4.89 1.56	0.91
20 °C	k = 21.7 1/n = 0.31	2.35 0.61	0.96
30 °C	k = 20.1 1/n = 0.45	2.69 0.41	0.96

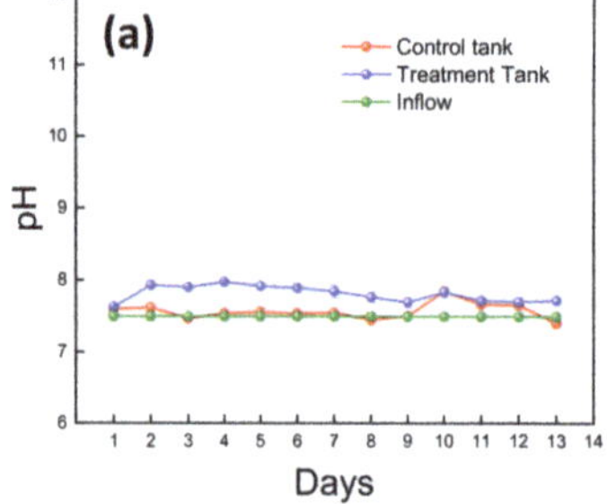

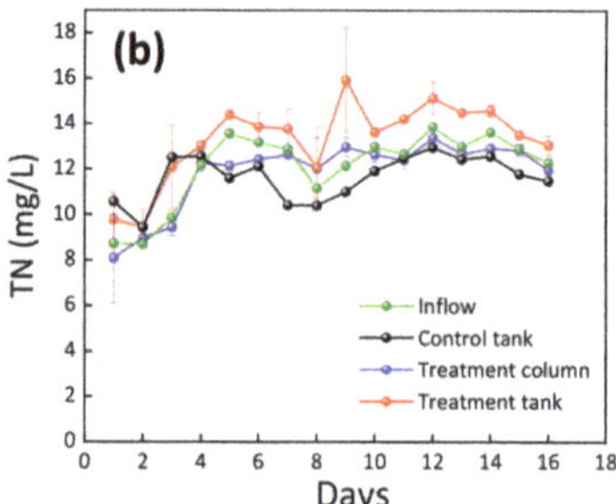

Figure 3. The dynamics of pH (**a**) and total nitrogen (TN) (**b**) in the inflow and outflow from the adsorption columns and tanks.

Notably, in real applications adsorbents will always be affected by the comprehensive surrounding factors, e.g., sediment, organism, vegetation, etc. As the first scaled-up study, this investigation focused on stoichiometric capacity, regardless of the complexity mentioned before. The results revealed the stable and high stoichiometric P-adsorption capacity of the LAH in a pilot study, which will be the foundation for further development into real applications. Synthetic investigations including sediment, organisms, and vegetation in lakes should also be conducted in further research.

3.3. Implementation Feasibility

The current study demonstrated the high P adsorption capacity of the LAH adsorbent under a realistic inflow rate and P concentration. Such adsorption ability could, in turn, decrease the applied dosage, which might contribute to a reduced cost for practice and La release potential. Currently, P adsorbents are often directly applied into water bodies for in situ P recapture. This study demonstrated that inducing runoff or river water flow through a P-adsorbent column can achieve significant P removal externally on land prior to the internal P recapture. Moreover, the P desorption of the LAH was reported in a previous study with a rate of 0.22%–4.05% when the pH changed within 4 to 8.5 [26]. The coated material improved the LAH from powder shape into granules, which makes the recycling and reutilisation of the adsorbent after desorption possible.

4. Conclusions

A pilot-scale system was used to simulate a river–lake system from the stoichiometric standpoint and to evaluate the approach for eutrophication control regardless of sediment, biology, and other environmental factors, based on the concept of combined external P removal from river water and internal P recapture from receiving lakes. After the treatment with the P-adsorption column equipped with the LAH adsorbent and direct application of LAH adsorbent into receiving tanks, the final P concentration could be reduced from 230 µg/L to below 10 µg/L. The synergy of external and internal phosphorus recapture proved to be an effective strategy for potential eutrophication management under mesotrophic P waters input. The results also demonstrated that the application of LAH adsorbent could reduce the cost for geoengineering for eutrophication management. Further research before application should be conducted, considering the synthetic effect of sediment, organisms, and vegetation in lakes.

Author Contributions: Conceptualization and supervision, G.P.; Investigation, L.B., L.W., H.Z., M.Z., J.C., C.Y., J.A.; Writing—Original Draft, M.P.; Writing—Review&Editing, T.L., S.B., N.R. and G.P. All authors have read and agreed to the published version of the manuscript.

Funding: This research was funded by the National Key Research and Development Program of China, grant number [2017YFA020724] and the Proof of Concept (POC) project at Nottingham Trent University. The PhD fellowship of Minmin Pan was supported by the China Scholarship Council (CSC).

Acknowledgments: Chen Liu, Ying Tang, Qing Xu, Xiaonan Ji, Xiaoguang Jin, Jing Su, Wanchun Zhang, Kaiqi Wang, Chen Wang and Mick Cooper are acknowledged for their valuable assistance with the experiment conduction, data analysis, and language improvement.

Conflicts of Interest: The authors declare no conflict of interest.

References

1. Elser, J.; Bennett, E. Phosphorus cycle: A broken biogeochemical cycle. *Nature* **2011**, *478*, 29–31. [CrossRef] [PubMed]

2. Van Vuuren, D.P.; Bouwman, A.F.; Beusen, A.H.W. Phosphorus demand for the 1970–2100 period: A scenario analysis of resource depletion. *Glob. Environ. Chang.* **2010**, *20*, 428–439. [CrossRef]

3. Peñuelas, J.; Poulter, B.; Sardans, J.; Ciais, P.; Velde, M.; Bopp, L.; Boucher, O.; Godderis, Y.; Hinsinger, P.; Llusia, J.; et al. Human-induced nitrogen-phosphorus imbalances alter natural and managed ecosystems across the globe. *Nat. Commun.* **2013**, *4*, 2934. [CrossRef] [PubMed]

4. Pan, G.; Lyu, T.; Mortimer, R. Comment: Closing phosphorus cycle from natural waters: Re-capturing phosphorus through an integrated water-energy-food strategy. *J. Environ. Sci.-China* **2018**, *65*, 375–376. [CrossRef] [PubMed]

5. Richardson, C.J.; King, R.S.; Qian, S.S.; Vaithiyanathan, P.; Qualls, R.; Stow, C.A. Estimating ecological thresholds for phosphours in the everglades. *Environ. Sci. Technol.* **2007**, *41*, 8084–8091. [CrossRef]

6. Carvalho, L.; McDonald, C.; Hoyos, C.D.; Mischke, U.; Phillips, G.; Borics, G.; Poikane, S.; Skjelbred, B.; Solheim, A.L.; Wichelen, J.V.; et al. Sustaining recreational quality of European lakes: Minimizing the health risks from algal blooms through phosphorus control. *J. Appl. Ecol.* **2013**, *50*, 315–332. [CrossRef]

7. Khare, Y.; Naja, G.M.; Stainback, G.A.; Martinez, C.J.; Paudel, R.; van Lent, T. A Phased Assessment of Restoration Alternatives to Achieve Phosphorus Water Quality Targets for Lake Okeechobee, Florida, USA. *Water* **2019**, *11*, 327. [CrossRef]

8. Dierberg, F.E.; DeBusk, T.A.; Jackson, S.D.; Chimney, M.J.; Pietro, K. Submerged aquatic vegetation-based treatment wetlands for removing phosphorus from agricultural runoff: Response to hydraulic and nutrient loading. *Water Res.* **2002**, *36*, 1409–1422. [CrossRef]

9. Grenon, G.; Madramootoo, C.A.; Singh, B.; Gaj, N. *Water Quality Management in the Holland Marsh, Ontario*; ASABE Annual International Meeting; American Society of Agricultural and Biological Engineers: St. Joseph, MI, USA, 2016; p. 1.

10. United States Environmental Protection Agency (U.S. EPA). *Nutrient Criteria Technical Guidance Manual, Lakes and Reservoirs*; Office of Science and Technology: Washington, DC, USA, 2000; p. 822. Available online: https://www.epa.gov/nutrient-policy-data/criteria-development-guidance-lakes-and-reservoirs (accessed on 20 June 2000).

11. Genz, A.; Kornmüller, A.; Jekel, M. Advanced phosphorus removal from membrane filtrates by adsorption on activated aluminium oxide and granulated ferric hydorxide. *Water Res.* **2004**, *38*, 3523–3530. [CrossRef]

12. Zhu, X.P.; Jyo, A. Column-mode phosphate removal by a novel highly selective adsorbent. *Water Res.* **2005**, *39*, 2301–2308. [CrossRef]

13. Kumar, P.S.; Korving, L.; van Loosdrecht, M.C.M.; Witkamp, G.J. Adsorption as a technology to achieve ultra-low concentrations of phosphate: Research gaps and economic analysis. *Water Res. X* **2019**, *4*, 100029. [CrossRef] [PubMed]

14. Bhadha, J.H.; Jennewein, S.P.; Khatiwada, R. Phosphorus Sorption Behavior of Torrefied Agricultural Byproducts under Sonicated Versus Non-Sonicated Conditions. *Sustain. Agric. Res.* **2017**, *6*, 4. [CrossRef]

15. Zhou, M.F.; Li, Y.C. Phosphorus-sorption characteristics of calcareous soils and limestone from the southern everglades and adjacent farmlands. *Soil Sci. Soc. Am. J.* **2000**, *65*, 1404–1412. [CrossRef]

16. Proctor, D.; Fehling, K.; Shay, E.; Wittenborn, J.; Green, J.; Avent, C.; Bigham, R.; Connolly, M.; Lee, B.; Shepker, T. Physical and chemical characteristics of blast furnace, basic oxygen furnace, and electric arc furnace steel industry slags. *Environ. Sci. Technol.* **2000**, *34*, 1576–1582. [CrossRef]

17. Lürling, M.; Waajen, G.; van Oosterhout, F. Humic substances interfere with phosphate removal by lanthanum modified clay in controlling eutrophication. *Water Res.* **2014**, *54*, 78–88. [CrossRef]

18. Reitzel, K.; Andersen, F.; Egemose, S.; Jensen, H.S. Phosphate adsorption by lanthanum modified bentonite clay in fresh and brackish water. *Water Res.* **2013**, *47*, 2787–2796. [CrossRef]

19. Kim, D.; Ryoo, K.S.; Hong, Y.P.; Choi, J.H. Evaluation of loess capability for adsorption of total nitrogen (T-N) and total phophorous (T-P) in aqueous solution. *Bull Korean Chem. Soc.* **2014**, *35*, 2471–2476. [CrossRef]

20. Molle, P.; Lienard, A.; Grasmick, A.; Iwema, A. Phosphorus retention in subsurface constructed wetlands: Investigations focused on calcareous materials and their chemical reactions. *Water Sci. Technol.* **2003**, *48*, 75–83. [CrossRef]

21. Drizo, A.; Frost, C.A.; Grace, J.; Smith, K.A. Physico-chemical screening of phosphate-removing substrates for use in constructed wetland systems. *Water Res.* **1999**, *33*, 3595–3602. [CrossRef]

22. Drizo, A.; Forget, C.; Chapuis, R.P.; Comeau, Y. Phosphorus removal by electric arc furnace steel slag and serpentinite. *Water Res.* **2006**, *40*, 1547–1554. [CrossRef]

23. Wang, Y.; Yu, Y.G.; Li, H.Y.; Shen, C.C. Comparison study of phosphorus adsorption on different waste solids: Fly ash, red mud and ferric-alum water treatment residues. *J. Environ. Sci.-China.* **2016**, *50*, 79–86. [CrossRef] [PubMed]

24. Svatos, K.B. Commercial silicate phosphate sequestration and desorption leads to a gradual decline of aquatic systems. *Environ. Sci. Pollut. Res.* **2018**, *25*, 5386–5392. [CrossRef] [PubMed]

25. Gonçalves, M.A.S.C. *New Eco-Efficient Polymers for Phosphorus Recovery*; Universidade do Minho-Campus de Gualtar: Braga, Portugal, 2012; Available online: https://search.proquest.com/docview/1900972696?pq-origsite=gscholar (accessed on 1 April 2012).

26. Xu, R.; Zhang, M.; Mortimer, R.J.; Pan, G. Enhanced phosphorus locking by novel lanthanum/aluminum-hydroxide composite: Implications for eutrophication control. *Environ. Sci. Technol.* **2017**, *51*, 3418–3425. [CrossRef]

27. Iizuka, A.; Sasaki, T.; Hongo, T.; Honma, M.; Hayakawa, Y.; Yamasaki, A.; Yanagisawa, Y. Phosphorus adsorbent derived from concrete sludge (PAdeCS) and its phosphorus recovery performance. *Ind. Eng. Chem. Res.* **2012**, *51*, 11266–11273. [CrossRef]

28. Yin, H.; Kong, M. Simultaneous removal of ammonium and phosphate from eutrophic waters using natural calcium-rich attapulgite-based versatile adsorbent. *Desalination* **2014**, *351*, 128–137. [CrossRef]

29. Hsieh, C.H.; Davis, A.P.; Needelman, B.A. Bioretention column studies of phosphorus removal from urban stormwater runoff. *Water Environ. Res.* **2007**, *79*, 177–184. [CrossRef]

30. Fu, X.Y.; Cui, G.Y.; Huang, K.; Chen, X.M.; Li, F.S.; Zhang, X.Y.; Li, F. Earthworms facilitate the stabilization of pelletized dewatered sludge through shaping microbial biomass and activity and community. *Environ. Sci. Pollut. Res.* **2016**, *23*, 4522–4530. [CrossRef]

31. Ding, Z.B.; Yu, Z.H.; Cheng, T.T.; Zhang, P.; Gao, X. Research on experimental condition of analysing total nitrogen of water by ultraviolet spectrophotometric method. *Water Purif. Technol.* **2008**, *27*, 61–64. Available online: http://en.cnki.com.cn/Article_en/CJFDTotal-ZSJS200801018.htm (accessed on 25 January 2008).

32. Pang, G.; Liss, P.S. Metastable-equilibrium adsorption theory. *J. Colloid Interface Sci.* **1998**, *201*, 71–76.

33. Xing, X.; Gao, B.; Wang, W.; Yue, Q.; Wang, Y.; Ni, S. Adsorption of phosphate from aqueous solutions onto modified wheat residue: Characteristics, kinetic and column studies. *Colloid Surface B* **2009**, *70*, 46–52.

34. Han, R.; Wang, Y.; Zou, W.; Wang, Y.; Shi, J. Comparison of linear and nonlinear analysis in estimating the Thomas model parameters for methylene blue adsorption onto natural zeolite in fixed-bed column. *J. Hazard. Mater.* **2007**, *145*, 331–335. [CrossRef]

35. Douglas, G.; Hamilton, D.; Robb, M.; Pan, G.; Spears, B.M.; Lurling, M. Guiding principles for the development and application of solid-phase phosphorus adsorbents for freshwater ecosystems. *Aquat. Ecol.* **2016**, *50*, 385–405. [CrossRef]

36. Robb, M.; Greenop, B.; Goss, Z.; Douglas, G.; Adeney, J. Application of Phoslock, an innovative phosphorus binding clay, to two Western Australian waterways: Preliminary findings. *Hydrobiologia* **2003**, *494*, 237–243. [CrossRef]

37. Ozkundakci, D.; Hamilton, D.P.; Scholes, P. Effect of intensive catchment and in-lake restoration procedures on phosphorus concentrations in a eutrophic lake. *Ecol. Eng.* **2010**, *36*, 396–405. [CrossRef]

38. Haghseresht, F.; Wang, S.; Do, D.D. A novel lanthanum-modified bentonite, Phoslock, for phosphate removal from wastewaters. *Appl. Clay Sci.* **2009**, *46*, 369–375. [CrossRef]

39. Conley, D.J.; Paerl, H.W.; Howarth, R.W.; Boesch, D.F.; Seitzinger, S.P.; Havens, K.E.; Lancelot, C.; Likens, G.E. Controlling eutrophication: Nitrogen and phosphorus. *Science* **2009**, *323*, 1014–1015. [CrossRef]

40. Langlois, J.L.; Johnson, D.W.; Mehuys, G.R. Adsorption and recovery of dissolved organic phosphorus and nitrogen by mixed-bed ion-exchange resin. *Soil. Sci. Soc. Am. J.* **2003**, *67*, 889–894. [CrossRef]

Article

Hyporheic Process Restoration: Design and Performance of an Engineered Streambed

Paul D. Bakke [1,*], Michael Hrachovec [2] and Katherine D. Lynch [3]

[1] The Science of Rivers, 4031 Wexford Loop SE, Olympia, WA 98501, USA
[2] Natural Systems Design, 1900 N. Northlake Way, Suite 211, Seattle, WA 98103, USA; rocky@naturaldes.com
[3] Seattle Public Utilities, 700 5th Ave, Suite 4900, Seattle, WA 98124, USA; katherine.lynch@seattle.gov
* Correspondence: thescienceofrivers@outlook.com; Tel.: +1-360-412-0220

Received: 13 December 2019; Accepted: 23 January 2020; Published: 5 February 2020

Abstract: Stream restoration designed specifically to enhance hyporheic processes has seldom been contemplated. To gain experience with hyporheic restoration, an engineered streambed was built using a gravel mixture formulated to mimic natural streambed composition, filling an over-excavated channel to a minimum depth of 90 cm. Specially designed plunge-pool structures, built with subsurface gravel extending down to 2.4 m, promoted greatly enhanced hyporheic circulation, path length, and residence time. Hyporheic process enhancement was verified using intra-gravel temperature mapping to document the distribution and strength of upwelling and downwelling zones, computation of vertical water flux using diurnal streambed temperature patterns, estimation of hyporheic zone cross section using sodium chloride tracer studies, and repeat measurements of streambed sand content to document evolution of the engineered streambed over time. Results showed that vertical water flux in the vicinity of plunge-pool structures was quite large, averaging 89 times the pre-construction rate, and 17 times larger than maximum rates measured in a pristine stream in Idaho. Upwelling and downwelling strengths in the constructed channel were larger and more spatially diverse than in the control. Streambed sand content showed a variety of response over time, indicating that rapid return to an embedded, impermeable state is not occurring.

Keywords: hyporheic; hyporheic zone restoration; stream restoration; urban watercourse restoration; streambed temperature; tracer studies; streambed grain-size distribution; vertical water flux

---

1. Introduction

A river's boundary does not end at the channel margins. Instead, the river system includes physical processes that extend laterally, into the riparian zone and floodplain, and vertically, into the substrate beneath the channel and floodplain [1–3]. This transitional zone between subterranean and surface aquatic environments, commonly referred to as the hyporheic zone [4], provides ecological functions/benefits that help sustain streambed and aquatic conditions. Among these functions are vertical water flux between the stream and subsurface, water temperature moderation, recycling of carbon, energy, and nutrients, natural attenuation of certain pollutants, a sink/source of sediment for the channel, and habitat for benthic and interstitial organisms [4]. The role of the hyporheic zone is increasingly recognized for its significance with regard to river management, conservation, and restoration [4–9], and as such, the restoration of hyporheic zone processes was one of the primary design goals of the City of Seattle's floodplain pilot projects.

Although improved hyporheic processes may be mentioned as a project goal [10], rarely do hyporheic processes drive restoration or design [11,12]. Usually, the argument is put forth that, by constructing a channel that mimics a natural alluvial morphology, and by reestablishing natural processes of sediment mobilization, transport, and deposition, the hyporheic functions will be reestablished or enhanced as an ancillary benefit. Even in cases of "stream simulation design" [13], as in road crossing structures, where continuity of the streambed through the structure without changes in grain size composition or width is a key design goal, features that explicitly enhance hyporheic processes are not included, and hyporheic zone size or function is not quantified.

Recently, however, interest in using hyporheic processes to achieve specific, quantifiable benefits has been growing. Herzog et al. [14], for example, showed how streambed composition and channel slope could be engineered to maximize hyporheic interchange. In urban streams, stormwater contaminants such as metals, petrochemical derivatives, excessive nutrients, and pathogens could be attenuated by designing hyporheic residence times to match contaminant transformation time [14]. Crispell and Endreny [15] mapped and quantified hyporheic exchange flow around J-hook vanes and rock cross vanes, two commonly applied techniques used in channel reconstruction, which extend hyporheic residence times. Engineered streambeds, which provide instream treatment, could conceivably supplement other emerging stormwater best management practices, such as "rain gardens," "biofiltration swales", and "compost-amended vegetated filter strips" [16], which improve the quality of stormwater runoff [17,18].

Other water quality benefits, such as temperature moderation, particularly reduction of excessive summer temperatures, could also be achieved, and would be a more easily measured benefit than contaminant mitigation, which can be difficult to quantify [19–26]. Creation of a large thermal "heat sink" in the streambed could result in net cooling of surface waters [27]. Even in cases where the thermal mass of the streambed becomes temperature-saturated [28], the time lag between when water enters versus reemerges from the hyporheic zone could effectively dampen diurnal temperature fluctuations, by causing cooler nighttime water to reemerge during mid-day, mixing with, and cooling, warm surface water [27].

Water quality problems are only beginning to be addressed in infrastructure with the same diligence as runoff timing and volume. In places like the Pacific Northwest of the U.S., where culturally and economically important fish still exist in streams of urban watersheds, there is substantial political will to reverse the declining trends in water quality that have, up until now, been largely ignored. Early spawner mortality of salmonid fishes, for example, has spurred efforts to identify the nature and sources of its chemical causes [29,30]. Early spawner die-off is a syndrome typically tracked in female coho salmon, and is observed when females, ready to spawn, migrate into a creek, but die before laying eggs. The exact cause is not known [31,32], but the phenomenon is more extensive in urban areas, and is correlated with stormwater runoff from impervious surfaces, particularly roads with high traffic volumes [33,34]. Even without information about the specific toxic chemicals, efforts are underway to devise engineered infrastructure, such as compost-amended vegetated filter strips, which might treat the sources of stormwater runoff adequately enough to reduce this source of mortality [16–18].

Nevertheless, current and developing engineered designs for stream restoration and stormwater treatment are unlikely to achieve the vision of a fully functioning, self-maintaining stream system, which has been described as the "river corridor" [7]. Because of the narrow scope of engineered solutions, and the piecemeal approach of addressing only one, or a narrow range of impacts at a time, these solutions alter the behavior of the river corridor in unforeseen ways. As is common to all complex systems, rivers have emergent behaviors, that is, behaviors that appear only due to the system functioning as a whole and cannot be predicted from individual parts of that system. In particular, rivers have self-organizing and self-maintaining properties that stormwater ponds, or compost-amended vegetated filter strips, do not. Rivers undergo threshold responses to change, even gradual change, that depend on the interconnections between the wetted channel, the floodplain, the riparian zone vegetation, the soil and streambed, and subsurface flows of water, both lateral and vertical [35]. The riverbed is the integrated product of hydrological patterns, sediment dynamics, aquatic biota, riparian inputs, channel structural complexity, etc. It maintains and renews itself through the interaction of all these things.

Common engineered solutions, by contrast, do not sustain themselves in a way that is coherent with the river corridor. They usually require maintenance, such as, dredging and material disposal, which can be expensive. Moreover, they do not create functional habitat, and the types of habitat that result are often attractive nuisances. For example, by using a stormwater pond that is saturated with toxic sediments, waterfowl may reduce their survival fitness. Stormwater ponds and other infrastructure may be physically dangerous to humans as well, requiring special measures to exclude them. Finally, they often take up large portions of the limited land surface area that could provide functional habitat, often, ironically, occupying former floodplain wetland sites.

The City of Seattle tried a different approach to manage stormwater by pilot testing and evaluating the two Thornton Creek floodplain reconnection projects, as a way to reduce flooding and erosion from peak flow discharges, and to restore hyporheic and stream processes to improve reach-scale channel morphology, streambed composition, and water quality. These latter two elements involved engineering the streambed.

Currently, streambed engineering to achieve water quality benefits can be considered a field in its infancy, and treatments are all experimental in nature [36]. There are no standard engineering practices to achieve it, and no widespread acceptance of it within the community of river managers and restoration practitioners. The authors know of no examples of such streambed engineering in urban settings other than the one to be described in this paper.

This study was designed to test the hypothesis that it is possible to improve reach scale streambed condition and hyporheic processes by using an engineered streambed that is designed to maximize hyporheic exchange and residence time. A modified Before-After-Control-Impact Design was used to compare a restored reach ("treatment reach/site") to its pre-project condition (Before Impact), and to an upstream reference reach ("control reach/site"). Streambed and hyporheic process improvements were measured using a combination of physical, biological, and water quality (temperature and chemistry) monitoring. This study focused on the physical and water temperature monitoring, and the biological and water chemistry were monitored by other researchers ([37] and [30], respectively). The streambed and physical hyporheic processes were monitored over a period of three years after project construction to evaluate the ability of the engineered streambed to continue performing over time, particularly in a highly urbanized environment, which is characterized by large inputs of sediment, especially sand.

2. Thornton Creek Watershed and Study Site Description

Thornton Creek drains nearly 29 km^2 (11.1 square miles) of urban Seattle and Shoreline, Washington, USA, flowing through two second order tributaries and a 2.1 km (1.3 mile) mainstem into Lake Washington, just north of the Lake Washington Ship Canal [38] (see Figure 1). The watershed lies entirely within Pleistocene glacial drift, consisting of till and outwash deposits overlying compacted glaciolacustrine clay, and ranges in elevation from about 150 m in its headwater areas to 2.5 m at its mouth (490 to 8.2 feet, respectively) [38].

Originally heavily forested, the watershed receives nearly 89 cm (35 inches) of precipitation annually, which falls primarily as rain between October and May [39]. Currently, roads, buildings, and other impervious surfaces comprise 59% of the watershed area [38], and most of the remainder has been converted to non-native vegetation such as lawns and landscaping with different runoff, soil stability, and erodibility properties than originally existed.

The stream channel has been straightened and its banks hardened over much of its length. In most places, the streambed has been simplified by channel incision, deliberate channelization, and removal of its structural framework of large wood, such that alluvial gravels and sands exist only in a thin veneer, 15–25 cm thick (6–10 inches), over compacted till or clay. Structural complexity, which formerly would have retained much deeper alluvial deposits, is largely missing. The pre-project streambed surface layer median (D_{50}) grain sizes ranged from 15.7 to 58.2 mm (0.6–2.3 inches), and widespread embedded conditions reduce permeability to water movement.

The glacial-drift stratigraphy of the Thornton Creek watershed is conducive to development of springs where permeable sands overlie till or clay. This groundwater emerges on the edge of the ravines through which creek channel flows. Normally, this water feeds floodplain wetlands, or enters the stream channel along its banks in places where the channel lies adjacent to ravine walls.

Several species of salmonid fish are present. Cutthroat trout (*Oncorhynchus clarki*) are the most common, but coho salmon (*O. kisutch*), and on occasions, Chinook salmon (*O. tshawytscha*), sockeye salmon (*O. nerka*), and rainbow trout (*O. mykiss*) are seen [38]. Rarely has the anadromous form of rainbow trout, the steelhead, been documented [38].

Typical of many urban streams [40], Thornton Creek suffers from excessive stormwater runoff and poor water quality. As a consequence, average early spawner mortality for coho salmon in Thornton Creek is the highest (79%) among Seattle's creeks [38,39].

In order to address the effects of stormwater runoff, Seattle Public Utilities (SPU) embarked on two channel and floodplain reconstruction projects: The Kingfisher site, located on the South Fork of Thornton Creek, and the Forks Confluence site, located at the confluence of the South and North Forks of Thornton Creek (see Figure 1). The Kingfisher site has a drainage area of 6.6 km^2 (2.5 square miles) and the Forks Confluence site drains 9.8 km^2 (3.8 square miles).

Control sites were established immediately upstream of each of the restoration sites in sections of the channel that were morphologically similar to pre-restoration conditions. For the purposes of discussion in this paper, the restoration sites shall be referred to as the treatment sites.

This paper will focus almost exclusively on the Kingfisher Site, as at this site the full suite of hyporheic restoration techniques were implemented, whereas at the Forks Confluence site, only a portion of the hyporheic restoration techniques were implemented. At both sites, the control reach was located as close as possible upstream from the treatment sites; at Kingfisher, the control reach is approximately 90 m (300 feet) upstream of the treatment reach.

The Kingfisher reach is a gaining reach, with significant amounts of groundwater entering the channel from the surrounding hillsides (particular the south side) in a mixture of surface flow at discrete locations and dispersed subsurface flow [41]. Figure 2 illustrates typical pre-construction site conditions in and adjacent to the channel at the Kingfisher Treatment site. Small seasonal wetlands and overbank sand deposits existed on top of compacted artificial fill, which covered the original (but highly disturbed) alluvial deposits. Impermeable compacted clay underlies the site, forming a barrier to vertical water movement and a partial control on channel incision.

Figure 3 presents pre- and post-restoration photos of the Kingfisher Control and Treatment reaches, as well as the relative locations of each. Flow arrows on Figure 3a document direction of flow.

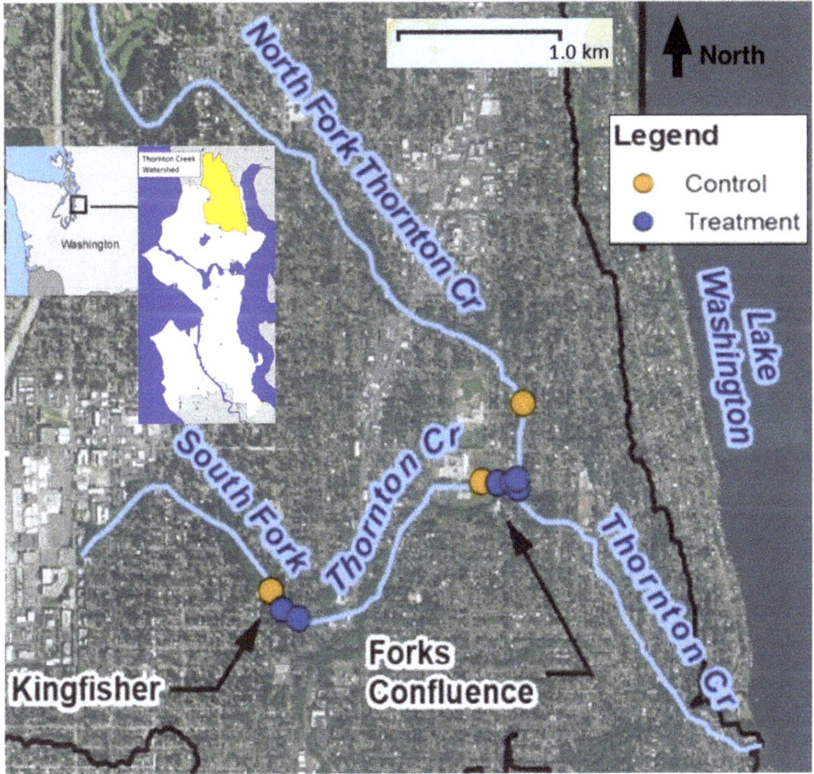

Figure 1. Locations of control and treatment sites for both Seattle Public Utilities (SPU) Thornton Floodplain projects. Right inset shows the watershed (yellow) within the City of Seattle (white). Left inset is a portion of Washington State, USA. Scale shows 1 km = 0.6 mile.

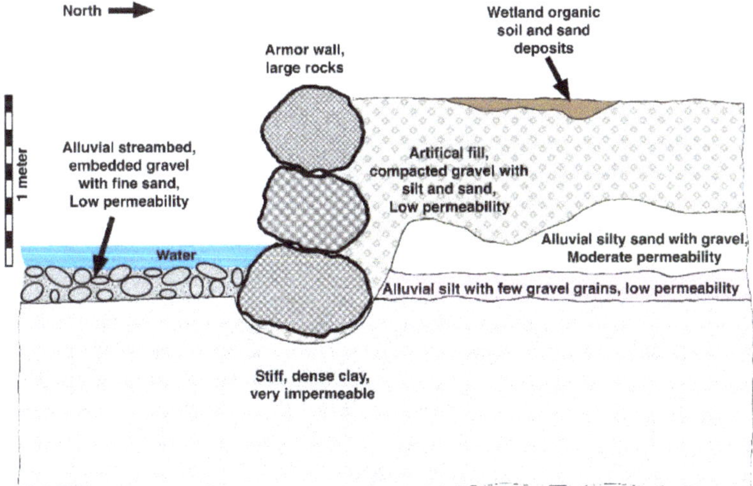

Figure 2. Schematic pre-construction substrate cross section at the Kingfisher Treatment site, looking upstream with channel center at left.

(a)

(b)

(c)

(d)

Figure 3. Views of Kingfisher Control and Treatment reach, including: (**a**) Aerial photo showing locations of the Kingfisher Control and Treatment reaches, (**b**) photo of the control reach, and photos of treatment reach (**c**) before, and (**d**) after, construction, taken from close to same vantage point (the "before" is taken closer to the south slope, on the right side of this photo).

Wherever possible, monitoring techniques were implemented both on the treatment and control reaches. Both the treatment and the control reaches have been heavily impacted by human activities, including complete removal of historic forest cover, removal of instream wood, channel straightening, installation of major sewer lines through and along the channel, encroachment by residential yards, bridges, and culverts. The control site is in a more natural condition than was the pre-project restored site, having a small but functioning floodplain, an established forest cover (albeit with numerous non-native species), small amounts of in-channel large wood, and a wider channel, being about 4.6–5.8 m (15–19 feet) in width. The pre-project treatment site had a channel width of only 2.4 m (8 feet), was lined with large rocks, and had no large wood, and the floodplain had been replaced by several feet of fill to accommodate the five houses, and their outbuildings, that were removed from the site. Of critical importance for this study, both the pre-project treatment site and the control site had experienced similar channel incision due to human activities resulting in extremely similar streambed conditions. Both treatment and control streambeds consisted of a thin veneer, 15–25 cm (6–10 inches) thick, of moderately embedded alluvial gravel over a dense, impermeable clay substrate. As such, hyporheic exchange and flow were extremely low [42].

3. Design of Hyporheic Restoration Features

Design of the hyporheic restoration features as a part of the overall stream restoration project was completed in 2013. Project construction occurred during the summer of 2014 and was essentially complete by October of that year. Monitoring of hyporheic processes began in the summer of 2015, after the constructed streambed had experienced one winter season of adjustment to the hydraulic conditions and sediment load, and then continued through October 2017.

Urban stream restoration typically consists of channel reconstruction with hardened elements, such as large-wood or rock structures, to control channel gradient and channel location. The constructed morphology, as much as is possible, is designed to mimic morphology and habitat features found in more pristine settings. Often, clean, highly permeable gravel mixtures are imported to improve streambed conditions at shallow depths of 15–30 cm (0.5–1 feet), but only directly along the channel alignment.

Key design features of both treatment sites consisted of construction of a longer, wider, more sinuous channel, with numerous large-wood structures and creation of an inset floodplain designed to engage at the 1.5-year annual peak discharge. A primary goal of this floodplain restoration project was to store and detain floodwater temporarily, reducing water velocities, moderating flood impacts downstream, and reactivating soil water storage and deeper infiltration.

A unique aspect of this restoration design was incorporation of an engineered streambed for improved hyporheic function. Flow of water into, and out of, the hyporheic zone was very small under pre-project conditions, on the order of 0.018–0.019 m/day (0.7 inch/day) [42], and hyporheic cross section was quite small, due to the thinness and embedded condition of the alluvial streambed [39]. By building a streambed specifically engineered to enhance hyporheic exchange, increase hyporheic residence times, and increase hyporheic volume, improvements to temperature and chemical water quality were anticipated.

Since this engineered streambed was innovative and experimental, an intensive effort was made to monitor its effectiveness in improving hyporheic function. This paper both describes the new design elements used in hyporheic restoration, as well as the monitoring protocol and subsequent results for both treatment and control reaches. The Kingfisher site was selected for this presentation since it has a thicker engineered streambed, and therefore was expected to exhibit a larger response, than the Forks Confluence site. A flowchart illustrating the project design process appears in Figure 4.

Four new techniques were implemented to augment hyporheic processes. These included: (1) Over-excavation and backfilling with clean gravel; (2) plunge-pool structures; (3) subsurface water drains; and (4) hyporheic logs and pocket-water logs.

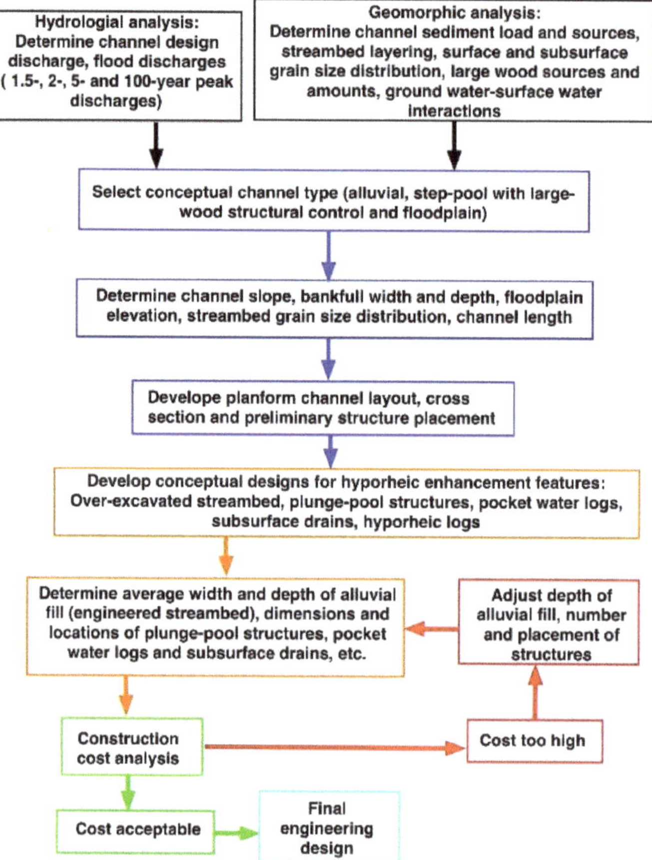

Figure 4. Project design process flowchart.

3.1. Over-Excavation and Backfilling with Clean Gravel

The primary hyporheic restoration action was installing a deep (90–240 cm or 3–8 feet), wide (up to 9.1 m or 30 feet) alluvial gravel corridor into which the active channel was constructed. The goal was to greatly increase the hyporheic cross section over pre-project conditions, and to make the hyporheic cross section significantly larger than is usually attained in urban restoration projects. At the Kingfisher site, a 90 cm (3 feet) thickness was specified, tapering to a smaller depth at the edges of the active channel. Meanwhile, average active channel width was increased from about 2.6 to 6.8 m (8.6 to 22.2 feet), which considerably increased the hyporheic cross section. The alluvial layer target width was 9.1 m (30 feet), varied to accommodate site constraints and was approximately twice the average bankfull channel width.

A critical design element was sizing of the gravel to maximize hyporheic function. The design specifications were set to approximate the surface gravel size at the control site as shown in Figure 5, but the fine materials (<2 mm) were omitted to maximize hydraulic conductivity. In the native (subsurface) material, this fine material fraction comprises approximately 15% of the volume. The goal was to create a mixture that would be similar enough to the original streambed alluvial material so that it would be maintained over time by natural scour, transport, and fill processes, and yet would not be occluded with fine sand and silt. A thin layer, 10–20 cm (4–12 inches) of native soil material was placed over the top of the alluvial layer outside of the boundary of the active channel to facilitate plant growth on the floodplain.

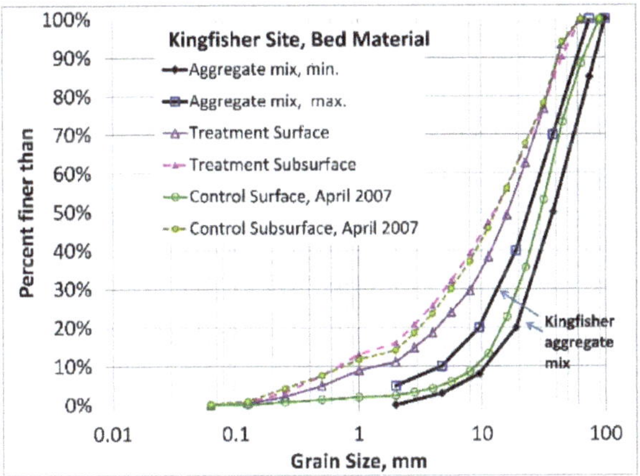

Figure 5. Design aggregate gravel size distribution relative to pre-project streambed gravel sizes.

3.2. Plunge-Pool Structures

Plunge-pool structures, constructed with channel-spanning sill logs to create a 24 cm (0.8 feet) step in the longitudinal profile of both the water surface and streambed, are a common design feature in stream restoration projects [43]. This step size is consistent with criteria for juvenile salmonid fish passage. The basic habitat enhancement goals for this structure are to generate plunging flow to maintain a scour pool for deep-water fish rearing habitat, with a resulting riffle at the tail-out of the pool; the upstream end of the riffle is typically a preferential location for salmonid spawning. For Thornton Creek, plunge-pool structures were designed as illustrated in Figure 6 below, with a deeply over-excavated placement hole, as much as 2.4 m (8 feet) deep, with 90 cm (3 feet) of clean gravel mixture placed in the hole before placing the channel-spanning sill log and footer log. In addition, a layer of compacted sand/silt/crushed-gravel mixture was placed on top of this gravel foundation to create an occluding layer, ranging in thickness from 1.5 m (5 feet) upstream to 30 cm (1 foot) downstream, which forces downwelling water to travel deeply beneath the structure before emerging in the tail out zone of the pool downstream. This increases the residence time and path length for water in the downwelling zone on the upstream side, in response to the hydraulic head differential created by the sill logs. The plunge-pool structure drives local hyporheic circulation, on the scale of one habitat unit, but the circulation below the pool is much deeper, and has a longer path length and residence time, than that associated with a simple channel-spanning log. At the Kingfisher site, six plunge-pool structures were installed, one of which was studied intensively.

A primary habitat benefit of this enhanced hyporheic circulation is cooling of the water as the heat from the incoming water is transferred to the subsurface gravel, and to the cooler native material below. Because the water is forced down more deeply, it encounters a cooler mineral substrate and interacts with a greater thermal mass for heat exchange. This cooler water then emerges into the pool at the upstream end of the riffle, benefiting both juvenile salmon rearing in the pool, as well as those eggs laid in the riffle [44]. The cooler water maintains a lower-stress condition for the eggs during development, and will improve overall survival [45]. Though not explored in this study, it is anticipated that in winter, the hyporheic circulation will warm the incoming water, with heat transferring from the native material into the hyporheic water [42]. On the broadest scale, this enhanced hyporheic circulation acts as a thermal buffer, moderating diurnal and seasonal water temperatures through greater heat exchange with the native parent material. This local water flux is superimposed upon the reach-scale hyporheic circulation induced by the site-wide over-excavation and backfilling with gravel. That is, some fraction of the water, which percolates down into the hyporheic zone at the upstream end of the reach, will stay in the subsurface, flowing downstream until forced to re-emerge at the end of the treatment project site.

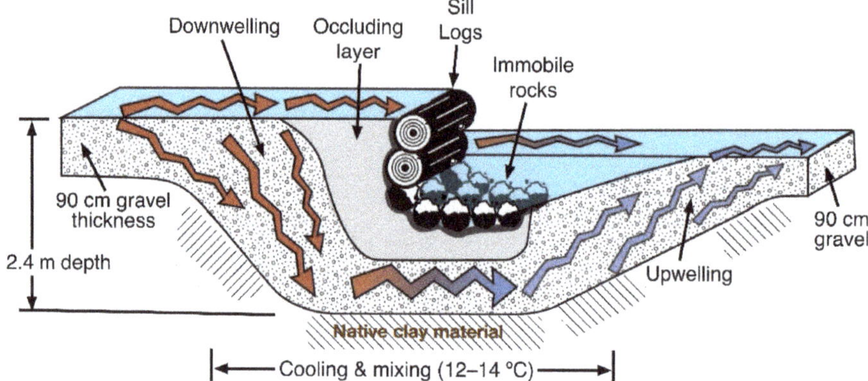

Figure 6. Longitudinal cutaway view of a plunge-pool structure, showing direction of surface and hyporheic flow.

3.3. Subsurface Drains

In pre-project conditions, water from hillslope seeps percolated through the wet soil or flowed across the soil surface to enter the stream channel. To capture this clean, cool water prior to it being warmed through air exposure or used up by plants, in one location, a subsurface trench was filled with permeable gravel that routed directly to the stream channel under the floodplain surface to deliver the water directly into adjacent plunge-pools, where cooler temperature inflows would provide the greatest habitat benefit (Figure 7). The effects of subsurface drains would manifest in reduced instream temperatures where the subsurface drain enters the stream. The effects of individual subsurface drains were not studied, though they would have a localized effect similar to upwelling hyporheic water.

3.4. Hyporheic Logs

Hyporheic logs were installed as single logs buried in the streambed at the bottom of the gravel layer within the active channel, approximately perpendicular to the flow, and with 10 cm (4 inches) of gravel separating them from the native substrate. They were intended to force hyporheic flow to mix laterally and vertically and increase hyporheic residence time and flow path length. The individual effects of hyporheic logs were not studied, though they would have an overall influence in greater lateral mixing of hyporheic flow and increased hyporheic residence time.

3.5. Pocket-Water Logs

Pairs of logs were installed at several locations to stabilize streambed elevation and slope (grade control), generate small scour pools (pocket water), and encourage hyporheic exchange similar to a plunge-pool, but on a smaller (shallower) scale. Each log had 30–46 cm (1–1.5 feet) of a coarse cobble mixture between it and the native substrate, which limited scour depth while allowing hyporheic flow. This cobble mixture was in turn covered with 15 cm (0.5 feet) of streambed gravel. Structures similar to these are frequently used in stream restoration projects [43] so they were not studied but are called out as they are part of the overall stream restoration project and integrate with other the hyporheic enhancement features.

Figure 7 shows an overall plan view of the majority of the Kingfisher restoration project, particularly the elements that were evaluated as a part of this study.

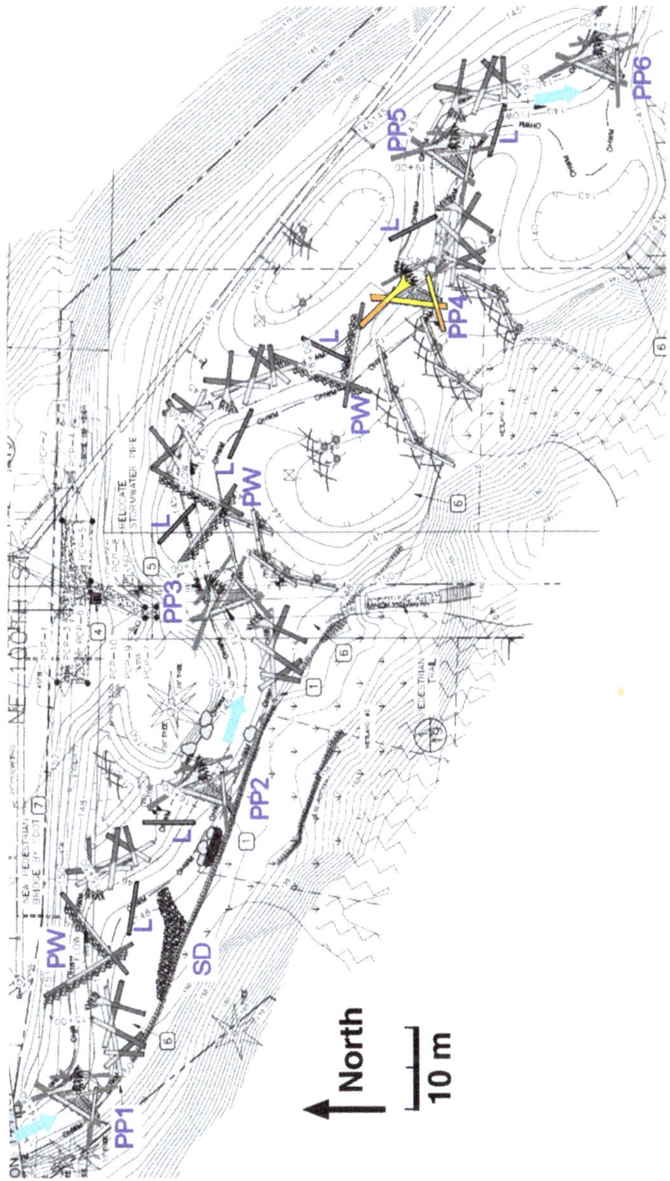

Figure 7. Plan view of the majority of the Kingfisher restoration project. Flow is from upper left to lower right, indicated by light blue arrows. Hyporheic enhancements include: PP 1–4 = plunge-pool structures; SD = subsurface water drain; PW = pocket-water logs; and, L = hyporheic log. Plunge-pool No. 4 (PP4) is highlighted in yellow (above ground portion) and orange (portion buried in streambed); this structure was singled out for intensive study.

4. Materials and Methods for Hyporheic Study

The physical condition and function of the hyporheic zone was monitored using five attributes, which are described below and summarized in Table 1. Techniques used to monitor each attribute differed in terms of their cost and effort, whether the outcome was precise and quantitative versus relative or qualitative, whether the scope was locally intensive versus spatially extensive, and their utility for detecting change over time. Taken together, this suite of techniques provides a basis for inferences about the degree to which hyporheic function has been achieved and its persistence. All tests were completed during stable, baseflow hydrological conditions, during dry weather when diurnal air temperature patterns were not changing in order to minimize confounding effects from changes in water discharge or weather-induced temperature shifts.

Table 1. Summary of evaluation attributes and techniques used to measure them.

Attribute of Evaluation	Technique Used	Cost/Effort	Inference Type	Spatial Extent	Detect Change over Time
Vertical water flux	Temperature piezometers with data loggers	High	Quantitative	Localized	Yes
Mapping upwelling and downwelling zones	Shallow tube array for intra-gravel temperatures	Low	Relative	Extensive	No
Immobile versus mobile cross-section ratio	NaCl tracer studies	Low	Relative	Average for entire reach	No
Lateral subsurface hydraulic gradient	Floodplain wells and stilling wells	Medium	Relative	Localized	Yes
Streambed surface and subsurface sand content	Volumetric surface and subsurface streambed samples	High	Quantitative	Localized	Yes

4.1. Measurement of Vertical Water Flux

Hyporheic exchange rate was quantified by estimating the vertical water flux at discrete points on the streambed from the 10-minute-interval temperature time series measured at the surface and at two depths in the gravel, 10 and 20 cm (4 and 8 inches). To calculate the vertical hyporheic flux in $m^3/m^2/day$ (which becomes m/d) between two elevations (e.g., streambed surface and 10 cm below the surface), we used the method of Garuglio et al., 2013 [46,47], which allows calculation of vertical water flux from temperature time-series data alone, without knowing pieziometric head or hydraulic conductivity.

At the Kingfisher restoration site, plunge-pool structure No. 4 was selected for vertical water flux instrumentation, as it was located in the middle of the reach in an area with floodplains on both sides of the channel, away from the influence of valley sides, and was installed with a full width of hyporheic gravel. At each of five locations upstream and five downstream of plunge-pool No. 4, stainless-steel piezometers 38 mm (1.5 inches) in diameter were driven into the streambed to a depth of 61 cm (2 feet) with their tops capped and flush to the streambed surface. Piezometers were custom fabricated from stainless-steel water pipe by crimping shut the lower end to act as a drive point. Each piezometer was equipped with a ring of 6 mm (0.25 inch) screen holes surrounding temperature data loggers (TidbiT v2 water temperature data loggers, model UBTI-001, Onset Computer Corp., Bourne, MA, USA) held at precise depths of 10 cm and 20 cm beneath the streambed surface. Foam gaskets (made from short sections of water pipe insulation) prevented water from moving vertically between each datalogger with its piezometer. The piezometers were placed in a T-shaped array of five units upstream of the plunge-pool structure, allowing for both a cross-channel and a longitudinal transect to be created. Downstream of the structure, only one of the five units in the array could be placed outside of the channel midline due to a more complex streambed topography and more vigorous gravel deposition/scour patterns. Each of the two arrays was also equipped with one data logger to record surface water temperature. Temperatures were measured every 10 min. In addition to the 10 piezometers bracketing plunge-pool No. 4, a piezometer was installed at the upstream end of the reach, just upstream of the sill log of plunge-pool No. 1, and two were placed at the far downstream end, in the pool tail-out zone for plunge-pool No. 6 (see Figure 7). These are zones where, on a reach scale, water is expected to first enter and then to finally exit the hyporheic zone, respectively [44].

To calculate water flux from the temperature data, we determined the ratio of the diurnal temperature sine wave amplitudes and the amount of phase shift between the two temperature sine waves bracketing the depth interval chosen. Amplitudes and phases were determined from the temperature time series by fitting the data to a sine function using non-linear least squares (NLS) regression. Calculations were performed for five time periods of 3–10-day durations, with stable (unchanging) weather and relatively constant discharge, and which were close enough to a piezometer maintenance visit to assure accurate knowledge of any streambed deposition or scour over the piezometer top. Flux over each of three possible depth intervals was determined: 0–10 cm, 0–20 cm, and 10–20 cm. The shallowest (0–10 cm) interval is likely to be the most representative of vertical exchange, since the assumption of 1-dimensional (vertical) flow holds best near the streambed surface. Large discrepancies between flux calculated for different depth intervals at the same piezometer location can occur if significant lateral flux is present.

Since there was frequent sediment deposition in this reach, the streambed elevation on top of each piezometer was surveyed with a total station theodolite (Nikon model DTM-520, Nikon-Trimble Co., Ltd., Tokyo, Japan) accurate to 3 mm (0.1 inch), to determine the actual depth to each data logger, before servicing (downloading and resealing). These actual depths were used in the calculations.

4.2. Mapping of Upwelling and Downwelling Zones

The locations and relative strengths of upwelling and downwelling zones were mapped out on a reach-wide scale by means of a comparison between the surface water temperature and the temperature at a depth of 10 cm (4 inches) into the streambed [48]. To record the temperature at the appropriate depths, an array of approximately 60 plastic tubes, of 9 mm (0.4 inch) inner diameter, were inserted into the streambed in equally spaced (approximately) transects of two to six tubes each. Tubes were cut from rolled flexible polyethylene water pipe. Figure 8 shows tube and data-logger locations for the Kingfisher Treatment reach.

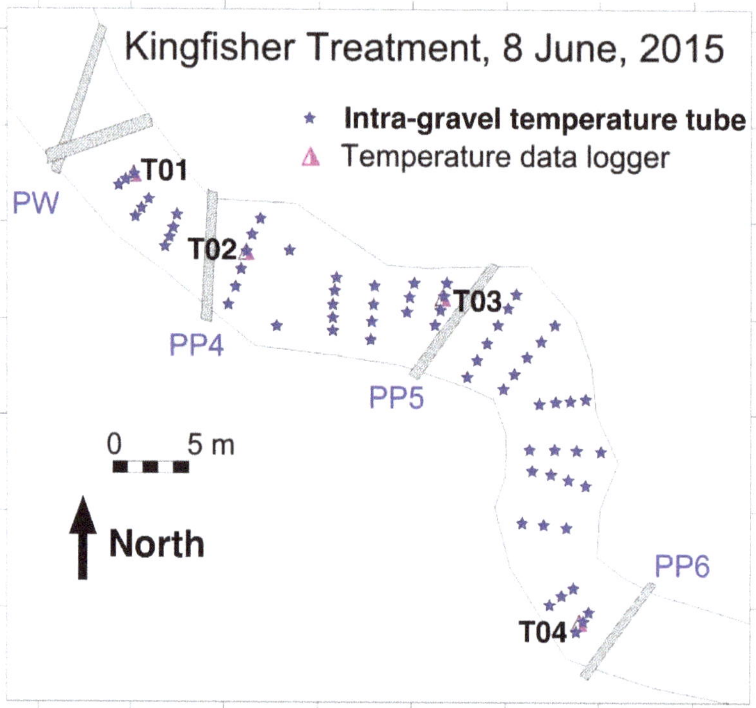

Figure 8. Locations of intra-gravel temperature measurement tubes and the four surface water temperature data loggers, T01–T04. PP4–6 are sill logs for plunge-pool structures No. 4–6, and PW is a pocket-water log structure.

The streambed temperature at the bottom of each tube was measured three times over the course of a few hours using a precision hand-held thermister probe accurate to 0.015 °C (model HH41 ultra-high accuracy hand held thermistor thermometer, Omega Engineering, Inc., Norwalk, CT, USA). The surface water temperature was measured at four locations in the reach at 10- or 15-minute intervals using programmable temperature data loggers accurate to 0.21 °C (TidbiT v2 water temperature data loggers, model UBTI-001, Onset Computer Corp., Bourne, MA, USA), and surface water temperature at each transect of tubes estimated by spatial and temporal interpolation. The average difference in temperature between the surface water and the interstitial water 10 cm (4 inches) into the streambed is an indicator of relative upwelling versus downwelling, and the relative strength of that process [48]. The temperature differences between surface water and subsurface were normalized by subtracting each difference from the median of all the differences, and then the average of the three normalized differences for each tube was computed for comparison. These averages are reported in the results.

Since the four surface water temperature data loggers were placed at locations uniformly spaced along the reach, as shown in Figure 8, they were also used to examine the longitudinal temperature patterns in the reach. The test was conducted during warm, sunny weather, before the vegetation at the Kingfisher Treatment site had re-established any shade. The hypothesis is that, under these conditions, steady warming of surface water in the downstream direction would be expected in the absence of influx of cool subsurface water. A different or less uniform pattern of temperature change would be strong evidence for hyporheic mixing and/or substantial groundwater influence. This longitudinal temperature pattern will be interpreted in the results separately.

4.3. Immobile versus Mobile Cross Section Ratio

Sodium chloride (NaCl) tracer studies are well-established as a method to characterize hyporheic exchange processes, using the relative size of the estimated "immobile" (i.e., long residence time) water cross-sectional area compared with that of the surface water flow ("mobile," short residence time) cross-sectional area as a surrogate for the strength of hyporheic water flux processes [49].

In a tracer study, concentrated sodium chloride solution is injected into the stream at a constant rate, and the time series of electrical conductivity (a surrogate for chloride ion concentration) in the surface water is measured. Measurement at a point where the water is thoroughly mixed immediately downstream of the injection point constitutes the "input" pulse, and measurement at the far downstream end of the reach documents how this input pulse is modified by longitudinal dispersion and exchange and mixing with hyporheic and other relatively "immobile" water. Once the pulse of increased conductivity at the end of the reach has stabilized to a constant "plateau" value, the tracer injection is turned off, and the conductivity time series measurements continued as the conductivity declines towards the background level.

A one-dimensional dispersion-advection model with immobile storage elements (OTIS-P, [50]) is then calibrated to maximize fit of the measured conductivity pulse at the end of the reach to the model predictions, given the measured mixing zone input pulse. Calibration gives an estimate of the effective "immobile" and "mobile" zone water cross sections [50,51]. The ratio of these cross sections is then used to infer hyporheic function [51].

At the Kingfisher Treatment and Control sites, electrical conductivity was measured at 30-second intervals using submersible data loggers (Hobo fresh water conductivity data logger, model U24-001, Onset Computer Corp., Bourne, MA, USA). Conductivity was normalized by subtracting the upstream background conductivity, which was fairly high (average 218–221 microsiemens/cm). Surface water discharge was 8.9 L/s (0.32 ft^3/s) and 23.8 L/s (0.84 ft^3/s) at the treatment and control sites, respectively. At the treatment site, tracer solution containing 205 g/L NaCl was injected at a rate of 0.920 L/min, with a 6-min interruption (76–84 min) due to a pump reservoir blockage. At the control site, 205 g/L NaCl solution was injected at a rate of 0.710 L/min.

Model calibration focused on optimizing model parameters to minimize sum of squared error over the time representing the rapidly increasing limb of the pulse, the plateau, and the rapidly decreasing portion of the trailing limb, but not the long-term recovery to background levels. This strategy is consistent with standard practice, given that the assumptions of a one-dimensional model approximation become less valid over longer time intervals [51].

4.4. Lateral Subsurface Hydraulic Gradient

Lateral hyporheic flow potential was estimated using an array of five floodplain monitoring wells installed outside the limits of the constructed streambed gravel, and stilling wells equipped with data loggers to record water surface levels within the channel (Hobo 13-foot fresh water level data logger, model U20-001-04, Onset Computer Corp., Bourne, MA, USA, measurement accuracy 3 mm). The wells were aligned in three distinct transects representing the subsurface hydraulic gradient perpendicular to in-channel flow. Two transects were defined upstream of plunge-pool structure No. 4 and one downstream of plunge-pool No. 4. Measurements of water surface elevation in the floodplain wells was done by hand on eight occasions during steady, baseflow conditions, allowing inferences about lateral subsurface flow throughout different seasons. Hand measurements were done with a custom-made (by the authors) electronic water surface detector attached to a stiff measuring tape, accurate to 1 mm.

4.5. Streambed Surface and Subsurface Sand Content

Streambed sediment texture (grain-size composition), through its effect on hydraulic conductivity and porosity, controls hyporheic processes. In particular, evolution of the streambed grain size composition over time, as the project adjusts to its hydraulic and sediment regime, will determine whether hyporheic processes in the engineered streambed persist. The pre-project streambed was highly embedded with fine sand, effectively cutting off water movement. Thus, one consideration of over-excavation and backfilling with clean gravel was the possibility that embedded conditions might reestablish themselves in the surface layers, blocking hyporheic exchange, and that this could potentially happen within only a year or two.

Sample sites were selected to be representative of active alluvial portions of the streambed. Where possible, these comprised pool tail-out zones or glides. To collect the sample for use in volumetric analysis, the site was first isolated with a 3-sided plywood shield sealed to the streambed to prevent flushing of fine sediment by water currents as the sample was excavated but open on the downstream side to allow easy access. First, the surface layer was excavated using gloved hands to the embedded depth of the larger particles comprising it. Finally, the subsurface layer was separately excavated beneath this level, to the depth necessary to obtain roughly the same amount of material as was taken for the surface layer. These methods are described in McNamara et al. [52] and Bunte and Abt [53].

Samples were taken to the laboratory, dried, sieved, and weighed according to standard methods [53], using sieve sizes based on a powers of two (Wentworth scale) sequence. Although a complete particle grain-size distribution was obtained by this process, detailed interpretation of the overall distribution was beyond the scope of this report. For inference purposes, comparisons were based on percent sand content by weight, with sand being defined as all grain sizes passing through the 2-mm sieve.

Sampling was performed in 2015 after the streambed had one winter season of post-construction adjustment, and then again in 2017 at the Kingfisher sites. At the Confluence sites, post-project sampling was performed in 2016 and 2017. For this study attribute, results from both the Kingfisher and Forks Confluence sites are reported in the results, to give a more comprehensive picture of sediment dynamics.

5. Results

Physical monitoring results are presented separately for each technique.

5.1. Vertical Water Flux

Prior to restoration, vertical (upward) water flux at the Kingfisher site was estimated at 0.019 m/d (0.7 in/d) or 2.20×10^{-7} m/s, using a deeper piezometer placement (1.4 m or 4.5 feet) and a modeling approach that utilized both temperature and piezometric head data [39,42]. After restoration, the site demonstrated an average downwelling of 1.69 m/d (5.5 ft/d) and maximum of 8.10 m/d (25.6 ft/d), an 89-fold increase over pre-project rates. The comparison in rates is indirect because two different methods were used; the thin veneer of alluvial substrate present during pre-project conditions necessitated modeling vertical flux as an average over a much larger thicknesses (1.4 m), which included non-alluvial substrate beneath this thin veneer [39,42]. The vertical water flux values at Kingfisher after restoration are significantly greater than those reported in pristine streams (a maximum of 0.48 m/d or 1.6 ft/d, in Bear Valley Creek, Idaho, USA, reported by Gariglio et al. [46]).

Upwelling was monitored with only 1/20 of the intensity of sampling wells and ranged from none observed to as high as 6.59 m/d (21.6 ft/d). Most upwelling zones in the vicinity of plunge-pool No. 4 were located in areas of very dynamic streambed scour and fill, which were unsuitable for piezometer placement. The computed vertical water flux is summarized below in Table 2, which reports the maximum and average values for each layer and time period.

Table 2. Summary of vertical water flux measurements for the Kingfisher restoration site.

| | | Layer | | | | | |
| | | 0–10 cm | | 0–20 cm | | 10–20 cm | |
Start Date	Value	Up-Welling, m/d	Down-Welling, m/d	Up-Welling m/d	Down-Welling, m/d	Up-Welling m/d	Down-Welling, m/d
27 September 2015	Max	None obs.	3.39	None obs.	9.91	6.59	8.07
	Mean	-	1.40	-	3.57	6.59	2.91
5 May 2016	Max	0.40	8.10	0.37	8.76	3.78	9.36
	Mean	0.40	3.08	0.37	2.34	1.65	2.64
10 August 2016	Max	0.31	3.40	0.49	3.39	0.88	3.27
	Mean	0.31	1.19	0.49	1.51	0.47	1.93
24 August 2016	Max	None obs.	5.60	None obs.	7.08	0.05	8.92
	Mean	-	1.82	-	2.22	0.05	2.55
7 May 2017	Max	None obs.	3.24	None obs.	3.07	5.68	6.07
	Mean	-	1.46	-	0.91	2.09	1.18
No. of measurements		2	44	3	42	2	44
Maximum		0.40	8.10	0.88	9.91	6.59	9.36
Mean		0.35	1.69	0.44	2.13	2.00	2.24
Range		No–0.40	0.01–8.10	No–0.49	0.08–9.91	0.05–6.59	0.01–9.36

Figure 9 is a graphical depiction of the vertical water flux in m/d for the different depth layers at plunge-pool No. 4 of the Kingfisher site for a range of dates, to illustrate the variation found between various dates and locations near the plunge-pool. Red or blue dots indicate downwelling versus upwelling, respectively, and are sized by magnitude. For scale, red and blue dots in map legend represent 1.0 m/d (3.3 ft/d) flux. The complete set of values plotted on this figure are provided in the Supplementary Materials, Spreadsheet S1.

Floodplain well measurements (red and blue diamonds) are part of the discussion of lateral subsurface hydraulic gradient (Section 3.5, below) but are included in these figures to assist in interpretation of the piezometer analysis. Where the floodplain well has a higher water surface elevation than the nearby surface water stilling well, it is shown in blue (the stream is gaining water); where it is lower (stream losing water), it is shown in red. The size of the diamonds indicates the magnitude of gaining or losing during the time period being represented. In this time period, these wells show that lateral movement is generally from bottom to top across the channel (south to north) and is stronger on the downstream side of plunge-pool No. 4.

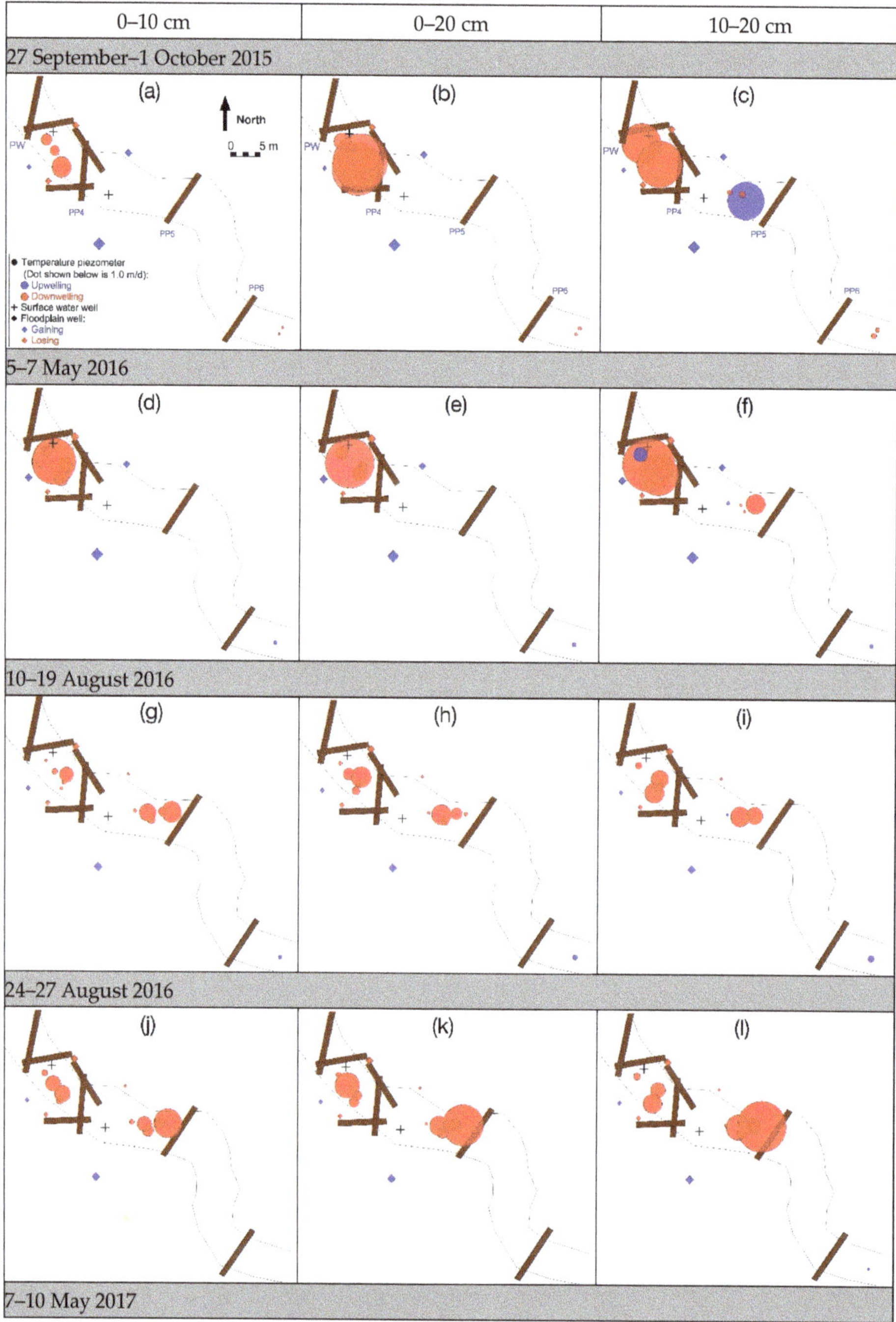

Figure 9. *Cont.*

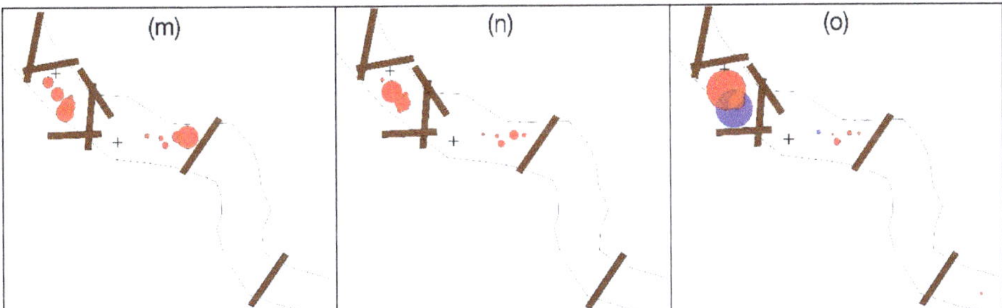

Figure 9. Vertical Flux (m/d) at the Kingfisher Treatment site by depth layer and date. Maps in (**a–o**) are arranged chronologically in horizontal rows by date, with each column representing a layer (0–10 cm on the left, 0–20 cm in the middle, and 10–20 cm on the right). For scale, red and blue dots in map legend, in upper left panel (**a**), represent 1.0 m/d flux. Plunge-pools 4–6 (PP4–PP6) and a pocket-water structure (PW) are labelled in (**a**).

Lateral flow will manifest itself as patterns of difference in magnitude, or even direction, between vertical flux calculated in one layer as compared with another layer. For example, comparing the panels for 7–10 May 2017 in Figure 9, and focusing attention on the bottom-most piezometer in the left-most array of five, it can be seen that the vertical flux went from a moderately large downwelling magnitude (2.62 m/d or 8.6 ft/d) in the 0–10 cm layer, to a much smaller apparent downwelling (0.38 m/d or 1.2 ft/d) in the 0–20 cm layer to a large apparent upwelling (5.68 m/d or 18.6 ft/d) in the 10–20 cm layer. This piezometer is known to be located in a zone where lateral flow moves from bottom left to top right in the figures, as shown by floodplain well measurements.

The distribution and magnitude of upwelling and downwelling is more complicated than what would be expected from simple flow around and beneath a log sill. For example, focusing on the 10–19 August 2016 time period, the strongest values of downwelling were in the downstream array, within the influence of the sill for plunge-pool No. 5 (right-most log sill shown). Upwelling did not occur anywhere in the upstream or downstream arrays but did appear, as expected, in the single piezometer at the far downstream (right) end of the channel (10–19 August, 0–10 cm). The original intent was to place the downstream array of five piezometers closer to the pool tail-out zone for plunge-pool No. 4, which is presumably a zone of upwelling, but this area turned out to have a very unstable streambed, with ongoing sediment deposition and bar formation rendering it unsuitable for piezometer installation.

The picture for the 0–20 cm layer for the this same time period is similar, in terms of the variety of downwelling magnitudes in close proximity to each other, and the maximum downwelling flux measured (3.39 m/d or 11.1 ft/d versus 4.40 m/d or 14.4 ft/d in the 0–10 cm depth range). Downstream, the sole piezometer at the end of the reach was estimated to have a somewhat larger upwelling magnitude, 0.49 m/d or 1.6 ft/d (10–19 August 2018, 0–20 cm), in contrast to 0.31 m/d or 1 ft/d in the 0–10 cm layer during this same date range.

The situation for the 10–20 cm during 10–19 August 2016 is more interesting, with similar maximum downwelling magnitude, 3.27 m/d (1.1 ft/d), a larger upwelling magnitude, 0.88 m/d (2.9 ft/d), at the downstream end of the reach, but with one of the piezometers in the array between plunge-pools No. 4 and No. 5 now showing up as slightly upwelling (0.06 m/d or 0.2 ft/d). This could be attributable to the effects of lateral subsurface water movement, which is generally bottom to top in this portion of the reach and would likely be carrying cooler water. However, it is difficult to explain why this influence would only affect one, and not all five, of the piezometers in that array.

The single piezometer at the upstream end of the reach is not shown in these figures, as it was inadvertently installed adjacent to a buried pipe which affected its temperature readings. Likewise, one of the two piezometers at the downstream end of the reach had to be abandoned due to persistent, deep scour caused by high velocity water from plunge-pool structure No. 6.

A complete listing of the computed vertical flux values and location coordinates in provided in spreadsheet format in the Suplementary Materials, Spreadsheet S1: Summary of computed vertical water flux.

5.2. Mapping of Upwelling and Downwelling Zones

Normalized differences between surface water and intra-gravel temperatures for the Kingfisher Treatment site are shown in Figure 10 below as circles, with the diameters scaled according to the temperature difference magnitude. The test was conducted during a period of stable, sunny weather, with constant baseflow discharge of about 0.017 m³/s (0.6 ft³/s). The positive normalized differences are zones of relative upwelling (shown as blue circles) while the negative normalized differences are zones of relative downwelling (shown as red circles). In this figure, the circle size in the legend represents a temperature difference of 1.0 °C. The maximum and minimum actual temperature differences (in degrees C, not normalized) were 2.89 °C (upwelling) and −0.03 °C (downwelling), but these numbers, like the size of the circles, only indicate relative tendencies, and are not meaningful in an absolute or quantitative sense. That is, a location mapped as "upwelling" might actually be physically downwelling, but at a weaker rate than those "downwelling" locations with temperature differences less than the overall median.

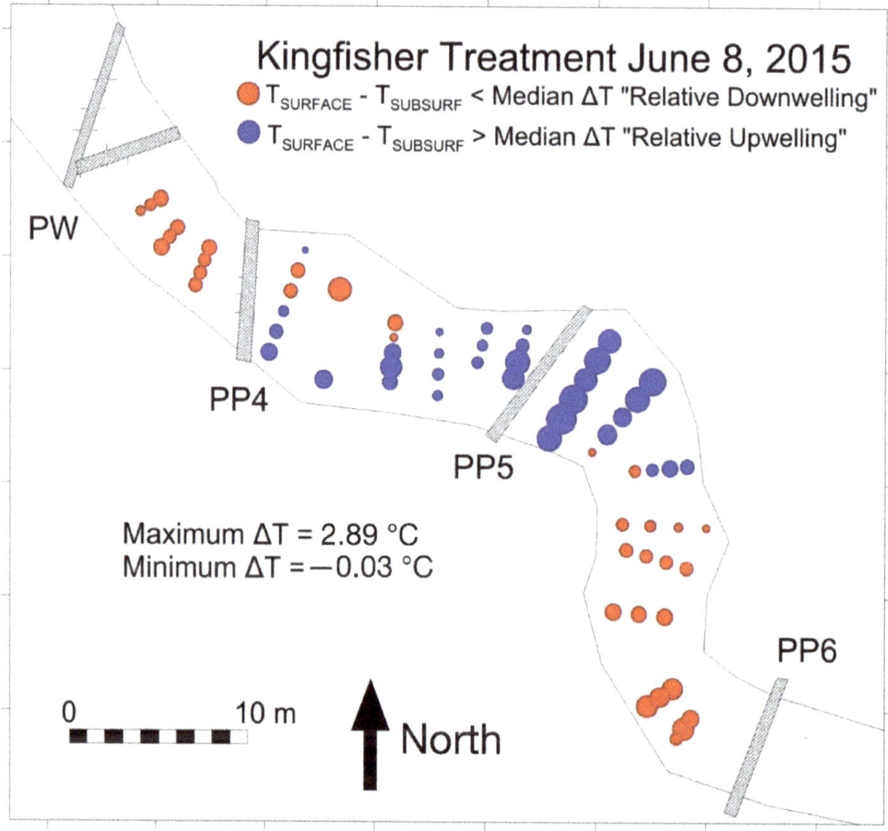

Figure 10. Map showing relative downwelling (red) and upwelling (blue) at the post-project Kingfisher Treatment site. The dots in the legend represent 1.0 °C temperature difference. PP4–6 are sill logs for plunge-pool structures No. 4–6, and PW is a pocket-water log structure.

The results indicate a large diversity of shallow hyporheic interactions across the study reach. Note that several transects showed relative downwelling zones on one side of the channel, with upwelling zones on the other. The middle portion of the reach, between plunge-pool No. 4 and No. 5, was an area of strong influence from lateral subsurface flow entering from the south (bottom edge of the channel in the figure). Although this subsurface flow feeds floodplain wetlands before continuing to flow subsurface to enter the stream channel, it still entered the channel cooler than the surface water during the time of year this test was conducted (8 June 2015). There were fairly consistent subsurface temperature differences from the south (cooler) to north (warmer) sides in this area. This lateral input of cool water probably explains why the upstream (left) side of plunge-pool No. 5 is mapped as an upwelling zone, even though this would be expected to be a downwelling area. Looking at the strength of the upwelling, as indicated by the larger circles on the downstream side, we can still conclude that the structure is functioning as expected, with water entering the streambed on the upstream side and reemerging downstream.

Figure 11 illustrates the relative upwelling and downwelling found at the Kingfisher Control reach during identical discharge and weather conditions.

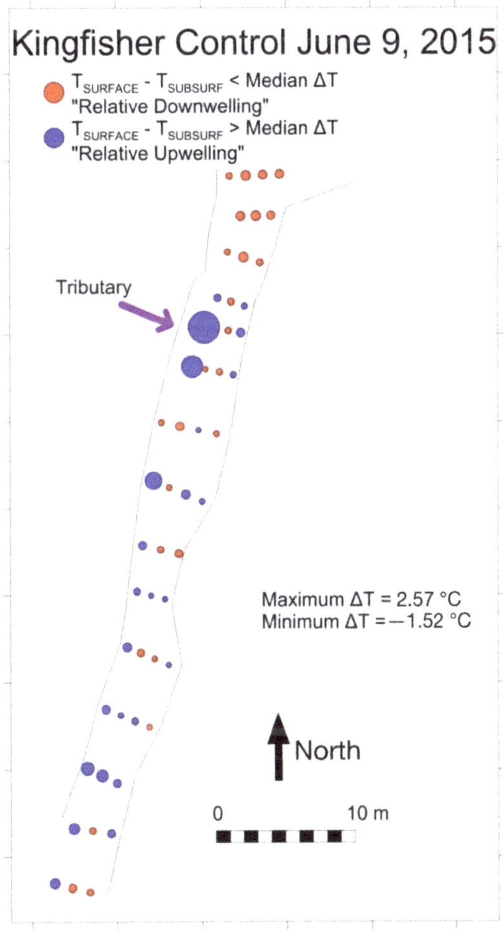

Figure 11. Map showing relative downwelling (red) and upwelling (blue) locations in the Kingfisher Control reach on 9 June 2015. The dots in the legend represent 1.0 °C temperature difference.

Note the large apparent upwelling zones occurring in the vicinity of the alluvial fan of a tributary, which enters in the left upper side of the figure. Other than this zone, most of the points in the Kingfisher Control had weaker upwelling or downwelling tendencies than those observed in the restored (treatment) reach, and less diversity in relative intensity. The dots in the legend represent 1.0 °C temperature difference. Median and average (absolute) temperature differences between the surface and intra-gravel temperatures were larger in the treatment reach (Figure 10) than in the control reach (Figure 11, although the maximum (2.57 °C) and minimum (−1.52 °C) were somewhat more extreme in control reach. Median and average (absolute) temperature differences in the treatment reach were larger, and the maximum and minimum more extreme, post-construction than in the pre-construction channel in 2006 [39].

This subdued intensity of hyporheic interchange at the control reach was expected, due to the limited thickness of the alluvial streambed in the control reach. Table 3 summarizes the intra-gravel temperature measurements, where temperature difference refers to (surface water)–(intra-gravel) temperature. Intra-gravel temperature was measured 10 cm below the streambed surface. The data reveal that the restored treatment site yielded a mean temperature difference 1.5 times greater than control and nearly 4 times greater than the pre-restoration site. Variability, as indicated by standard deviation (S.D.) of temperature differences, shows a similar pattern. The complete set of intragravel temperature differences is provided in spreadsheet format in the Supplementary Materials, Spreadsheet S2: Summary of surface-intragravel temperature differences.

Table 3. Summary of intra-gravel streambed temperature measurements.

Site/Date	Time, PDT		No. of Tubes	Temperature Difference, Degrees C				
	Begin	Duration		Median	Mean	S.D.	Min	Max
Kingfisher Treatment, 8 June 2015	14:15	3:15h	66	1.18	1.25	0.72	−0.03	2.89
Kingfisher Treatment, 9 June 2015	10:33	3:09h	52	0.73	0.81	0.50	−1.52	2.97
Kingfisher Treatment pre-restoration, 27 September 2006	13:03	3:05h	51	0.35	0.31	0.22	−0.26	0.78

5.3. Surface Water Longitudinal Temperature Pattern

This section presents a portion of the time series of surface water temperatures collected during the intra-gravel temperature difference study. Locations of temperature data loggers are shown in Figure 8, and the temperature measurements are shown in Figure 12. Cooling occurred during part of the day between T01 and T02, even though this was a completely open, unshaded portion of the reach during the study. The only explanation for this is some combination of hyporheic cooling due to plunge-pool No. 4, which lies between these two data loggers, and lateral input of cool subsurface water. By contrast, there was considerable solar heating evident between T02 and T03, which is similarly situated relative to groundwater influx, has similar slope and morphology, and is also without shade, but has no log sills to promote hyporheic flow.

Even more surprising is the significant (>1.5 °C) cooling that occurred between T03 and T04, the most downstream data logger. This portion of the reach is characterized by deep pools and sluggish water velocities, so in the absence of lateral subsurface flow or hyporheic cooling due to plunge-pool No. 5, we would expect substantial heating between these two points. The only suitable explanation for this significant temperature decrease is due to the cooling provided via hyporheic exchange as water passes under and through plunge-pool No. 5.

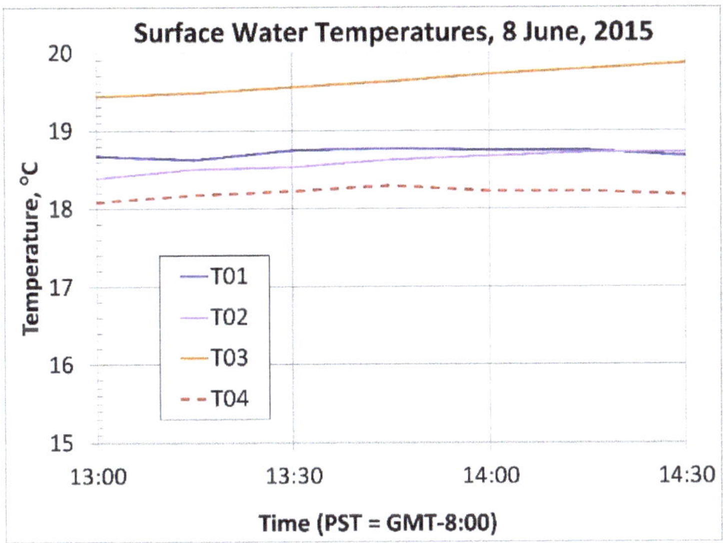

Figure 12. Surface water temperature longitudinal changes for temperature data loggers T01–T04 (see Figure 8 for locations).

5.4. Immobile versus Mobile Cross-Section Ratio

Figure 13 shows end-of-reach water electrical conductivity data compared to optimized OTIS-P model predictions for the Kingfisher Treatment site to illustrate the degree of model fit to data. The model optimization excluded data beyond 219 min in order to focus parameter optimization on the periods of rapid change, which included the rising limb, plateau, and steep portions of the falling limbs of the tracer pulse, but not the long gradual recovery to background levels. This strategy is consistent with standard procedures [50,51]. The test reach was 80.5 m long, with water discharge of approximately 0.0089 m^3/s (0.32 ft^3/s).

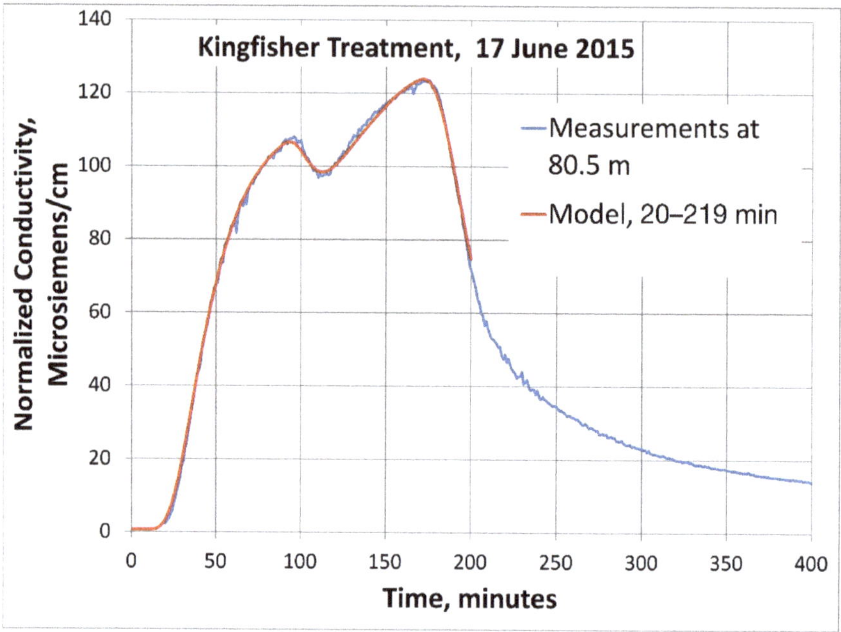

Figure 13. NaCl tracer test comparison of normalized measured values versus model predictions, Kingfisher Treatment site.

Table 4 summarizes the hydraulic conditions and optimized model parameters for the Kingfisher Treatment and Control site tracer tests. Estimates of model parameters are within the ranges expected. The discharge measured in the restoration site was only 37% as large as that measured immediately upstream in the control site (0.0089 vs 0.0238 m^3/s or 0.32 vs 0.84 ft^3/s). This is an indication of the large volume of water that flows subsurface in the restoration site.

Both tracer tests resulted in DamKohler number, DaI, values that are within the acceptable range of about 0.2–2.0, indicating that reliable parameter estimates can be obtained via model optimization [54].

Table 4. Summary of NaCl tracer test parameters and results.

Site Characteristics	Site:	
	Kingfisher Treatment	Kingfisher Control
Reach length, m	80.5	100.3
Discharge, m^3/s	0.0089	0.0238
Model Output		
Average velocity, m/s	0.027	0.118
Mobile cross section, m^2	0.336	0.201
Immobile zone cross section, m^2	1.095	0.266
Exchange coefficient, alpha, 1/s	0.000191	0.000230
DamKohler number, DaI	0.76	0.34
Immobile/Mobile ratio, I/M	3.26	1.33

The Kingfisher Treatment site had an I/M ratio of 3.26, which is 2.5 times as large as I/M for the corresponding control site. Both sites had noticeable amounts of stagnant water along their margins, as well as several constrictions and expansions in channel cross section that generated slackwater areas. According to the model, immobile cross sections were 1.095 and 0.266 m^2, respectively, for the treatment and control sites. One potentially confounding issue is that the restoration site had an increased surface morphological complexity compared to control.

To ascertain the distinction between hyporheic function as separate from surface morphological complexity, it is observed that the control site had a predicted immobile cross section that was 1.33 times the predicted mobile cross section, and virtually all of this was surface marginal zone, as this reach had a very thin alluvial streambed (15–30 cm thick, or 6–12 inches) with little hyporheic component [39]. Thus, most of the 0.266 m^2 (2.9 ft^2) of immobile zone consists of channel margin area. If, by extrapolation, the channel margin area is represented by a similar percentage in the treatment site, we would expect a channel margin immobile zone of $0.366 \times 1.33 = 0.446$ m^2 (4.8 ft^2). The remainder of the total immobile cross section, 0.650 m^2 (7.0 ft^2), would then represent hyporheic zone, giving a ratio of hyporheic to mobile cross section of 1.93. Thus, the post-construction effective hyporheic cross section was almost two times larger than the mobile cross section, and about 3.2 times larger than the control reach mobile cross section.

5.5. Lateral Subsurface Hydraulic Gradient

Lateral hyporheic flow potential is illustrated through measurements of water surface elevations along the two upstream and one downstream lateral transects shown in Figure 14. Locations of the transects are shown in Figure 14d. The central point on each data series is the instream surface-water measurement well (stilling well), and distance was measured from north (top) to south (left to right, as viewed facing downstream): (a) Upstream transect closest to plunge-pool No. 4; (b) upstream floodplain transect No. 2, furthest from plunge-pool; (c) downstream transect.

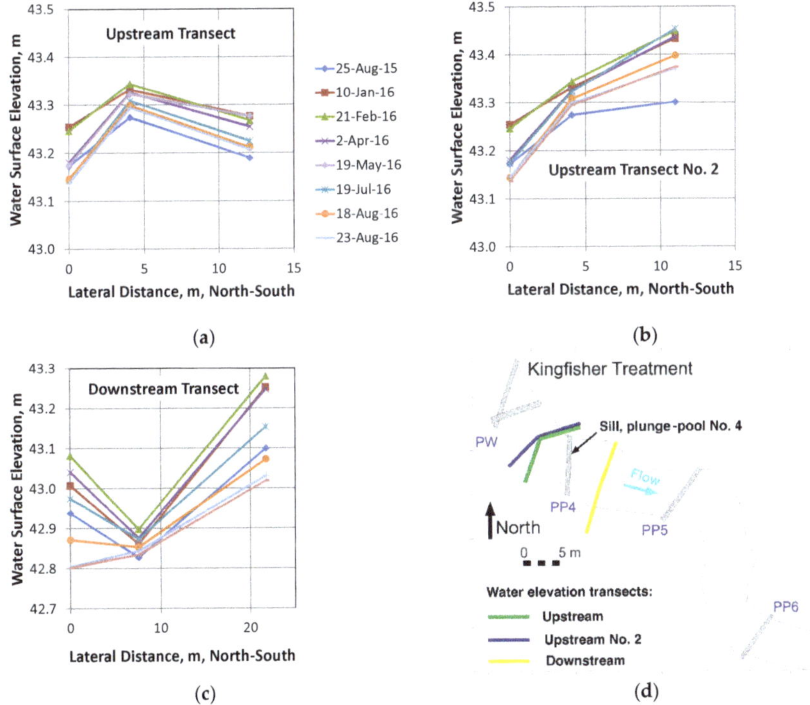

Figure 14. Water surface elevations along lateral transects, oriented as viewed from upstream. First upstream transect is shown in (**a**), second upstream transect in (**b**) and downstream transect in (**c**). Transect locations shown in (**d**).

The upstream transect closest to plunge-pool No. 4 (Figure 14a) shows the classic pattern of lateral hyporheic flow around a channel spanning step [55,56]. Subsurface flow moves laterally outward, away from the channel center at both sides. Likewise, the downstream transect (Figure 14c) shows how this water, after having briefly entered the floodplain, flows back into the channel from both sides. Superimposed on this local lateral hyporheic circulation, is the regional flow of subsurface water from the valley sides on the south in a northerly direction [41], encountering the right side of the stream channel (as facing downstream). This is the cause of higher surface water elevations on the right side of Figure 14a, and steeper piezometric gradients on the right side of the downstream transect, Figure 14c. The effect of this regional subsurface water flow is even more evident in Figure 14b, where the orientation of the transect is such that this regional flow overwhelms the local circulation, producing a situation of through flow from southwest to northeast. Moreover, during certain times of the year, notably, in August, when surface water is at its lowest, this regional trend overwhelms the local circulation along the downstream transect as well.

5.6. Streambed Surface and Subsurface Sand Content

Summary statistics for the surface and subsurface layers pre-project and post project for both Kingfisher and Confluence Forks sites are summarized in Table 5, with additional sample data and locations shown in Appendix A.

In order to investigate streambed evolution in terms of grain size, changes to the percentage of sand (grain sizes less than 2 mm) were computed. Data from all nine sample locations at Kingfisher and Forks Confluence are shown in Table 5, in order to provide context for the discussion of variability observed.

Table 5. 2017 median grain size, D_{50}, (in mm) and percent change (all grain sizes < 2mm) 2016–2017 (2015–2017 for the three Kingfisher sites).

	Treatment				Control	
	Site 1 (Upstream)		Site 2 (Downstream)			
Location (Reach)	D_{50} (mm)	Change in Sand Content (%)	D_{50} (mm)	Change in Sand Content (%)	D_{50} (mm)	Change in Sand Content (%)
Surface						
Kingfisher	32.2	2.9	28.9	16.3	45.7	4.0
Forks Confluence, North Fork	23.2	2.0	n/a	n/a	34.0	−0.9
Forks Confluence, South Fork	28.4	−1.1	16.9	−1.9	34.2	−1.0
Forks Confluence, Mainstem	30.1	−9.1	n/a	n/a	n/a	n/a
Subsurface						
Kingfisher	30.0	−2.0	14.8	4.7	15.4	7.1
Forks Confluence, North Fork	18.2	3.8	n/a	−1.6	31.0	−6.3
Forks Confluence, South Fork	20.3	0.5	6.8	12.7	14.3	−3.2
Forks Confluence, Mainstem	30.1	−1.6	n/a	n/a	n/a	n/a

At Kingfisher, the median surface layer grain size, D_{50}, ranged from 45.7 to 28.9 mm (1.8 to 1.1 inches) in 2017, as compared to 58.2 to 35.5 mm (2.3 to 1.4 inches) in 2015. Similarly, the median subsurface layer grain size ranged from 30.0 to 14.8 mm (1.1 to 0.6 inches) in 2017, as compared to 27.0 to 17.2 mm (1.06 to 0.7 inches) in 2015. Data from pre-project monitoring in 2007 are shown in Appendix A for the control site, to provide some perspective on inter-annual variability. These numbers, D_{50} of 42.3 mm (1.7 inches) for the surface layer and 18.4 mm (0.72 inches) for the subsurface, are not radically different, but are at the finer and of the observed range. Values for the upstream-most site at the constructed reach taken in 2007 is of historic interest, but cannot be used in for tracking change, because this location was replaced with the new engineered design for channel and hyporheic restoration.

Interpreting trends in streambed grain size evolution is difficult from only two points in time, given the apparent variability in grain size distribution and sand content, and given the episodic nature of sediment input to the stream system, to which the streambed must adjust. Generally, the dominant sources of coarse sediment and sand are the walls of the ravine immediately upstream from the Kingfisher Control site. Thus, it is not surprising that sand content generally increased in the surface and subsurface layers in the Kingfisher sites, which are closest to this source. An exception is the most upstream sample on the Kingfisher Treatment (KFT1), which diminished by 2% in sand content, but only in its subsurface layer. So, even in these sites closest to the source, it is possible for the natural processes of streambed scour and fill to clear fine sediment from streambed, as would happen in a more pristine watershed setting. It also bears mentioning that these changes in sand content are based on volumetric samples, and thus appear small due to the relatively greater influence of the larger gravel -sized particles on overall sample mass [53,57]. That is, the streambed surface visual appearance will change from slightly to moderately "sandy" over a fairly small range of sand percentage as measured in a volumetric sample [57]. Nevertheless, volumetric sampling is a more accurate way to quantify fine particle content than surface particle counts [57,58] when tracking processes of sediment deposition and transport, since substantial infilling of surface layer voids can occur before these fine sediments are evident in particle counts.

In order to more confidently answer the question of whether the engineered streambed is destined to steadily increase its sand content over time, it is useful to bring into this discussion the three sediment sampling sites on the South Fork, near the Forks Confluence restoration site, 1.7 km (1.1 miles) downstream from Kingfisher. These are shown in Table 5, in upstream to downstream order. Here, all three sites decreased their surface sand content over the time period from 2016–2017. This was in spite of the fact that, according to repeat channel cross-section surveys, the South Fork Confluence constructed site was gaining streambed sediment (aggrading) in some locations over this time, as it adjusted to its new hydraulic and sediment load conditions. Magnitudes of decrease in surface sand content were larger than at the control site, which also decreased its sand content in the subsurface layer.

In general, the surface layer responds more rapidly to pulses of sediment and peak flows than does the subsurface, which tends to better represent long-term average bedload size distribution [59]. However, the fact that all of the South Fork Confluence sites decreased in surface sand content is indicative that scour and fill processes that periodically mobilize, and clean, the streambed of excessive fine sediment can function in this stream system. This study time interval, 2015–2017 in the case of Kingfisher and 2016–2017 in the case of the South Fork Confluence, is too short to make definitive conclusions about long-term trends in streambed condition. The data collected to date indicate the restored engineered channel has been resilient, has continued to maintain its significant hyporheic performance, and has not become embedded despite continuous inputs of fine sediment.

6. Discussion

Watersheds and river systems are complex, even those urban streams that have been simplified by eliminating much of that complexity. Variability on small spatial and temporal scales is an inherent property of river systems, and therefore measurement uncertainty and complex responses are expected [35]. There are limits to what we can measure or predict. Specific metrics, which can be measured in a way that temporal and spatial variability are adequately known, may be prohibitively costly to undertake, or may require longer study duration than is practical. Evaluating the effectiveness of an individual restoration project is also made difficult by the vast difference in spatial scale between the project footprint (a few hectares) and the watershed itself (29 km^2 or 11.1 square miles), which establishes controls on hydrology, sediment dynamics, solutes, pollutants, and energy. The "signal" from a restoration treatment must overwhelm the background variability established by watershed-scale processes in order to be detected. In this situation, restoration effectiveness is best indicated not by any single attribute (such as stream temperature), but by a "weight of evidence" approach, in which a suite of attributes, taken together, provide evidence for detection of change. In this monitoring effort, five different attributes were measured. Each one was found to exhibit considerable variability in the results. However, taken together, they create a coherent picture of increased hyporheic processes, both in terms of flow and energy exchange.

Moreover, treating a restoration project as an experiment, in which the design includes a control, and a post- and pre-treatment period can be difficult. True control reaches, which resemble the treatment reach in every respect except restoration, may be impossible to find. The Kingfisher Control reach, though morphologically different in terms of cross section and floodplain occurrence, did have hyporheic conditions very similar to the pre-project treatment reach, including lateral subsurface water influx, and thus was a good control for the purposes of this study.

Hyporheic process restoration by means of an engineered streambed is very new to the community of stream restoration practitioners [11,12,14,36]. As such, it brings up three important questions. First, does it increase the magnitude and extent of hyporheic flow? Second, does it improve surface water quality? Finally, how long will it continue to function? It is to answer these questions that the weight of the evidence approach followed in this study is directed.

6.1. Vertical Water Flux

Vertical water flux measurements reveal that this hyporheic restoration design significantly increases the magnitude and extent of hyporheic process relative to both control sites and natural sites. This study documented improvements in the strength and diversity of upwelling and downwelling zones, which indicates improved hyporheic circulation. Vertical water flux was quantified, and found to be quite large, about 89 times as large, on average, as measured in the pre-project channel, with a maximum 16 times as large as maximum values measured in a pristine stream in Idaho, which, presumably, has a fully functional, self-maintaining streambed.

The large variability of vertical flux over time, and over short spatial distances, could lead one to suspect methodological imprecision. However, Peter et al. [30] confirmed by direct observation that hyporheic pathways, and presumably fluxes, do change radically over short time periods. These authors conducted tests using visible die tracers to map specific hyporheic pathways through plunge-pool No. 4 as part of their study to quantify hyporheic residence times and identify potential chemical transformations in the streambed. They found that the exit point for water entering the same injection location changed substantially over time periods as short as a few weeks.

Although the method assumes that the hyporheic water flux in the streambed is vertical, and does not explicitly account for lateral movement, it has been shown to be fairly robust even when there are moderate amounts of lateral flow. In general, the shallower the layer (e.g., 0–10 cm as opposed to 0–20 cm), the more closely to vertical the flux will be, and the better the method fits its assumptions [46]. Lateral flux may confound vertical flux estimates, especially for deeper layers in streambed and or in locations proximate to subsurface flow inputs.

6.2. Mapping of Upwelling and Downwelling Zones

The primary method to measure how well the restored hyporheic design benefits water quality is to quantify water temperature effects. Immediate effects on hyporheic process was observed by measuring the difference between the temperature of surface and subsurface water under stable, baseflow conditions in sunny weather. The restored reach had a temperature difference 1.5 times that of the control reach and 4 times that of the pre-restoration reach. The temperature difference is more significant than absolute temperature, as the temperature difference indicates the cooling potential of a system, where the absolute temperature simply reflects local, immediate ambient conditions. Any hyporheic mixing achieved on the project site with this increased temperature difference will yield an overall cooler surface water in summer conditions.

Though it was not measured here, it is also feasible that this temperature difference could operate in reverse during winter months, with the ability to warm incoming surface waters through mixing. This warming could potentially improve conditions for spawning and rearing during the cold season [42]. Overall, a system with a large intra-gravel to surface water temperature difference and robust hyporheic mixing can act as to moderate thermal changes within the stream seasonally, cooling the surface waters in the summer and warming surface waters in the winter.

This attribute does not give us a quantitative measure of vertical water flux but does provide the relative strength from one point to another, extensively, over the entire reach. The results indicated much stronger upwelling and downwelling within the restoration reach compared to the control reach, demonstrating the relative uplift in hyporheic function in the restored reach. Results also showed greater upwelling and downwelling strength in the treatment reach than had been observed in pre-project conditions.

The normalized temperature-difference measurements obtained through this method do not tell us the actual threshold between upwelling and downwelling, however. Theoretically, in the absence of significant reach-scale groundwater influx or efflux, local upwelling and downwelling should balance, and this method would give a realistic picture of the distribution of these opposing processes. However, in a reach such as Kingfisher Treatment or Control that gains substantial amounts of groundwater, it is conceivable that upwelling could occur everywhere, in which case this method would only indicate the relative strength of that upwelling.

6.3. Surface Water Longitudinal Temperature Pattern

Longitudinal surface water temperature measurements indicated consistent temperature decreases during warm, sunny conditions through the zone of channel occupied by plunge-pool No. 5. On an established restoration site with full vegetation, this cooling would be attributable, at least in part, to shade from vegetation. However, this explanation is not sufficient here, as the site had only recently been replanted and there was no shade over the stream channel. In the absence of hyporheic water exchange or lateral groundwater input, we would expect steady increases in surface water temperature in the upstream to downstream direction, due to solar heating. Instead, the data loggers recorded a cooling as water flowed downstream, which can only be attributed to thermal exchange with the subsurface as driven by hyporheic flow under plunge-pool No. 5.

6.4. Immobile versus Mobile Cross-Section Ratio

A critical element to determining how well the restored hyporheic design can restore water quality is by considering the retention time of the surface water versus the hyporheic water. Longer retention times can yield greater reduction of incoming chemicals through adsorption, biological, or chemical transformation. Using NaCl tracer studies, this study documented large improvements in hyporheic volume compared to the control reach, as represented by a more than 3-fold increase in immobile cross section found in the tracer test modeling. A larger hyporheic cross section (and volume) translates into slower subsurface flow, through hydraulic continuity, and thus into longer subsurface residence times.

6.5. Lateral Subsurface Hydraulic Gradient

Subsurface water elevations along a transect perpendicular to the stream channel is an indicator of potential for lateral or floodplain hyporheic flow. In this study, evidence for local hyporheic flow around a plunge-pool structure, latterly, into the floodplain was demonstrated, as well as a regional flow of subsurface water from south to north across the site. The greatest hydraulic gradient was evident in the winter when there is greater in-channel flow, though the pattern of lateral flow was fairly consistent between seasons. In low flow summer conditions, reach-scale conditions such as lateral inputs from groundwater can overwhelm local gradients produced by the plunge-pool structures.

6.6. Streambed Surface and Subsurface Sand Content

Determining whether the design will continue to perform successfully over longer time spans was evaluated by sampling changes in streambed sediment grain size composition, in particular, its sand content. Sampling of streambed sediment has demonstrated continued ability of the stream to maintain its engineered bed as loose, mobile, alluvial gravel, without becoming embedded with fine sand, to date. This condition is associated with good hyporheic function. The literature is vague regarding the amount of time needed to characterize a trend in evolution of streambed grain size composition (e.g., [53,58]), but judging from the fact that there were significant observed changes to some of the permanent cross sections used to monitor morphological changes in the control reaches over the period 2006–2017 [41], 10 years would not be an unreasonable estimate. Subsequent sediment composition sampling is recommended to document any future changes, which could be influenced by changes in incoming substrate size, quantities, and/or hydraulics due to changes in channel cross section.

In this, and other channel reconstruction projects, the exposed portions of large wood structures have a finite lifespan in their original designed configuration. Eventually, as the wood decays, and as the channel adjusts to sediment pulses, fallen trees, floods and other disturbances, it will attain a configuration that is self-maintaining over time, if allowed to do so. Wood remaining partly or wholly buried in the streambed (e.g., the hyporheic logs), however, is expected to have a very long lifespan, potentially on the order of a thousand years [60]. As the system retains incoming wood (falling trees and/or washed in logs) it likely will retain additional alluvium, burying the wood as it builds its channel and floodplain over time. This configuration will likely have a hyporheic zone that is much larger and higher functioning than the pre-project condition long into the future.

6.7. Relation to Other Studies and Observations

Using mass spectroscopy, Peter et al. [30] found significant reductions in the number and concentration of organic chemicals, some of which are potentially associated with coho mortality (as identified in [29]) both under both base flow and stormflow conditions through plunge-pool No. 4 at the Kingfisher site. For example, polypropylene glycols (PPGs) were reduced in concentration by 46%–100% in a hyporheic pathway with residence time of only 32 min and estimated path length 4.4 m (14.4 feet) beneath plunge-pool No. 4 (called hyporheic design element or HDE No. 4 by these authors). In a nearby hyporheic pathway of residence time 3.75 h and path length about 5.0 m (18.4 feet), the PPGs were reduced by 92%–100%. Reductions in chemical concentrations were found to be significantly larger in the hyporheic flow paths, even though these were short, than reductions in the surface flow paths. During stormflow conditions, less than 17% of the 1900 non-target chemicals (i.e., those chemicals analyzed as a group without knowing the exact identity of all of them) were reduced in the surface flow by more than 50%, compared to 59% and 78% reductions in the short and long hyporheic pathways, respectively. These researchers attribute the contaminant reduction to sorption by biofilms on hyporheic sand and gravel particles. At base flow, the majority of the flow was found to circulate through the hyporheic zone, about 20% at each of the six structures. During storm flow conditions, 20%–60% of the flow was found to circulate through the hyporheic zone, indicating substantial water quality treatment potential associated with the plunge-pool structures.

Researchers documenting the biological response to these projects also detected differences between constructed and control sites. They sampled biota from the hyporheic zone at a depth of 15–25 cm (6–10 inches) beneath the streambed and found greater macroinvertebrate density and taxa richness in the treatment than in the control reaches [37]. Even though there was substantial variability from 2015 to 2016 (i.e., one to two years, post construction), the treatment reach consistently was higher than the control in both of these metrics. They also documented increased microbial metabolic activity as indicated by carbon metabolism, and changes to microbial and invertebrate taxonomic structure.

Part of the biological response study involved experimental inoculation of the hyporheic zone with macro- and micro-invertebrates from a more pristine reference stream, the Cedar River, located in forested setting in the municipal watershed for the City of Seattle. Mesh baskets of gravel were placed on the Cedar River streambed for a period of several months, then taken to Thornton Creek and placed into vertical perforated pipes embedded into the engineered streambed. However, only small, transient changes to the microbial taxonomic community structure were observed, and no significant changes to macroinvertebrate density or structure [37].

In support of this result, during autumn 2018, Chinook salmon were found spawning in the restored Forks Confluence [61]. This is the first documented spawning in this area ever since spawning surveys began in 1999, and is attributed to the improved streambed condition (which was visibly apparent in 2018, four years post-construction), the wider, more complex channel morphology, and restored hyporheic function providing the needed conditions to match spawning requirements.

6.8. Future Prospects of Hyporheic Restoration

Significant water quality benefits have been documented on a local scale by evaluating in detail a handful of parameters on an individual plunge-pool structure among the six plunge-pool structures in the Kingfisher restoration site and eight on the Confluence restoration site. Water quality within a stream is degraded by diffuse interactions with impervious surfaces, unshaded open water, and exposure to environmental chemicals occurring throughout the watershed. The next step is to explore how to correlate the amount of stream and hyporheic restoration, which is required to provide significant water quality benefits to treat that incoming pollution from a watershed, and how this work can be integrated into a larger vision which also incorporates current input-water engineered treatment solutions such as vegetated filter strips and stormwater retention structures. In the broadest sense, it is worth considering whether stream and hyporheic restoration, when implemented at scale throughout a watershed, serve as a superior replacement to current engineering practices and provide water quality enhancements with combined habitat uplift benefiting instream fauna as well as riparian functions including improvements to air quality, avian habitat, and soil moisture retention.

7. Conclusions

This new approach to enhanced hyporheic design successfully restored hyporheic processes and yielded significant water quality improvements compared to control and pre-project conditions. The engineered streambed and large-wood structures central to this design have yielded vertical water flux averaging 89 times greater than pre-project conditions, and with a maximum 16 times greater than that found in a pristine watercourse. Mapping of upwelling and downwelling zones in the restored reach yielded a surface-water to intra-gravel temperature difference 1.5 times that of the control reach and 4 times that of the pre-restoration reach, demonstrating the ability of the restoration design to yield an improved hyporheic circulation in summer conditions. Surface water longitudinal temperature patterns documented greater than 1.5 degrees of cooling as the water passed through a plunge-pool structure in full sun in the middle of the day, which can only be attributed to hyporheic exchange driving flows subsurface and subsequent cooling. Effective hyporheic zone cross section, as estimated by the immobile to mobile cross-section ratio, improved by a factor of 3.

The engineered streambed has been resilient to the influence of its sediment load. Surface and subsurface sand content, an indicator of hyporheic performance, has shown no observable tendency for the streambed to revert to the embedded, impermeable state that existed before project construction, at least over 2–3 years of sampling. The design has thus proven capable of maintaining natural scour and fill processes which slow or reverse embedded conditions and may actually improve over time through natural stream evolution, including processes of large-wood recruitment and associated accumulation of more diverse and extensive alluvial streambed material.

Other researchers studying the water chemistry and biology have shown that this hyporheic restoration has also been effective at significantly reducing concentrations of incoming pollutants [29], and at enhancing both hyporheic microbial heterotrophic production and macroinvertebrate taxa diversity and richness [37].

Supplementary Materials: The following are available online at http://www.mdpi.com/2073-4441/12/2/425/s1: Spreadsheet S1: Summary of computed vertical water flux, and Spreadsheet S2: Summary of surface-intragravel temperature differences.

Author Contributions: Conceptualization of engineered streambed design, P.D.B., M.H., and K.D.L.; formal project engineering design, M.H.; methodology, investigation, data curation, formal analysis, validation, visualization, and original draft preparation, P.D.B.; review and editing, P.D.B., K.D.L. and M.H.; resources, P.D.B. and K.D.L.; funding acquisition, K.D.L. All authors have read and agreed to the published version of the manuscript.

Funding: This research was funded by the City of Seattle, contract No. 14-133-A.

Acknowledgments: We thank Chapin Pier and Steve Damm (City of Seattle), Tracy Leavy, Kennith King, Alex Bell, Zach Moore, Michael Carlson and Michael Elam (U.S. Fish and Wildlife Service), Lila Neal and Corrine Hoffman (volunteers), Michaela Lowe, Jawad Frangieh and Jeffrey Lee (Americorps) for their diligent field assistance. We thank Jeffery Johnson (U.S. Fish and Wildlife Service), Michaela Lowe, Laurel Low, and Jeffery Lee (Americorps) for generous assistance with lab analysis of sediment. We thank Jeffrey Godden for software coding assistance. We also thank Jennifer Hrachovec for her thoughtful edits, Colin Riordan (Natural Systems Design) for help with map graphics and Danielle Devier (Natural Systems Design) for illustrations. We thank the U.S. Fish and Wildlife Service, under whose employment Paul D. Bakke (author) performed all of his work except for manuscript preparation. Finally, we give special thanks to Katherine D. Lynch, for her tenacious advocacy of this project to the City of Seattle leadership over many years, without which none of this would have been possible.

Conflicts of Interest: The authors declare no conflict of interest. Opinions expressed in this article represent those of the authors and not of the U.S. Fish and Wildlife Service, the City of Seattle, The Science of Rivers, or Natural Systems Design.

Appendix A

Table A1. Streambed surface layer grain size statistics for all 9 Thornton Creek sites (see Figure A1).

Site	Year	Surface Grain Size Distribution (mm):				
		D_{15}	D_{35}	D_{50}	D_{84}	D_{95}
Kingfisher Control, KFC	2007	17.0	31.4	42.3	81.2	109.7
	2015	15.5	39.4	58.2	87.0	111.4
	2017	6.7	24.3	45.7	96.7	117.3
Kingfisher Treatment, KFT1	2007	4.1	14.0	23.0	52.3	69.4
	2015	20.2	30.0	35.5	51.1	59.7
	2017	14.6	26.5	32.2	51.4	61.3
Kingfisher Treatment, KFT2	2015	11.2	27.3	35.5	52.8	60.2
	2017	0.6	18.9	28.9	50.4	65.7
South Fork Control, SFC	2007	7.5	24.0	39.3	98.2	117.8
	2016	8.5	25.4	36.6	68.7	82.7
	2017	7.3	21.8	34.2	68.1	82.5
South Fork Treatment, SFT1	2016	12.4	22.6	29.6	76.5	111.2
	2017	13.4	23.4	28.4	54.3	76.0
South Fork Treatment, SFT2	2007	5.7	11.1	15.7	37.4	96.6
	2016	2.5	10.3	17.8	39.5	59.7
	2017	3.2	10.7	16.9	29.9	39.9
North Fork Control, NFC	2007	9.7	25.9	44.4	112.7	151.7
	2016	4.5	15.3	23.2	46.5	57.9
	2017	7.4	25.3	34.0	53.0	61.5
North Fork Treatment, NFT	2007	4.1	17.5	29.9	63.7	94.1
	2016	9.0	20.8	28.6	90.7	114.9
	2017	7.9	17.6	23.2	34.0	42.3
Mainstem Treatment, MST	2016	2.9	22.1	26.3	44.0	59.0
	2017	11.0	24.5	30.1	55.7	69.9

Table A2. Streambed subsurface layer grain size statistics for all 9 Thornton Creek sites (see Figure A1).

Site	Year	Subsurface Grain Size Distribution (mm)				
		D_{15}	D_{35}	D_{50}	D_{84}	D_{95}
Kingfisher Control, KFC	2007	3.0	10.2	18.4	51.1	67.3
	2015	2.8	13.1	27.0	81.4	110.8
	2017	1.1	7.1	15.4	62.6	88.7
Kingfisher Treatment, KFT1	2007	2.5	9.2	17.7	53.6	75.4
	2015	2.9	13.7	23.5	36.6	47.0
	2017	6.1	24.5	30.0	50.4	61.2
Kingfisher Treatment, KFT2	2015	4.4	8.7	17.2	33.7	49.0
	2017	2.6	7.9	14.8	35.4	43.4
South Fork Control, SFC	2007	2.6	8.8	16.3	42.8	62.0
	2016	1.0	6.2	12.9	45.9	62.5
	2017	1.5	7.0	14.3	53.6	77.7
South Fork Treatment, SFT1	2016	3.1	15.3	24.2	47.7	60.2
	2017	2.7	12.6	20.3	36.5	44.9
South Fork Treatment, SFT2	2007	1.8	7.4	12.7	30.4	41.6
	2016	1.9	14.3	25.0	52.4	74.5
	2017	0.6	3.3	6.8	23.1	30.2
North Fork Control, NFC	2007	3.3	10.3	18.1	48.5	68.7
	2016	1.0	7.0	14.7	38.9	56.4
	2017	1.9	20.3	31.0	55.6	75.5
North Fork Treatment, NFT	2007	1.7	7.7	14.5	38.6	54.6
	2016	2.5	12.9	22.7	44.6	60.0
	2017	1.5	10.3	18.2	46.4	77.8
Mainstem Treatment, MST	2016	1.0	16.6	23.7	40.2	103.5
	2017	11.0	24.5	30.1	55.7	69.9

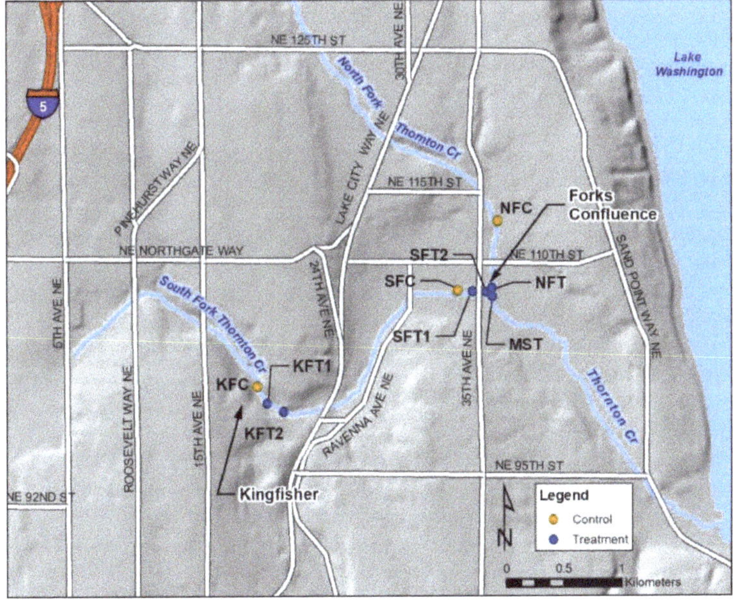

Figure A1. Locations of streambed sediment sampling sites referenced in Tables A1 and A2.

References

1. Federal Interagency Stream Restoration Working Group (FISRWG). *Stream Corridor Restoration: Principles, Processes, and Practices*; U.S. Government: Washington, DC, USA, 1998; ISBN-0-934213-59-3.
2. Budd, W.W.; Cohen, P.L.; Saunders, P.R.; Steiner, F.R. Stream corridor management in the Pacific Northwest: I. Determination of stream-corridor widths. *Environ. Mgmnt.* **1987**, *11*, 587–597. [CrossRef]
3. Stanford, J.A.; Ward, J.V. An ecosystem perspective of alluvial rivers: Connectivity and the hyporheic corridor. *Jour. N. Am. Benth. Soc.* **1993**, *12*, 48–60. [CrossRef]
4. Buss, S.; Cai, Z.; Cardenas, B.; Fleckenstein, J.; Hannah, D.; Heppell, K.; Wood, P. *The Hyporheic Handbook: A Handbook on the Groundwater-Surface Water Interface and Hyporheic Zone for Environment Managers*; Integrated Catchment Science Programme: Bristol, UK, 2009; ISBN 9781849111317.
5. Edwards, R.T. The Hyporheic Zone. In *River Ecology and Management: Lessons from the Pacific Coastal Ecosystem*; Naiman, R.J., Bilby, R.E., Eds.; Springer: New York, NY, USA, 1998; pp. 399–429, ISBN 978-0-387-95246-8.
6. Cardenas, M.B. Hyporheic zone hydrological science: A historical account of its emergence and a prospectus. *Water Resour. Res.* **2015**, *51*, 3601–3616. [CrossRef]
7. Harvey, J.; Gooseff, M. River corridor science: Hydrologic exchange and ecological consequences from bedforms to basins. *Water Resour. Res.* **2015**, *51*, 6893–6922. [CrossRef]
8. Fischer, H.; Kloep, F.; Wilzcek, S.; Pusch, M.T. A river's liver: Microbial processes within the hyporheic zone of a large lowland river. *Biogeochemistry* **2005**, *76*, 349–371. [CrossRef]
9. Stanford, J.A.; Ward, J.V. The hyporheic habitat of river of river ecosystems. *Nature* **1988**, *335*, 64–66. [CrossRef]
10. Hester, E.T.; Doyle, M.W. In-stream geomorphic structures as drivers of hyporheic exchange. *Water Resour. Res.* **2008**, *44*. [CrossRef]
11. Ward, A.S.; Gooseff, M.N.; Johnson, P.A. How can subsurface modifications to hydraulic conductivity be designed as stream restoration structures? Analysis of Vaux's conceptual models to enhance hyporheic exchange. *Water Resour. Res.* **2011**, *47*. [CrossRef]
12. Boulton, A.J. Hyporheic rehabilitation in rivers: Restoring vertical connectivity. *Freshwater Biol.* **2007**, *52*, 632–650. [CrossRef]
13. Barnard, R.J.; Johnson, J.; Brooks, P.; Bates, K.M.; Heiner, B.; Klavas, J.P.; Ponder, D.C.; Smith, P.D.; Powers, P.D. *Water Crossings Design Guidelines*; Washington Department of Fish and Wildlife: Olympia, WA, USA, 2013; pp. 28–69. Available online: https://wdfw.wa.gov/publications/01501 (accessed on 29 January 2020).
14. Herzog, S.P.; Higgins, C.P.; McCray, J.E. Engineered streambeds for induced hyporheic flow: Enhanced removal of nutrients, pathogens, and metals from urban streams. *J. Environ. Eng.* **2015**, *142*, 04015053-1–04015053-10. [CrossRef]
15. Crispell, J.K.; Endreny, T.A. Hyporheic exchange flow around constructed in-channel structures and implications for restoration design. *Hydrol. Process.* **2009**, *23*, 1158–1168. [CrossRef]
16. Washington State Dept. of Transportation (WSDOT). Stormwater best management practices. In *Highway Runoff Manual, M31-16.04, Suppl. February 2016*; Washington State Dept. of Transportation: Olympia, WA, USA, 2016; pp. 5-1–5-227. Available online: https://www.wsdot.wa.gov/Publications/Manuals/M31-16.htm (accessed on 29 January 2020).
17. McIntyre, J.K.; Davis, J.W.; Hinman, C.; Macneale, K.H.; Anulacion, B.F.; Scholz, N.L.; Stark, J.D. Soil bioretention protects juvenile salmon and their prey from the toxic impacts of urban stormwater runoff. *Chemosphere* **2015**, *132*, 213–219. [CrossRef] [PubMed]
18. Spromberg, J.A.; Baldwin, D.H.; Damm, S.E.; McIntyre, J.K.; Huff, M.; Sloan, C.A.; Anulacion, B.F.; Davis, J.W.; Scholz, N.L. Coho salmon spawner mortality in western US urban watersheds: Bioinfiltration prevents lethal storm water impacts. *J. Appl. Ecol.* **2016**, *53*, 398–407. [CrossRef] [PubMed]
19. Böhlke, J.K.; Antweiler, R.C.; Harvey, J.W.; Laursen, A.E.; Smith, L.K.; Smith, R.L.; Voytek, M.A. Multi-scale measurements and modeling of denitrification in streams with varying flow and nitrate concentration in the upper Mississippi river basin, USA. *Biogeochemical* **2009**, *93*, 117–141. [CrossRef]
20. Böhlke, J.K.; Harvey, J.W.; Voytek, M.A. Reach-scale isotope tracer experiment to quantify denitrification and related processes in a nitrate-rich stream, midcontinent United States. *Limnol. Oceanogr.* **2004**, *49*, 821–838. [CrossRef]

21. Butturini, A.; Sabater, F. Importance of transient storage zones for ammonium and phosphate retention in a sandy-bottom Mediterranean stream. *Freshwater Biol.* **1999**, *41*, 593–603. [CrossRef]

22. Groffman, P.M.; Altabet, M.A.; Böhlke, J.K.; Butterbach-Bahl, K.; David, M.B.; Firestone, M.K.; Giblin, A.E.; Kana, T.M.; Nielsen, L.P.; Voytek, M.A. Methods for measuring denitrification: Diverse approaches to a difficult problem. *Ecol. Appl.* **2006**, *16*, 2091–2122. [CrossRef]

23. Lewandowski, J.; Putschew, A.; Schwesig, D.; Neumann, C.; Radke, M. Fate of organic micropollutants in the hyporheic zone of a eutrophic lowland stream: Results of a preliminary field study. *Sci. Total Environ.* **2011**, *409*, 1824–1835. [CrossRef]

24. Mulholland, P.J.; Valett, H.M.; Webster, J.R.; Thomas, S.A.; Cooper, L.W.; Hamilton, S.K.; Peterson, B.J. Stream denitrification and total nitrate uptake rates measured using a field 15N tracer addition approach. *Limnol. Oceanogr.* **2004**, *49*, 809–820. [CrossRef]

25. Mulholland, P.J.; Helton, A.M.; Poole, G.C.; Hall, R.O.; Hamilton, S.K.; Peterson, B.J.; Tank, J.L.; Ashkenas, L.R.; Cooper, L.W.; Dahm, C.N.; et al. Stream denitrification across biomes and its response to anthropogenic nitrate loading. *Nature* **2008**, *452*, 202–205. [CrossRef]

26. Orr, C.H.; Predick, K.I.; Stanley, E.H.; Rogers, K.L. Spatial autocorrelation of denitrification in a restored and a natural floodplain. *Wetlands* **2014**, *34*, 89–100. [CrossRef]

27. Poole, G.C.; Berman, C.H. An ecological perspective on in-stream temperature: Natural heat dynamics and mechanisms of human-caused thermal degradation. *Environ. Mgmnt.* **2000**, *27*, 787–802. [CrossRef] [PubMed]

28. Burkholder, B.K.; Grant, G.E.; Haggerty, R.; Khangaonkar, T.; Wampler, P.J. Influence of hyporheic flow and geomorphology on temperature of a large, gravel-bed river, Clackamas River, Oregon, USA. *Hydrol. Process.* **2008**, *22*, 941–953. [CrossRef]

29. Peter, K.T.; Tian, Z.; Wu, C.; Lin, P.; White, S.; Du, B.; McIntyre, J.K.; Scholz, N.L.; Kolodziej, E.P. Using high-resolution mass spectrometry to identify organic contaminants linked to urban stormwater mortality syndrome in Coho Salmon. *Environ. Sci. Technol.* **2018**. [CrossRef]

30. Peter, K.T.; Herzog, S.; Tian, Z.; McCray, J.; Lynch, K.; Kolodziej, E.P. Evaluating emerging organic contaminant removal in an engineered hyporheic zone using high resolution mass spectrometry. *Water Res.* **2019**, *150*, 140–152. [CrossRef]

31. Spromberg, J.A.; Scholz, N.L. Estimating the future decline of wild Coho salmon populations resulting from early spawner die-offs in urbanizing watersheds of the Pacific Northwest, USA. *Integr. Environ. Assess. Manag.* **2011**, *7*, 648–656. [CrossRef]

32. Feist, B.E.; Buhle, E.R.; Arnold, P.; Davis, J.W.; Scholz, N.L. Landscape ecotoxicology of Coho salmon spawner mortality in urban streams. *PLoS ONE* **2011**, *6*, e23424. [CrossRef]

33. Feist, B.E.; Buhle, E.R.; Baldwin, D.H.; Spromberg, J.A.; Damm, S.E.; Davis, J.W.; Scholz, N.L. Roads to ruin: Conservation threats to a sentinel species across an urban gradient. *Ecol. Appl.* **2017**, *27*, 2382–2396. [CrossRef]

34. Scholz, N.L.; Myers, M.S.; McCarthy, S.G.; Labenia, J.S.; McIntyre, J.K.; Ylitalo, G.M.; Rhodes, L.D.; Laetz, C.A.; Stehr, C.M.; French, B.L.; et al. Recurrent die-offs of adult Coho salmon returning to spawn in Puget Sound lowland urban streams. *PLoS ONE* **2011**, *6*, e28013. [CrossRef]

35. Harris, G.P.; Heathwaite, A.L. Why is achieving good ecological outcomes in rivers so difficult? *Freshwater Biol.* **2011**, *57*, 91–107. [CrossRef]

36. Lawrence, J.E.; Skold, M.E.; Hussain, F.; Silverman, D.R.; Resh, V.H.; Sedlak, D.L.; Luthy, R.; Mccray, J.E. Hyporheic zone in urban streams: A review and opportunities for enhancing water quality and improving aquatic habitat by active management. *Environ. Engr. Sci.* **2013**, *30*, 480–501. [CrossRef]

37. Morley, S.; Rhodes, L.; Baxter, A.; Goetz, G. Biological outcomes of hyporheic zone restoration in an urban floodplain. In Proceedings of the 2018 Annual Stream Restoration Symposium, Stevenson, WA, USA, 5–8 February 2018; River Restoration Northwest: Portland, OR, USA, 2018. Available online: http://www.rrnw.org/wp-content/uploads/1.4_2018-02-River-Restoration-NW-Talk-Morley.pdf (accessed on 7 December 2018).

38. Hall, J.; Basketfield, S.; Cooksey, M.; Zisette, R. *City of Seattle State of the Waters 2007, Vol. 1: Seattle Water Courses*; City of Seattle: Seattle, WA, USA, 2007; pp. 150–187. Available online: https://www.seattle.gov/util/cs/groups/public/@spu/@conservation/documents/webcontent/spu01_003413.pdf (accessed on 7 December 2018).

39. Leavy, T.R.; Bakke, P.; Peters, R.J.; Morley, S. *Influence of Habitat Complexity and Floodplain Reconnection Projects on the Physical and Biological Conditions of Seattle Urban Streams: Pre-Project Monitoring;* Final Draft Report, 2005–2009; U.S. Fish and Wildlife Service: Lacey, WA, USA, 2010.

40. Walsh, C.J.; Roy, A.H.; Feminella, J.W.; Cottingham, P.D.; Groffman, P.M.; Morgan, R.P. The urban stream syndrome: Current knowledge and the search for a cure. *J. N. Am. Benthol. Soc.* **2005**, *24*, 706–723. [CrossRef]

41. Bakke, P.D. *Thornton Creek Floodplain Restoration: Effectiveness Monitoring of Physical Processes;* Report to City of Seattle, Vol. 1, Contract 14-133-A; U.S. Fish and Wildlife Service: Lacey, DC, USA, 2018; pp. 1–56.

42. Bakke, P.D.; Lynch, K.; Leavy, T.; Peters, R. Monitoring of groundwater-surface water interactions in support of restoration of hyporheic processes in an urban stream, Thornton Creek, Washington (poster). In Proceedings of the Geological Society of America Annual Meeting, Portland, OR, USA, 20 October 2009.

43. Washington Dept. of Fish and Wildlife (WDFW); U.S. Fish and Wildlife Service (USFWS). Technique 6, Instream structures. In *Stream Habitat Restoration Guidelines;* Cramer, M.L., Ed.; Co-published by the Washington Departments of Fish and Wildlife, Natural Resources, Transportation and Ecology, Washington State Recreation and Conservation Office, Puget Sound Partnership, and the U.S. Fish and Wildlife Service: Olympia, DC, USA, 2012; pp. T6-1–T6-36.

44. Baxter, C.V.; Hauer, F.R. Geomorphology, hyporheic exchange, and selection of spawning habitat by bull trout. *Can. J. Fish. Aquat. Sci.* **2000**, *57*, 1470–1481. [CrossRef]

45. Bjornn, T.C.; Reiser, D.W. Habitat requirements of salmonids in streams. In *Influences of Forest and Rangeland Management on Salmonids Fishes and Their Habitats, American Fisheries Society Special Publication 19;* Meehan, W.R., Ed.; American Fisheries Society: Bethesda, MD, USA, 1991; pp. 83–138.

46. Gariglio, F.P.; Tonina, D.; Luce, C.H. Spatiotemporal variability of hyporheic exchange through a pool-riffle sequence. *Water Resour. Res.* **2009**, *49*, 1–20. [CrossRef]

47. Luce, C.H.; Tonina, D.; Gariglio, F.; Applebee, R. Solutions for the diurnally forced advection-diffusion equation to estimate bulk fluid velocity and diffusivity in streambeds from temperature time series. *Water Resour. Res.* **2013**, *49*, 488–506. [CrossRef]

48. Conant Jr., B. Delineating and quantifying ground water discharge zones using streambed temperatures. *Groundwater* **2004**, *42*, 243–257. [CrossRef]

49. Webster, J.R.; Ehrman, T.P. Solute dynamics. In *Methods in Stream Ecology;* Hauer, F.R., Lambert, G.A., Eds.; Academic Press: New York, NY, USA, 1996; pp. 145–160, ISBN 0-12-332905-1.

50. Runkel, R.L. *One-Dimensional Transport with Inflow and Storage (OTIS): A Solute Transport Model for Streams and Rivers;* Water-Resour. Invest. Rept. 98–4018; U.S. Geological Survey: Denver, CO, USA, 1998. [CrossRef]

51. Bencala, K.E.; Walters, R.A. Simulation of solute transport in a mountain pool-and-riffle stream–a transient storage model. *Water Resour. Res.* **1983**, *19*, 718–724. [CrossRef]

52. McNamara, M.L.; Klingeman, P.C.; Bakke, P.D.; Sullivan, R.L. Channel maintenance flows in the upper Klamath Basin, Oregon. In *Riparian Ecology and Management in Multi-Land Use Watersheds;* Wigington, P.J., Beschta, R.L., Eds.; TPS-00-2; American Water Resources Association: Middleberg, VA, USA, 2000; pp. 53–58, ISBN 978-1882132515.

53. Bunte, K.; Abt, S.R. *Sampling Surface and Subsurface Particle-Size Distributions in Wadable Gravel-and Cobble-Bed Streams for Analyses in Sediment Transport, Hydraulics, and Streambed Monitoring;* General Technical Report RMRS-GTR-74; U.S. Department of Agriculture, Forest Service, Rocky Mountain Research Station: Fort Collins, CO, USA, 2001. [CrossRef]

54. Wagner, B.J.; Harvey, J.W. Experimental design for estimating parameters of rate-limited mass transfer: Analysis of stream tracer studies. *Water Resour. Res.* **1997**, *33*, 1731–1741. [CrossRef]

55. Buffington, J.M.; Tonina, D. Hyporheic exchange in mountain rivers I: Mechanics and environmental effects. *Geogr. Compass* **2009**, *3*, 1063–1086. [CrossRef]

56. Buffington, J.M.; Tonina, D. Hyporheic exchange in mountain rivers II: Effects of channel morphology on mechanics, scales, and rates of exchange. *Geogr. Compass* **2009**, *3*, 1038–1062. [CrossRef]

57. Sambrook Smith, G.H.; Nicholas, A.P.; Ferguson, R.I. Measuring and defining bimodal sediments. *Water Resour. Res.* **1997**, *33*, 1179–1185. [CrossRef]

58. Kondolf, G.M.; Li, S. The pebble count technique for quantifying surface bed material size in instream flow studies. *Rivers* **1992**, *3*, 80–87.

59. Parker, G.; Toro-Escobar, C.M. Equal mobility of gravel in streams: The remains of the day. *Water Resour. Res.* **2002**, *38*, 1–8. [CrossRef]

60. Hyatt, T.L.; Naiman, R.J. The residence time of large wood debris in the Queets River, Washington, USA. *Ecol. Apps.* **2001**, *11*, 191–202. [CrossRef]
61. Lynch, K. (Seattle Public Utilities, Seattle, WA, USA); Pier, C. (Seattle Public Utilities, Seattle, WA, USA). Personal communication, 2018.

Article

Reducing the Phytoplankton Biomass to Promote the Growth of Submerged Macrophytes by Introducing Artificial Aquatic Plants in Shallow Eutrophic Waters

Yue Wu [1,2], Licheng Huang [1], Yalin Wang [1], Lin Li [1,*], Genbao Li [1], Bangding Xiao [1] and Lirong Song [1]

[1] State Key Laboratory of Freshwater Ecology and Biotechnology, Key Laboratory of Algae Biology, Institute of Hydrobiology, Chinese Academy of Sciences, Wuhan 430072, China
[2] University of Chinese Academy of Sciences, Beijing 100049, China
* Correspondence: lilin@ihb.ac.cn; Tel.: +86-27-6878-0037

Received: 23 April 2019; Accepted: 29 June 2019; Published: 2 July 2019

Abstract: Harmful cyanobacterial blooms frequently occur in shallow eutrophic lakes and usually cause the decline of submerged vegetation. Therefore, artificial aquatic plants (AAPs) were introduced into enclosures in the eutrophic Dianchi Lake to investigate whether or not they could reduce cyanobacterial blooms and promote the growth of submerged macrophytes. On the 60th day after the AAPs were installed, the turbidity, total nitrogen (TN), total phosphorous (TP), and the cell density of phytoplankton (especially cyanobacteria) of the treated enclosures were significantly reduced as compared with the control enclosures. The adsorption and absorption of the subsequently formed periphyton biofilms attached to the AAPs effectively decreased nutrient levels in the water. Moreover, the microbial diversity and structure in the water changed with the development of periphyton biofilms, showing that the dominant planktonic algae shifted from Cyanophyta to Chlorophyta. The biodiversity of both planktonic and attached bacterial communities in the periphyton biofilm also gradually increased with time, and were higher than those of the control enclosures. The transplanted submerged macrophyte (*Elodea nuttallii*) in treated enclosures recovered effectively and reached 50% coverage in one month while those in the control enclosures failed to grow. The application of AAPs with incubated periphyton presents an environmentally-friendly and effective solution for reducing nutrients and controlling the biomass of phytoplankton, thereby promoting the restoration of submerged macrophytes in shallow eutrophic waters.

Keywords: artificial aquatic plants; periphyton biofilm; cyanobacteria; eutrophication; Dianchi Lake

---

1. Introduction

The principal characteristics of aquatic macrophytes are their ability to accumulate nutrients and sustain biogeochemical cycles in aquatic environments [1,2]. However, shallow lake eutrophication with the concomitant occurrence of cyanobacterial blooms has led to the decline of submerged macrophytes and the degeneration of water functions [3,4], which have transformed from a macrophyte-dominated state (clear water state) to a phytoplankton-dominated state (turbid water state) [5,6]. Environmental stressors, such as nutrient enrichment [7], high temperatures [8], and water level changes [9], favor cyanobacterial communities. Additionally, previous studies have reported that the low light intensity in many fertile lakes was associated with low submerged macrophyte productivity [10,11]. Therefore, the determination of an appropriate approach as a transitional means of creating beneficial conditions that control cyanobacterial blooms and restore macrophytes is urgently needed.

Many strategies have been proposed to alleviate the effects of cyanobacterial blooms, including the application of aluminum sulphate to decrease phosphorus concentrations through the reaction of co-precipitated phosphorus into the sediment [12]. Inorganic compounds in marl have been used to immobilize nutrients on the sediment surface, thus reducing nutrient concentrations in the water [13]. However, the aforementioned methodologies often have some effect on the original aquatic ecosystem, including changes in pH or salinity, which may threaten life in the lake [14]. Other measures have been applied to control the cyanobacterial blooms directly. Algaecides such as organic bromide, copper sulphate, and hydrogen peroxide have mostly been used to control large-scale cyanobacterial bloom as an emergency method, but these may result in an unsafe aquatic ecosystem and easily induce secondary pollution [15].

Periphyton is an assemblage of freshwater organisms mainly composed of photoautotrophic algae, heterotrophic, and chemoautotrophic bacteria, fungi, protozoans, metazoans and viruses, which attach to submerged surfaces [16]. Periphyton plays a crucial role in nutrient cycling and in supporting food webs and it has an excellent ability to degrade pollutants in aquatic environments [17]. Previous studies have utilized biofilms to remove nitrogen and phosphorus from wastewater and to biodegrade other hazardous contaminants [18,19]. Moreover, microbial communities are easily incorporated into bioreactors, which result in efficient bioremediation operations [20]. Additionally, the microbial compositions of periphyton are derived from original habitats, which prevent changes to the original community from occurring. On the basis of these properties, immobilizing the periphyton on artificial substrates and suspending it between the sediment and overlying water allows periphyton to serve as an advantageous approach for controlling contaminants in aquatic environments [21]. The artificial aquatic plant (AAP) was chosen as the most efficient for improving water quality among four different substrates (aquamats, biocompatibility carbon fiber, eco-carbon fiber, and AAP) based on our previous studies. The results demonstrated that both biocompatibility carbon fiber and the AAP improve water quality by reducing the concentration of nutrients of experiment water which restrained the reproduction of algae in the water column. The removal rates of TN, TP, and Chl-*a* by AAP and biocompatibility carbon fiber were: 66.81%, 61.85%; 57.89%, 8.26%; and 88.84%, 94.03% respectively [22]. Comprehensively, taking in to consideration the cost performance and construction convenience, the AAP was a more favorable material, with longevity over 15 years, for the application in the ecological restoration of shallow eutrophic waters. Due to the structure of plentiful micropores between the microfibers (8.5–9.0 cm per microfiber), the periphyton is provided with adhesive surfaces of 8000 m^2/m^3, and results in strong adsorption and absorption.

Research concerning periphyton has predominantly focused upon the purifying capacity of periphyton in bioreactors and rivers [23,24], while studies in eutrophic lakes remain rare [25,26]. Taking the periphyton biofilm as an assemblage, there is a need to understand the performance and processes involved with algal-bacterial biofilm-based wastewater treatment projects. In this study, we employ periphyton biofilms with artificial substrates to alleviate the negative effects of phytoplankton biomass on aquatic conditions in shallow eutrophic waters, and we explore whether or not this biological measure provides a transition for the restoration of submerged macrophytes. The main objectives of this study are (i) evaluation of nutrient removal performance and impacts on the microbial community, (ii) investigation of the dynamics of periphyton attached on artificial substrates during the experiment, and (iii) discussion of the possible role of periphyton to control cyanobacterial blooms.

2. Materials and Methods

2.1. Study Site

Dianchi Lake (24°51′ N, 102°42′ E, Figure 1) located on a plateau in southwestern China, has a water surface of 300 km^2, watershed area of 2920 km^2, and an altitude of 1886.5 m above sea level [27]. Since the establishment of Haigeng Dam in 1996, Lake Dianchi has been separated into two parts: Caohai in the north with a water surface of 11 km^2 and average water depth of 2.5 m and Waihai in the south with a water surface of 299 km^2 and average water depth of 5 m. The map of Dianchi Lake watershed is modified from Wang et al. [28]. Due to the rapid economic development and intensive use of the water resources, the water in this lake has become seriously polluted, and cyanobacterial blooms have occurred frequently in the past 20 years [29]. In this study, the experimental enclosures were built in Caohai Bay (24°58′ N, 102°38′ E) which were located in the north of Lake Dianchi.

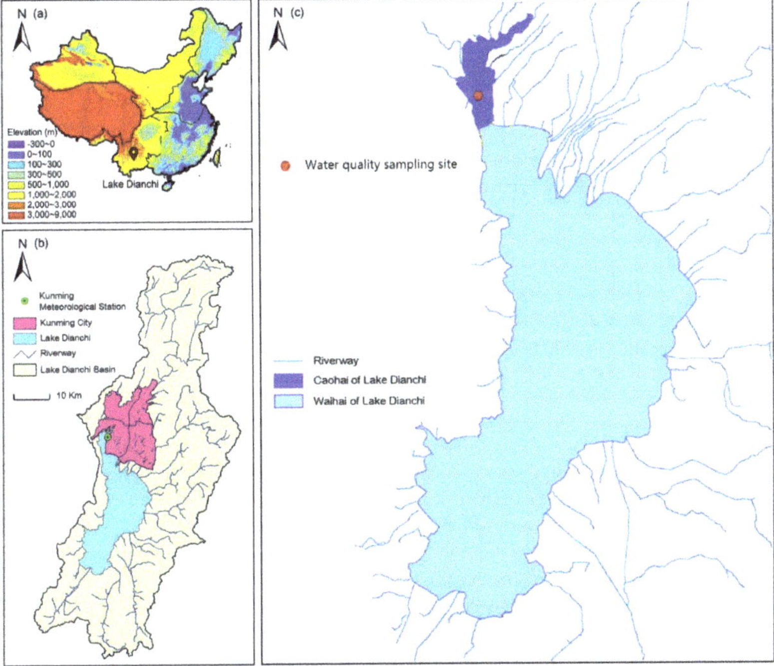

Figure 1. Map of Dianchi Lake watershed.

2.2. Experimental Design

Four enclosures (each 6 m × 6 m × 2 m, length × width × height, Figure 2) were constructed with rubber cloth (around all sides of the water column) and geotextile (underwater). They were open to the air and above the sediment at the bottom of Caohai Bay. Two enclosures were used as the control enclosure and the others employed fresh artificial aquatic plants (AAPs). The AAP strand (Figure 2) was composed of thousands of polypropylene microfibers, which imitate the structure of natural aquatic plants. A uniform weight was used for cement balls tied at the bottom of every strand of the AAP to maintain a vertical state and set a density of 1 strands/m^3 in the treatment enclosures.

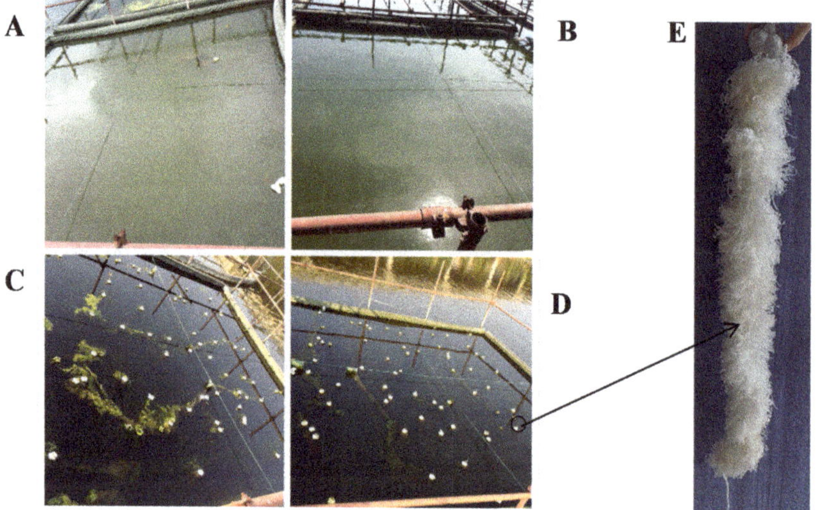

Figure 2. Experimental setup of the enclosures and structure of the (**E**) artificial aquatic plant (AAP); (**A,B**) control enclosures; (**C,D**) treated enclosures with 1 strand AAP /m^3 with length of 80–85 cm.

2.3. Sample Preparation

Representative microfibers, which contained periphyton biofilms, were removed from the AAP at different times. To assure vertical comparability, both the samples of water and microfibers of AAP were collected at a depth of 0.5 m during the experiment. For eliminating spatial heterogeneity, the water samples were collected at three sampling sites horizontally in the same enclosure and twice a month. Similarly, the microfiber samples were carefully cut from three different AAP strands in the same enclosure with sterile scalpels and then weighed for microbial analyses once per month.

2.4. Environmental Parameters

The performance of experimental enclosures was evaluated by comparing water quality indices during this experiment. Chemical oxygen demand (COD), total nitrogen (TN), and total phosphorus (TP) were determined according to the standard methods for water and wastewater monitoring and analysis [30]. The water temperature (WT), dissolved oxygen (DO), and pH were measured in situ via a multiparameter sonde (EXO 2, YSI Inc., Yellow Springs, OH, USA). Water transparency (SD) was measured with a 10 cm diameter black and white Secchi disk. The chlorophyll-a (Chl *a*) content of the water was determined after extraction of the filtered algal mat with 90% acetone as per AHPA [31]. Phytoplankton were identified and counted using the methods described by Hu et al. [32].

2.5. Phytoplankton Community and Total Phosphorous (TP) Associated with Periphyton Biofilms

After clearing microfibers of AAPs with distilled water, attached phytoplankton cells were extracted from the microfibers using a modified soil extraction method [33]. Briefly, 20 mL of a sterile saline solution was added to 0.5 g textile (dry weight) in 50 mL centrifugation tubes. After centrifuging for 2 min at a maximum speed (Vortex Genie2, Scientific Industries, Inc., Bohemia, NY, USA), the sample was shaken horizontally at 200 rpm for 30 min at room temperature. This process was repeated three times and the algal fluid volume was collected and used to measure the Chl *a* by spectrophotometry [34] and count the cell density of attached algae by microscopy [32].

To compare the dynamics of planktonic algae in the water column with algae attached to the AAP, the conversional relationship was examined using the volume of each enclosure (72 m^3) and the density of the AAPs (2 strands/m^2) in each enclosure (mean 72 strands/enclosure). The formula for the Chl *a* content adsorbed by the whole AAP in each enclosure, is given as Equation (1) and the cell density of the attached algae is defined by Equation (2). The content of the TP adsorbed by the periphyton biofilms attached to the microfibers of the AAP was measured by a modified method using the TP in the sediment. The periphyton biofilms were removed from the microfibers of the AAPs using distilled water and ash was formed at 450 °C in a muffle furnace with an acid dissolution [35].

To compare the concentration of the TP in the water column with that in the periphyton biofilms, the conversional relationship was examined using the TP content adsorbed by the periphyton biofilms attached to microfibers of AAPs in each enclosure, as defined by Equation (3).

$$A_1 = \frac{A_2 \times V_1 \times M_2 \times 72}{(72000 \times M_1)} \tag{1}$$

$$B_1 = \frac{B_2 \times V_1 \times M_2 \times 72}{(72000 \times M_1)} \tag{2}$$

$$C_1 = \frac{M_2 \times C_2 \times 72}{72000} \tag{3}$$

where, A_1 is the content of Chl *a* adsorbed by all strands of the AAPs in each enclosure; µg/L, A_2 is the content of Chl *a* adsorbed by microfibers of the AAP; µg/L, B_1 is the cell density of attached algae for all strands of the AAPs in each enclosure; and cells/L and B_2 is the cell density of attached algae by microfibers of the AAP, cells/L. Additionally, C_1 is the TP content adsorbed by periphyton biofilms attached to all strands of the AAPs in each enclosure; mg/L, C_2 is the TP content adsorbed by periphyton biofilms attached to the quantitative microfibers of the AAP, mg/g; M_1 is the dry weight of microfibers of the AAP, g; M_2 is the average dry weight per strand of the AAP, g; V_1 is the algal fluid volume collected from the microfibers of the AAP, L; 72 is the number of AAPs of each enclosure, and strand and 72,000 is the volume of each enclosure, L.

2.6. DNA Extraction and Sequencing

Quantitative microfibers that contained periphyton biofilms for the sequencing of 16S ribosomal RNA (rRNA) gene amplicons were first processed as follows: DNA was extracted, 16S rRNA gene polymerase chain reaction (PCR) amplification was performed, and PCR products were purified. For DNA extraction, 100 mL water samples were filtered by cellulose acetate filters (pore size = 0.2 mm, diameter = 45 mm, XingYa, Shanghai, China) and DNA was isolated from the filters, and directly from 0.2 g of biocarrier using the FastDNA kit (MP Biomedicals, Santa Ana, California, USA) and the FastPrep instrument (Savant Instruments, Inc., Holbrook, NY, USA) according to their protocols. The 16S rRNA genes were amplified using bacterial universal primers 27F (AGRGTTTGATCMTGGCTCAG) and 1492R (GGTTACCTTGTTACGACTT), DNA dilutions were used as a template, and PCR was conducted with a volume of 20 µL, with 10 µL of PCR master mix, 8 µL of sterile water, 0.5 µL each of forward and reverse primers, and 1 µL of the template DNA (~100 ng). Amplification used the following protocol: initial denaturation at 95 °C for 1 min, followed by 40 cycles of 95 °C for 25 s, 53 °C for 30 s, and 70 °C for 60 s; the samples were amplified in triplicate. The PCR products were then subjected to sequence analysis (Nuohe, Beijing, China).

2.7. Submerged Macrophyte—Elodea Nuttallii

In order to explore the feasibility of periphyton, biofilms attached on the AAPs to increase the transparency and incident light, thereby facilitating the growth of submerged macrophytes. Prepared 30 cm high seedlings of the submerged macrophyte, *Elodea nuttallii*, which can grow slightly even in winter if the daily mean water temperature is higher than 4 °C [36,37], were wrapped with loess clay and nonwoven fabrics and were arranged to transplant into all experimental enclosures at suitable conditions.

2.8. Data Analysis

Nutrient concentrations were expressed as the means ± standard deviation of duplicates (3 × 2). Because normality and homogeneity of variance assumptions were not satisfied, we used the nonparametric Kruskal–Wallis test followed by the Nemenyi test (post hoc test) to test whether the time phases or (interaction) installed AAPs have differences or similarities between two groups in each time phase, with PASW Statistics 20 (SPSS Inc., Hong Kong, China). The significance level was $p = 0.05$. Amplicon sequence analysis was performed using QIIME Pipeline Version 1.7.0 (Nuohe, Beijing, China) [38] and the weighted UniFrac distance for the unweighted pair group method with arithmetic mean (UPGMA), sample clustering tree, and principle coordinates analysis (PCoA).

3. Results and Discussion

3.1. Variation of Water Quality in the Enclosures

During the experiment, water temperature ranged between 13.1 °C and 15.6 °C. In control enclosures, turbidity decreased from 11.40 (±1.70) nephelometric turbidity units (NTU, Figure 3A) to 6.85 (±0.30) NTU by day 60. In treatment (i.e., within AAPs) enclosures, turbidity decreased from 10.80 (±2.04) NTU to 3.13 (±0.77) NTU by day 60. Water turbidity in control enclosures was significantly higher ($p < 0.05$) than that of treatment enclosures within the AAPs throughout the experiment after 60 days of operation. The concentration of COD (Figure 3B) declined from 12.06 (±1.35) mg/L to 4.04 (±0.037) mg/L and to 2.11 (±0.25) mg/L in the control and treatment enclosures, respectively; there was no significant difference between the enclosures ($p > 0.05$). Total nitrogen (Figure 3C) declined from 0.96 to 0.41 mg/L in treatment enclosures over 60 days, while TN slowly decreased from 0.96 to 0.90 mg/L in the first month and then gradually increased to 0.94 mg/L in the control enclosures. Jiang et al. measured total nitrogen removal of 92% by using a hybrid biofilm reactor [39]. Total phosphorous (Figure 3D) decreased from 0.067 (±0.002) mg/L to 0.051 mg/L and to 0.033 mg/L in the control and treatment enclosures, respectively. Water TP in control enclosures was significantly higher ($p < 0.05$) than that of treatment enclosures within the AAPs after 60 days of operation; the removal efficiency later exceeded 50%. Sukačová et al. reported a P removal efficiency of 99% within 24 hours of P addition when using an algal biofilm [40]. Chlorophyll a was used as an indicator of algal biomass (Figure 3E). In control enclosures, Chl *a* slightly declined from 285.46 (±22.66) to 177.65 (±2.42) µg/L by day 60, while the Chl *a* of treatment enclosures decreased from 283.17 (±36.01) µg/L to 43.82(±6.34) µg/L by day 60, which was significantly lower than that of control ($p < 0.05$) enclosures after 30 days of operation; the removal efficiency exceeded 80%.The results suggest the feasibility of using AAPs to reduce nutrient concentrations and algal biomass in eutrophic waters. The biophysical adsorption of phytoplankton by the microfibers and its adherent periphyton biofilms may contribute to the uptake, storage, and transformation of nutrients, as well as other chemicals, which play a key role in the self-purification of water bodies [41].

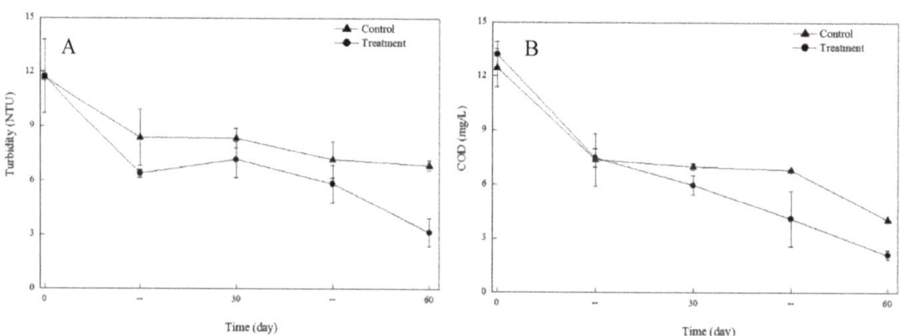

Figure 3. *Cont.*

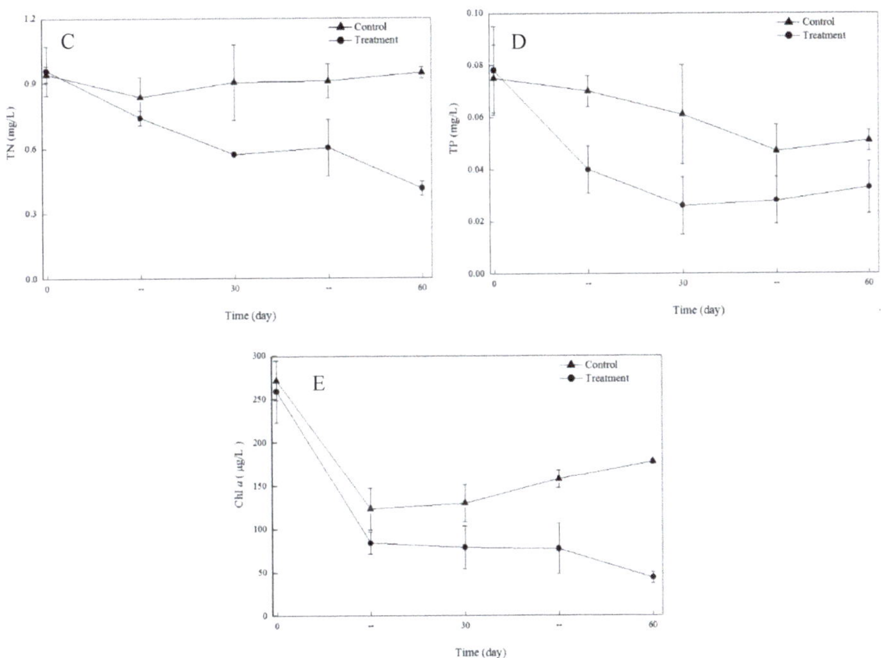

Figure 3. (**A**) turbidity, (**B**) chemical oxygen demand (COD), (**C**) total nitrogen (TN), (**D**) total phosphorous (TP) and (**E**) chlorophyll a (Chl *a*) content in the water column over 60 days. Error bars represent the standard deviation (n = 3 × 2).

3.2. Variation of TP Concentrations in Periphyton Biofilms Attached to Artificial Aquatic Plants (AAPs)

Compared to the average, dry weight per strand of AAP weighed 75 (±4.2) g. The mean dry weight per strand of AAP ranged from 87.3 (±5.1) g to 100.8 (±4.2) g from day 30 to day 60. According to the relationship expressed in Equation (3), the TP content adsorbed by periphyton biofilm attached to the AAPs ranged from 0.03 (±0.004) mg/L to 0.04 (±0.004) mg/L from day 30 to day 60 of the experiment. The resulting TP content in the water column was significantly decreased while that in the periphyton biofilms gradually increased in the same enclosure, which indicates that the cumulative adsorption and absorption by extensive periphyton biofilms attaching to AAPs played an important role in nutrient migration. Previous studies have reported that periphyton biofilms have a strong affinity for removing phosphorus, and therefore constitute an effective phosphorus sink [42]. This may be attributed to the uptake and transformation of biologically available soluble phosphorus for self-growth [43].

3.3. Dynamics of Phytoplankton in Water Column with the Development of Periphyton Biofilms Attached to AAPs

In this study, we investigated the biomass and composition of phytoplankton communities, which covered the water and microfibers of AAPs. Three types of samples were compared: (i) water samples from control enclosures without AAPs (CW), (ii) water samples from treatment enclosures within AAPs (TW), and (iii) microfiber samples containing periphyton biofilm attached to AAPs (TB) from treatment enclosures. The samples were then classified according to their different time phases.

The cell density of algae was driven mainly by the total phytoplankton and the cyanobacteria in all enclosures (Figure 4). Compared to the control enclosure, the mean cell density of total planktonic algae in the TW decreased rapidly from 1.09×10^8 (±0.32×10^8) cells/L to 0.15×10^8 (±0.04×10^8) cells/L, and the Chl *a* of TW declined at a rate of over 80% after 60 days of this experiment. Additionally, the mean cell density of cyanobacteria showed a similar trend and declined quickly from 6.0×10^7 ($\pm 1.91 \times 10^7$) cells/L to 0.5×10^7 ($\pm 0.22 \times 10^7$) cells/L. After 30 days of operation, the cell density of total planktonic algae in the TW1 was significantly lower than that of CW1 ($p < 0.05$). These results suggest that the phytoplankton were inhibited, obviously in the enclosures with AAP treatments, with respect to the control enclosure. Interestingly, it was found that the cell density of the total adherent algae in the TB amounted to 5.6×10^7 ($\pm 0.40 \times 10^7$) cells/L after 60 days of operation, and this included the cell density of cyanobacteria, which summed to 1.5×10^7 ($\pm 0.90 \times 10^7$) cells/L with the extension of incubation time. On the basis of the measurements of the microfibers of AAPs and the given conversion formula, the concentration of Chl *a* by periphyton biofilms attached to AAPs increased to 94.8 (± 5.48) µg/L. As substrata, the microfibers of AAPs provided a rough and porous surface that promoted higher planktonic micro-algal cell attachment. Therefore, with respect to adsorption, pioneer planktonic algal cells attached to the substrata and biological adhesion continued to form micro-colonies. In addition, the attachment of microalgal cells to a stable surface increased their abundance relative to the water column, which made them communicate and protect each other to adapt to environmental conditions [44].

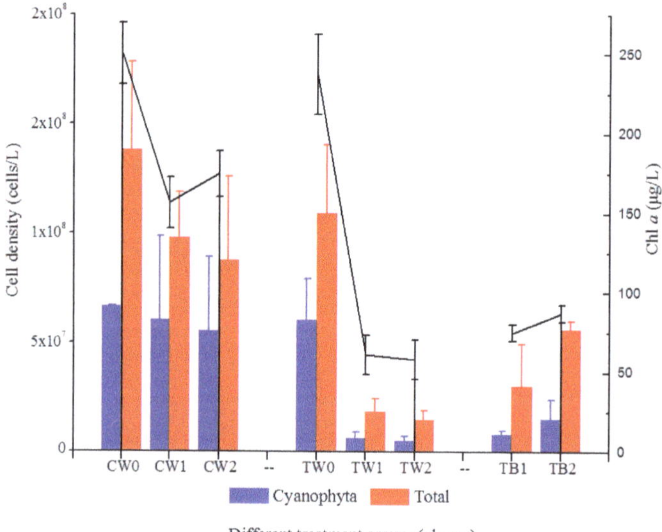

Figure 4. Variation of cell density relative to total algae (including cyanobacteria), and the corresponding concentration of Chl *a* (shown as black line) in all enclosures. Error bars represent standard deviation ($n = 3 \times 2$). Designations 0, 1, and 2 represent days 0, 30, and 60, respectively.

The aforementioned results, combined with the TP content resided in periphyton biofilms, may suggest that competition occurs between periphyton and planktonic algae, including cyanobacteria, for the phosphorus that algal cells require for growth. Some studies have shown that periphyton biofilms competed with planktonic algae for mutual resources, such as nutrients and space [45,46]. These results indicate the feasibility of the AAPs to reduce nutrient loads in water bodies and the phytoplankton biomass.

The structural composition of phylum-level algae was exhibited through the relative abundances (Figure 5). With the extension of experimental time, the proportion of cyanobacteria that consistently dominated and occupied more than 50% of the habitat reached 71.7% in the CW. In the TW, the dominant phylum was Cyanophyta at the initial stage and then shifted to Chlorophyta. Specifically, the proportion of Cyanophyta declined rapidly from 63.7% to 33.2%; simultaneously, the ratios of the Chlorophyta increased gradually from 17.5% to 49.2%. These trends may have been caused by the removal of nutrients, which agrees with the previous finding from Tlili et al. [47] that micropollutants from wastewater discharge were responsible for changes in community structure. An interesting result is that the ratios of Bacillariophyta in the periphyton biofilms that attached to the artificial aquatic plants (TB) accounted for a mean of 48.5%, and the dominant species were *Achnanthes exigua*, *Gomphonema acuminatum*, *Cymbella pusilla*, *Navicula placentula*, and *Navicula graciloides*. The success of diatoms implies that they have highly efficient and adaptable survival mechanisms and growth strategies. Previous studies have found that diatoms were generally regarded as the principal constituent of the primary colonizers in biofilms [27,48].

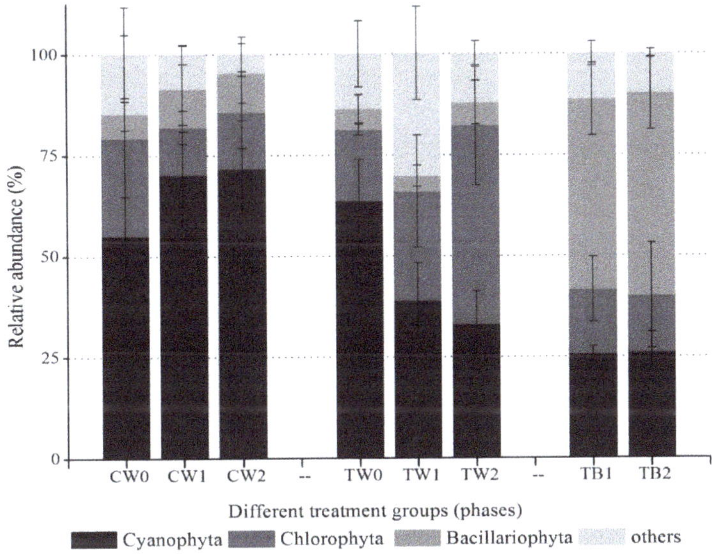

Figure 5. Variation of structural composition of phytoplankton in the enclosure.

3.4. Dynamics of Bacterial Communities with the Development of Periphyton Biofilms Attached to AAPs

The structure and succession of the bacterial communities were compared and analyzed during the intervention of the AAPs, which was regarded as an artificial micro-ecosystem where many microorganisms, including new prokaryotic and eukaryotic colonizers, interact in the periphyton biofilms [49]. Gene amplicon sequencing data of samples CW, TW, and TB were acquired based on the 16S rRNA. Due to the usage of fresh substrata (without any microbial community) in the experiment, TB0 (i.e., TB at day zero) was invalid.

In this study, PCoA (Figure 6) revealed that PC 1 and PC 2 explained 53.28% and 17.41% of the variation in the bacterial community, respectively. Importantly, TWs were clustered together and were effectively separated from the (CW) samples along PC 1, which was influenced by the decreased concentration of the nutrients of TW adsorbed by the periphyton biofilms attached to the AAPs. Carr et al. [50] reported that bacteria and algae in biofilms coexist in an association that offers space and resources to sustain production of both groups of organisms. Additionally, the heterotrophic bacterial community is an important component of periphyton biofilms and contributes to biodegrade pollutants, the changes in the microbial structure and function in periphyton biofilms exposed to different nutrient supplies [15]. For instance, Suberkropp et al. [51] reported a shift in the microbial structure by the continuous nutrient enrichment of a forested headwater stream. Moreover, TW and TB samples were slightly separated along PC2. This may indicate that the microbial communities of the periphyton biofilms preferred specific adaptable and functional microbes, although TB and TW shared the same source of microbial consortia. Therefore, to clarify the changes in the microbial structure in different samples, we considered and analyzed the diversity of the microbial community.

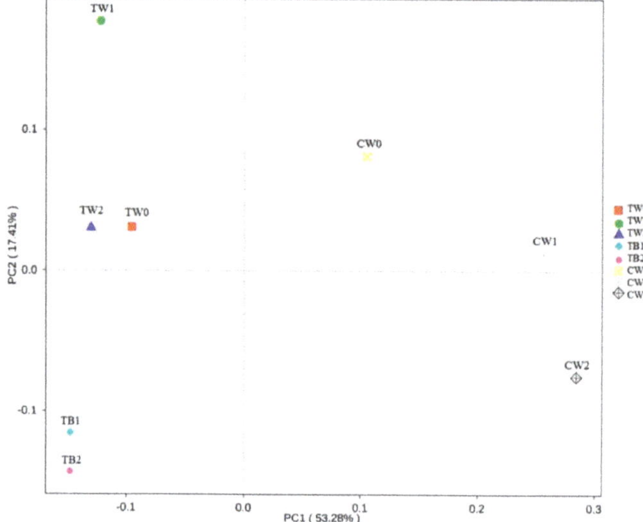

Figure 6. Coordinates analysis (PCoA) based on the weighted UniFrac distance of 16S rRNA genes.

This study chose to detect indices of operational taxonomic unit (OTU) and Shannon and Simpson indices for measuring species diversity and evenness, where a higher value represents greater diversity. The indices of Chao and ACE are richness estimators, where a higher value represents greater richness, and goods-coverage measures the reliability of the sequencing result.

The results indicate that the bacterial diversity and population richness of TW samples increased gradually, and this is revealed by the Shannon index and Simpson index. The microbial population richness of the TW sample was higher than that of the CW sample after the intervention of the AAPs. Meanwhile, it was indicated that bacterial diversity and population richness of sample TB were quite high and increased gradually as revealed by the Shannon (from 0 up 8.38) and Simpson (from 0 to 0.98) indices (Table 1). The results showed in Figure 7 illustrate the relative abundance of the 10 most abundant bacteria at the phylum level in the CW, TW, and TB samples.

Table 1. Statistics for different treatment samples.

Sample Name	Sequence Number	OTU Number	Shannon Index	Simpson Index	Chao Index	ACE Index	Goods-Coverage
CW0	50,000	1764	7.774	0.984	1701.511	1745.288	0.994
CW1	50,874	1445	6.370	0.925	1398.249	1420.827	0.995
CW2	44,922	1469	6.354	0.92	1512.567	2332.759	0.987
TW0	46,587	1672	6.764	0.969	1599.339	1649.734	0.992
TW1	61,178	2214	8.336	0.991	2174.538	2204.640	0.993
TW2	54,552	2014	8.273	0.992	1913.526	1992.917	0.993
TB1	52,842	2056	7.856	0.968	2009.426	2033.116	0.993
TB2	49,478	2099	8.377	0.984	2087.237	2097.015	0.993

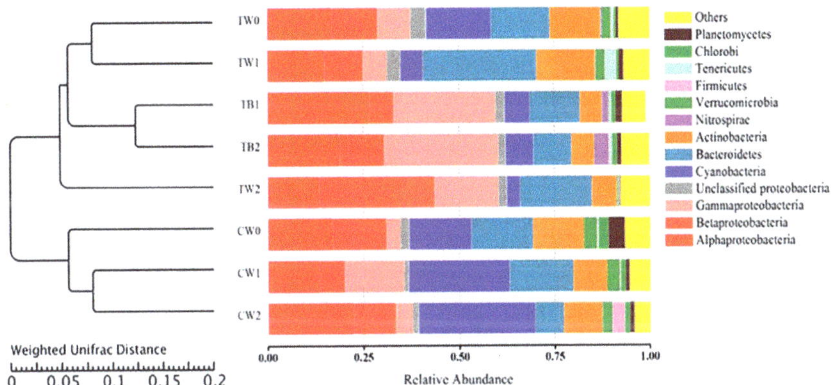

Figure 7. Succession of microbial community during biofilm development at different zones and stages.

Initially, cyanobacteria were found to be one of the dominant prokaryotes in the CW and TW samples. With the incubation of periphyton on AAPs, the relative abundance of cyanobacteria decreased gradually over time in TW, while it increased slightly in CW. This supported the cell fates of Cyanophyta detected by microscopy. Although Proteobacteria were frequently detected in all samples during this experiment, Proteobacteria of the α subdivision (Alphaproteobacteria) dominated in CW and TW, while the γ subdivision (Gammaproteobacteria) was dominant in TB. Members of Alphaproteobacteria are known for their biodegradation of COD [52], which explains the coexistence of the high efficiency of COD removal in CW and TW. Gammaproteobacteria were found mainly attached to the membrane module and easily led to biofouling [53,54], which made the microbial community that adhered to the surface of the artificial substrata and benefited from the development of periphyton biofilms. The relative abundance of *Nitrospirae* increased slowly in the TB, with ratios of 2.79%, 0.19%, and 0.02% in the samples TB, TW, and CW, respectively. *Nitrospirae* has an autotrophic metabolism and is associated with ammonia-oxidizing strains [55].

3.5. Restoration of Submerged Macrophytes with Possible Mechanisms

After 60 days of operation where AAPs were installed in enclosures within Dianchi Lake, the approximately 30 cm high seedlings of the submerged macrophyte (*Elodea nuttallii*), which were wrapped with loess clay and nonwoven fabrics, were transplanted into all experimental enclosures at a density of 4 strands/m². One month later, the planted *E. nuttallii* had matured and reached 50% coverage in treated enclosures while those in control enclosures did not grow. Previous studies have reported that the light compensation depth of submerged macrophytes in Caohai (northern Dianchi Lake) is much lower than the actual water depth, which leads to declines in the submerged macrophytes. The minimum transparency (0.79 ± 0.13 m) was required for the restoration of submerged macrophytes at depths of 2 m [56]. In this study, the transparency in the treatment enclosures with AAPs increased from the initial 0.6 ± 0.10 m to 1.2 ± 0.10 m after two months of the experiment, which attained the necessary transparency to restore *Elodea nuttallii*. Additionally, studies have demonstrated that the occurrence of aquatic macrophyte communities is also related to phytoplankton biomass and certain environmental variables (e.g., COD, nutrients) [57], which suggested the applicability of bioremediation for the restoration of submerged macrophytes.

Given the above, a schematic of mechanisms for improving water quality in shallow eutrophic waters by using AAPs is shown in Figure 8. When an AAP with a large specific surface area was introduced into the eutrophic lake water, large amounts of nutrients (e.g., N, P, and COD), phytoplankton, and planktonic bacteria in the water column were rapidly adsorbed in thousands of microfibers in the AAP to form the periphyton biofilm. Nutrients could then be gradually assimilated by adherent microalgae, and then biodegraded by attached bacteria with the amplification of periphyton biofilms. Thus, the reduced availability of nutrients and the competition between the planktonic and attached microbial community led to the restraint of cyanobacterial growth and the subsequent change in planktonic and attached microbial communities. Ultimately, the decrease of nutrients and cyanobacterial biomass in the water column increase the transparency and incident light, thereby facilitating the growth of submerged macrophytes.

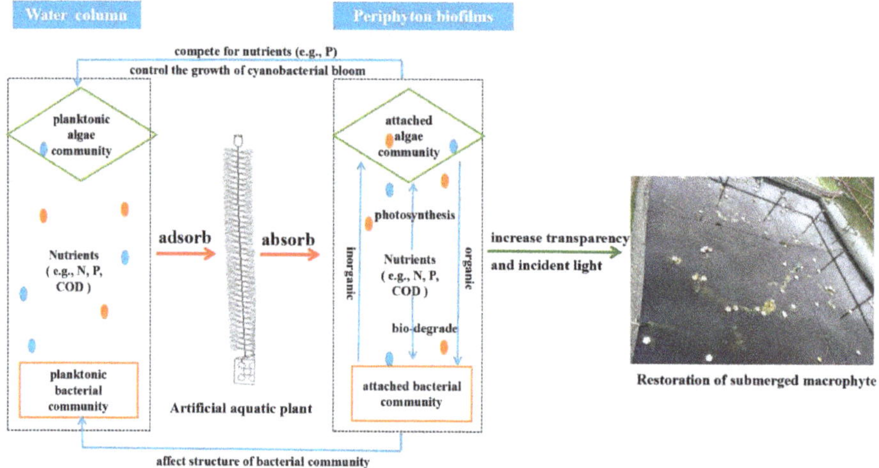

Figure 8. Mechanism of bioremediation for the restoration of submerged macrophytes.

4. Conclusions

Introducing the benign bioremediation of AAPs into in situ enclosures of the eutrophic Dianchi Lake for two months was found to reduce nutrient concentrations (both N and P) and phytoplankton biomass in the water column. The results of the increases in TP and Chl *a* and microbial communities in periphyton biofilms formed on AAPs greatly improved water quality and the subsequent restoration of submerged macrophytes. Management practices targeting periphyton could be potentially used to alleviate the eutrophic waters, which need to be explored further.

Finally, our results reveal the potential importance of AAPs for controlling phytoplankton biomass to promote the growth of submerged macrophytes in shallow eutrophic waters. We anticipate encountering some problems during the transfer of the small-scale experiments reported here to a whole lake situation. Technical problems of exposure and safe remove of thousands AAPs need to be solved. Human interference also needs to be taken into consideration. Long-term and large-scale studies would be of great interest and challenge to evaluate periphyton biofilms attached to AAPs as a transitional means of creating beneficial conditions that control cyanobacterial blooms and restore macrophytes under a complete annual cycle.

Author Contributions: Conceptualization, L.L.; Funding acquisition, L.L. and L.S.; Investigation, Y.W. (Yue Wu), L.H. and Y.W. (Yalin Wang); Methodology, Y.W. (Yue Wu), L.H. and L.L.; Discussion, Y.W. (Yue Wu), L.L., G.L., B.X. and L.S.; Writing—draft preparation, review & editing, Y.W. (Yue Wu), and L.L.; Project administration, L.L. and L.S.

Funding: This research was funded by the Major Science and Technology Program for Water Pollution Control and Treatment (2013ZX07102005, 2017ZX07203001) and the Joint NSFC-ISF Research Program (41561144008). The APC was funded by the National Key R&D Program of China (2017YFE0125700).

Acknowledgments: We are grateful to Chunbo Wang, Qichao Zhou, and Xingqiang Wu, for their efficient technical assistance during field and laboratory experiments and thank Eldon R. Rene (UNESCO-IHE) for his comments on the manuscript.

Conflicts of Interest: The authors declare no conflict of interest.

References

1. Carpenter, S.R.; Lodge, D.M. Effects of submersed macrophytes on ecosystem processes. *Aquat. Bot.* **1986**, *26*, 341–370. [CrossRef]
2. Maberly, S.C. *Ecology of Shallow Lakes*; Marten Scheffer, Chapman & Hall: London, UK, 1998; 357p, ISBN 0-412-74920-3.
3. Gulati, R.D.; Donk, E. Biomanipulation in the Netherlands: Applications in fresh water ecosystems and estuarine water—An introduction. *Hydrobiol. Bull.* **1989**, *23*, 1–4. [CrossRef]
4. Hosper, S.H. Biomanipulation, new perspective for restoring shallow, eutrophic lakes in The Netherlands. *Hydrobiol. Bull.* **1989**, *23*, 11–19. [CrossRef]
5. Scheffer, M.; Hosper, S.H.; Meijer, M.L.; Moss, B.; Jeppesen, E. Alternative equilibria in shallow lakes. *Trends Ecol. Evol.* **1993**, *8*, 275–279. [CrossRef]
6. Waters, M.N.; Schelske, C.L.; Brenner, M. Cyanobacterial dynamics in shallow Lake Apopka (Florida, USA) before and after the shift from a macrophyte-dominated to a phytoplankton-dominated state. *Freshwater Biol.* **2015**, *60*, 1571–1580. [CrossRef]
7. Smith, V.H. Low Nitrogen to phosphorus ratios favor dominance by blue-green algae in lake phytoplankton. *Science* **1983**, *221*, 669–671. [CrossRef]
8. Paerl, H.W.; Huisman, J. Climate change: A catalyst for global expansion of harmful cyanobacterial blooms. *Environ. Microbiol.* **2009**, *1*, 27–37. [CrossRef]
9. Waters, M.N.; Piehler, M.F.; Smoak, J.M.; Martens, C.S. The development and persistence of alternative ecosystem states in a large, shallow lake. *Freshwater Biol.* **2010**, *55*, 1249–1261. [CrossRef]
10. Brock, T.C.M.; Bongaerts, M.C.M.; Heijnen, G.J.M.A.; Heijthuijsen, J.H.F.G. Nitrogen and phosphorus accumulation and cycling by *Nymphoides peltata* (Gmel.) *O. Kuntze* (Menyanthaceae). *Aquat. Bot.* **1983**, *17*, 189–214. [CrossRef]
11. Spence, D.H.N. The zonation of plants in freshwater lakes. *Adv. Ecol. Res.* **1982**, *12*, 37–125.
12. Wu, Y.H.; Feng, M.Y.; Liu, J.T.; Zhao, Y. Effects of polyaluminium chloride and copper sulfate on phosphorus and UV254 under different anoxic levels. *Fresenius Environ. Bull.* **2005**, *14*, 406–412.
13. Stüben, D.; Walpersdorf, E.; Voss, K.; Rönick, H.; Schimmele, M.; Baborowski, M.; Luther, G.; Elsner, W. Application of lake marl at lake Arendsee, NE Germany: First results of a geochemical monitoring during the restoration experiment. *Sci. Total Environ.* **1998**, *218*, 33–44. [CrossRef]
14. Hullebusch, E.V.; Deluchat, V.; Chazal, P.M.; Baudu, M. Environmental impact of two successive chemical treatments in a small shallow eutrophied lake: Part I. Case of Aluminium Sulphate. *Environ. Pollut.* **2002**, *120*, 617–626. [CrossRef]
15. Chorus, I.; Bartram, J. *Toxic Cyanobacteria in Water. A Guide to their Public Health Consequences, Monitoring and Management*; F & FN Spon: London, UK, 1999; p. 416.
16. Larned, S.T. A prospectus for periphyton: Recent and future ecological research. *J. N. Am. Benthol. Soc.* **2010**, *29*, 182–206. [CrossRef]
17. Underwood, G.J.C.; Perkins, R.G.; Consalvey, M.C.; Hanlon, A.R.M.; Oxborough, K.; Baker, N.R.; Paterson, D.M. Patterns in Microphyto benthic primary productivity: Species-specific variation in migratory rhythms and photosynthetic efficiency in mixed-species biofilms. *Limnol. Oceanogr.* **2005**, *50*, 755–767. [CrossRef]
18. Sukačová, K.; Trtílek, M.; Rataj, T. Phosphorus removal using a microalgal biofilm in a new biofilm photobioreactor for tertiary wastewater treatment. *Water Res.* **2015**, *71*, 55–63. [CrossRef] [PubMed]
19. Wan, J.; Liu, X.; Wu, C.; Wu, Y. Nutrient capture and recycling by periphyton attached to modified agrowaste carriers. *Environ. Sci. Pollut. Res.* **2016**, *23*, 8035–8043. [CrossRef]
20. Wu, Y.; Zhang, S.; Zhao, H.; Yang, L. Environmentally benign periphyton bioreactors for controlling cyanobacterial growth. *Bioresour. Technol.* **2010**, *101*, 9681–9687. [CrossRef]
21. Wu, Y.; Xia, L.; Yu, Z.; Shabbir, S.; Kerr, P.G. In situ bioremediation of surface waters by periphytons. *Bioresour. Technol.* **2014**, *151*, 367–372. [CrossRef]
22. Huang, L.C. Ecological Restoration of Eutrophic Lake with Artificial Grass—Take Dianchi Lake for Example. Master's Thesis, Institute of Hydrobiology, Chinese Academy of Sciences, Wuhan, China, June 2016.
23. Prieto, D.M.; Rubinos, D.A.; Piñeiro, V.; Díaz-Fierros, F.; Barral, M.T. Influence of epipsammic biofilm on the biogeochemistry of arsenic in freshwater environments. *Biogeochemistry* **2016**, *129*, 291–306. [CrossRef]

24. Teng, Q.X.; Pang, Y.; Hu, X.Z.; Wang, Y.T.; Huang, T.Y. Application of Artificial Plants in Farmland Drainage Ditches of Boluo River. *J. Environ. Eng. Technol.* **2016**, *6*, 65–71.

25. Yang, Y.; Chen, W.; Yi, Z.; Pei, G. The integrative effect of periphyton biofilm and tape grass (*Vallisneria natans*) on internal loading of shallow eutrophic lakes. *Environ. Sci. Pollut. R.* **2017**, *25*, 1773–1783. [CrossRef] [PubMed]

26. He, H.; Luo, X.; Jin, H.; Gu, J.; Jeppesen, E.; Liu, Z.; Li, K. Effects of exposed artificial substrate on the competition between phytoplankton and benthic algae: Implications for shallow lake restoration. *Water* **2017**, *9*, 24. [CrossRef]

27. Liu, Y.; Chen, W.; Li, D.; Shen, Y.; Li, G.; Liu, Y. First report of aphantoxins in China—Water blooms of toxigenic *Aphanizomenon flos-aquae* in Lake Dianchi. *Ecotoxicol. Environ. Saf.* **2006**, *65*, 84–92. [CrossRef] [PubMed]

28. Wang, J.H.; Yang, C.; He, L.Q.S.; Dao, G.H.; Du, J.S.; Han, Y.P.; Hu, H.Y. Meteorological factors and water quality changes of Plateau Lake Dianchi in China (1990–2015) and their joint influences on cyanobacterial blooms. *Sci. Total Environ.* **2019**, *665*, 406–418. [CrossRef] [PubMed]

29. Liu, L.; Du, R.; Zhang, X.; Dong, S.; Sun, S. Succession and seasonal variation in epilithic biofilms on artificial reefs in culture waters of the sea cucumber *Apostichopus japonicus. Chin. J. Ocean. Limnol.* **2016**, *35*, 132–152. [CrossRef]

30. SEPA. *Standard Methods for Water and Wastewater Monitoring and Analysis*, 4th ed.; State Environmental Protection Administration: Beijing, China, 2002. (In Chinese)

31. APHA. *Standard Methods for the Examination of Water and Wastewater*, 20th ed.; American Public Health Association: Washington, DC, USA, 1998.

32. Hu, H.; Wei, Y. *The Freshwater Algae of China: Systematics, Taxonomy and Ecology*; Science Press: Beijing, China, 2006; 1023p.

33. Riis, V.; Lorbeer, H.; Babel, W. Extraction of Microorganisms from Soil: Evaluation of the Efficiency by Counting Methods and Activity Measurements. *Soil Biol. Biochem.* **1998**, *30*, 1573–1581. [CrossRef]

34. Nusch, E.A. Comparison of different methods for chlorophyll and phaeopigments determination. *Arch. Hydrobiol. Beih. Ergebn. Limnol.* **1980**, *14*, 14–36.

35. Rao, C.R.M.; Reddi, G.S. Decomposition procedure with aqua regia and hydrofluoric acid at room temperature for the spectrophotometric determination of phosphorus in rocks and minerals. *Analyt. Chim. Acta* **1990**, *237*, 251–252. [CrossRef]

36. Kunii, H. Characteristics of the winter growth of detached *Elodea nuttallii* (*planch.*) St John in Japan. *Aquat. Bot.* **1981**, *11*, 57–66. [CrossRef]

37. Kunii, H. Seasonal growth and profile structure development of *Elodea nuttallii* in pond Ojaga-Ike, Japan (Planch.) St John. *Aquat. Bot.* **1984**, *18*, 239–247. [CrossRef]

38. Caporaso, J.G.; Kuczynski, J.; Stombaugh, J.; Bittinger, K.; Bushman, F.D.; Costello, E.K.; Knight, R. QIIME allows analysis of high-throughput community sequencing data. *Nat. Med.* **2010**, *7*, 335–336. [CrossRef] [PubMed]

39. Jiang, L.L.; Zhou, L.; Li, S.M.; Fang, G.F. Nutrient removal and microbial community structure in hybrid biofilm reactor. *Chin. J. Ophthalmol.* **2012**, *48*, 24–27.

40. Sukačová, K.; Kočí, R.; Žídková, M.; Vítěz, T.; Trtílek, M. Novel insight into the process of nutrients removal using an algal biofilm: The evaluation of mechanism and efficiency. *Int. J. Phytoremediat.* **2017**, *19*, 909–914. [CrossRef] [PubMed]

41. Romaní, A.M.; Giorgi, A.; Acuña, V.; Sabater, S. The influence of substratum type and nutrient supply on biofilm organic matter utilization in streams. *Limnol. Oceanogr.* **2004**, *49*, 1713–1721. [CrossRef]

42. McCormick, P.V.; Shuford, R.B.E.; Chimney, M.J. Periphyton as a potential phosphorus sink in the Everglades Nutrient Removal Project. *Ecol. Eng.* **2006**, *27*, 279–289. [CrossRef]

43. Chiou, R.J.; Yang, Y.R. An evaluation of the phosphorus storage capacity of an anaerobic/aerobic sequential batch biofilm reactor. *Bioresour. Technol.* **2008**, *99*, 4408–4413. [CrossRef] [PubMed]

44. Hickman, J.W.; Tifrea, D.F.; Harwood, C.S. A chemosensory system that regulates biofilm formation through modulation of cyclic diguanylate levels. *Proc. Nat. Acad. Sci. USA* **2005**, *102*, 14422–14427. [CrossRef] [PubMed]

45. Li, S.; Wang, C.; Qin, H.; Li, Y.; Zheng, J.; Peng, C.; Li, D. Influence of phosphorus availability on the community structure and physiology of cultured biofilms. *J. Environ. Sci.* **2016**, *42*, 19–31. [CrossRef] [PubMed]

46. Spencer, D.F.; Linquist, B.A. Reducing rice field algae and cyanobacteria abundance by altering phosphorus fertilizer applications. *Paddy Water Environ.* **2013**, *12*, 147–154. [CrossRef]

47. Tlili, A.; Hollender, J.; Kienle, C.; Behra, R. Micropollutant-induced tolerance of in situ periphyton: Establishing causality in wastewater-impacted streams. *Water Res.* **2017**, *111*, 185–194. [CrossRef] [PubMed]

48. Artigas, J.; Fund, K.; Kirchen, S.; Morin, S.; Obst, U.; Romaní, A.M.; Schwartz, T. Patterns of biofilm formation in two streams from different bioclimatic regions: Analysis of microbial community structure and metabolism. *Hydrobiologia* **2012**, *695*, 83–96. [CrossRef]

49. Joint, I.; Tait, K.; Wheeler, G. Cross-kingdom signalling: Exploitation of bacterial quorum sensing molecules by the green seaweed Ulva. *Philos. Trans. R. Soc. Lond. Ser. B Biol. Sci.* **2007**, *362*, 1223–1233. [CrossRef] [PubMed]

50. Carr, G.M.; Morin, A.; Chambers, P.A. Bacteria and algae in stream periphyton along a nutrient gradient. *Freshwater Biol.* **2005**, *50*, 1337–1350. [CrossRef]

51. Suberkropp, K.; Gulis, V.; Rosemond, A.D.; Benstead, J.P. Ecosystem and physiological scales of microbial responses to nutrients in a detritus-based stream: Results of a 5-year continuous enrichment. *Limnol. Oceanogr.* **2009**, *55*, 149–160. [CrossRef]

52. Gao, X.Y.; Xu, Y.; Liu, Y.; Liu, Y.; Liu, Z.P. Bacterial diversity, community structure and function associated with biofilm development in a biological aerated filter in a recirculating marine aquaculture system. *Mar. Biol.* **2011**, *42*, 1–11. [CrossRef]

53. Atabek, A.; Camesano, T.A. Atomic force microscopy study of the effect of lipopolysaccharides and extracellular polymers on adhesion of *Pseudomonas aeruginosa*. *J. Bacteriol.* **2007**, *189*, 8503–8509. [CrossRef] [PubMed]

54. Gao, D.W.; Wang, X.L.; Xing, M. Dynamic variation of microbial metabolites and community involved in membrane fouling in A/O-MBR. *J. Memb. Sci.* **2014**, *458*, 157–163. [CrossRef]

55. Martiny, A.C.; Albrechtsen, H.J.; Arvin, E.; Molin, S. Identification of bacteria in biofilm and bulk water samples from a nonchlorinated model drinking water distribution system: Detection of a large nitrite-oxidizing population associated with *Nitrospira spp.*. *Appl. Environ. Microbiol.* **2005**, *71*, 8611–8617. [CrossRef]

56. Ren, J.C.; Zhou, H.; Sun, Y.T. Vertical distribution of light intensity and light compensation depth of submerged macrophyte in Lake Dianchi. *Acta Scicentiarum Nat. Univ. Pekin.* **1997**, *2*, 211–214. (In Chinese)

57. Feldmann, T.; Nõges, P. Factors controlling macrophyte distribution in large shallow Lake Võrtsjärv. *Aquat. Bot.* **2007**, *87*, 15–21. [CrossRef]

Article

Sorption Properties of the Bottom Sediment of a Lake Restored by Phosphorus Inactivation Method 15 Years after the Termination of Lake Restoration Procedures

Renata Augustyniak [1,*], Jolanta Grochowska [1], Michał Łopata [1], Katarzyna Parszuto [1], Renata Tandyrak [1] and Jacek Tunowski [2]

[1]   Department of Water Protection Engineering, University of Warmia and Mazury in Olsztyn, Prawocheńskiego St. 1, 10-720 Olsztyn, Poland; jgroch@uwm.edu.pl (J.G.); michal.lopata@uwm.edu.pl (M.Ł.); kasiapar@uwm.edu.pl (K.P.); renatat@uwm.edu.pl (R.T.)

[2]   Laboratory of Hydroacoustics, Inland Fisheries Institute, Oczapowskiego St. 10, 10-719 Olsztyn, Poland; j.tunowski@infish.com.pl

*   Correspondence: rbrzoza@uwm.edu.pl; Tel.: +48-89-523-37-24

Received: 3 September 2019; Accepted: 17 October 2019; Published: 19 October 2019

Abstract: Artificial mixing and phosphorus inactivation methods using aluminum compounds are among the most popular lake restoration methods. Długie Lake (Olsztyńskie Lakeland, Poland) was restored using these two methods. Primarily, P precipitation and inactivation methods significantly increased the sorption properties of Długie Lake bottom sediment. Fifteen years after the termination of the restoration procedure, the alum-modified "active" sediment layer still has higher P adsorption abilities, which can limit P internal loading. Relatively low amounts of phosphates in the near-bottom water of Długie Lake, even in anoxia, as well as the fact that the assessed maximum sediment P sorption capacity is still higher than NH_4Cl–P (labile P) and BD–P (Fe-bound P) sum ("native exchangeable P"), confirm that hypothesis. Among the tested P adsorption models for the sediment, the double Langmuir model showed the best fit to the experimental data (the highest R^2 values). This may indicate that phosphorus adsorption by the tested sediments most likely occurs through phosphate binding at two types of active sorption sites. P adsorption by the studied lake sediment during experiments was significantly connected to aluminum content in sediment. The research into the adsorption properties of sediment can be used as a tool for the evaluation of lake restoration effects.

Keywords: lake; sediment; phosphorus sorption; lake restoration; phosphorus inactivation; artificial mixing

1. Introduction

Currently, one of the biggest environmental problems is the constantly deteriorating surface water quality, resulting from the industrial revolution and a qualitative leap in improving the hygienic conditions of households and the growing human population [1,2]. Water bodies located in the direct vicinity of human settlements have very often been the victims of avalanche degradation resulting from the discharge of untreated municipal or industrial sewage. There have been many cases in the world of urban lakes transforming into hypertrophic and even saprotrophic water bodies [3–7].

In the case of the observed continuous deterioration of water quality in water bodies, the first absolutely necessary actions are to cut off the point sources of nutrients, the treatment of stormwater entering into lake, and the implementation of proper practices in land use (e.g., changes in cropland management practices and the promotion of particular crop types). Particularly, agriculture is a serious nutrient source for lakes and rivers, and in the case of water bodies with huge, agricultural catchments, it is very difficult to maintain good water quality. However, if these actions will be carried out, it is possible to observe a positive influence on lake water quality [8–11].

It should be emphasized that protective measures in the catchment should be performed before any other technical action directly in the lake bowl. Results should be monitored, and in the case, that such protective measures are not sufficient in the fight against negative eutrophication symptoms, technical solutions should be taken into consideration. This allows us to decrease the costs of the whole technical action [8,9,11]. In many cases, the only chance to stop or to reverse the eutrophication process in degraded lakes is to apply appropriate technical methods in the lake bowl [5–7,12,13]. These techniques are defined by Pan et al. [14] and Mackay et al. [15] as geoengineering. The main goal of geoengineering is to manipulate biogeochemical processes in order to obtain the improvement of a lake's ecological structure and function.

The phenomenon of internal nutrient loading in lakes is a serious barrier that can have a negative impact on the effects of restoration measures [16]. One of the popular methods used for the limitation of internal phosphorus loading is phosphorus inactivation [5,6,12,17].

The principle of this method is the addition of the factor eliminating phosphorus from the lake water and increasing the sediment P sorption capacity. Salts of iron, aluminum, calcium, zirconium or lanthanum are the coagulants used for this purpose. Also, a wide spectrum of solid materials (e.g., naturally occurring minerals or soils, synthetically produced materials, modified clays and soils or industrial by-products) can be used as P adsorbents [18,19]. These compounds can be dosed into lake water, or directly to the water-sediment interface [5,6,12,20–26]. In the case of degraded lakes in which there is thermal stratification, the anoxia occurring in hypolimnetic water promotes the intensification of the internal loading phenomenon. In such cases, aluminum salts are the most economically attractive option for lake treatment. Aluminum-phosphate complexes are insensitive to changes in redox potential in the anoxic bottom zone, while the main factor destroying aluminum-phosphate complexes is pH change [12,17]. Because of the fact that a phenomenon of sedimentation and the creation of a new layer of bottom sediment continuously occurs, the "active" sediment layer (a product of treatment) is covered with new sediment. Cooke et al. [17], Lewandowski et al. [23] and Rydin [24] maintain that phosphorus bound during inactivation remains permanently buried in the sediment. However, Reitzel et al. [27] and van Hullebush et al. [28] pay attention to the fact that, in shallow lakes, the sediment resuspension together with an increase in pH can dissolute aluminum-phosphate complexes and destroy the "active" layer. However, in stratified lakes the "active" layer should theoretically bind phosphorus under stable pH conditions until the sorption active sites are exhausted.

One of the most important questions in every lake restoration case are: how long its positive effects will persist and what factors can steer that persistence.

Thus, the aims of this study were as follows:

1. To determine the sorption characteristics of Długie Lake sediment;
2. To determine whether implemented lake restoration methods influenced P adsorption by sediment;
3. To determine whether the "active" layer of sediment (enriched with aluminum) produced during restoration (completed in 2003 year) still has a higher sorption capacity in relation to phosphorus.

2. Material and Methods

2.1. Lake Description

The object of the study was Długie Lake (area 0.268 km^2, max. depth 17.3 m), located in the western part of Olsztyn city (Olsztyńskie Lakeland, Poland). It is a seepage type of lake, without natural inflows and outflows. Długie Lake has been massively polluted by raw sewage input, which occurred in the second half of 20th century for twenty years (1956–1976). As the result of that pollution, the lake was transformed into a saprotrophic lake type [5,6,29]. The sewage inflow cut off (in 1976) did not result in the improvement of lake water quality, because of the internal loading phenomenon taking place in the lake. Then, it was obvious that the implementation of restoration techniques was the only way to obtain better water quality. Two technical restoration methods were successfully applied on Długie Lake: an artificial mixing method with the complete destratification of lake water (1987–1989 and

1991–2000 periods) and phosphorus inactivation method using an aluminum coagulant–polyaluminum chloride PAX 18 (2001–2003 period). The total pure aluminum dose per m^2 was equal 20.25 g m^{-2} [6]. The second restoration technique was supported with biomanipulation (the introduction of predatory fish species such as pike-perch (*Sander lucioperca* L.) and pike (*Esox lucius* L.)), which was implemented in cooperation with fishing users of the lake (Polish Angling Association).

2.2. Sampling

Dissolved oxygen content in the near-bottom water was measured using the probe ProOdo YSI Inc. (Yellow Springs, OH, USA).

The undisturbed bottom sediment cores were taken using Kajak's bottom sampler at November 2018 at three stations, localized at the deepest points of three separated lake parts (St. 1: the shallowest southern basin of the lake; St. 2: the deepest middle part of the lake; and St. 3: the northern basin of the lake) (Figure 1). At every station three cores were taken. Sediment cores (20 cm long) were immediately divided into four layers (0–5 cm, 5–10 cm, 10–15 cm and 15–20 cm). Lake water for further laboratory experiments was taken into 5 dm^3 tanks.

2.3. Water Analyses

After transportation to the laboratory sediment samples were subjected to centrifugation (3000 rpm, t = 20 min., MPW-351 centrifuge, MPW Med Instruments, Warsaw, Poland) for interstitial water separation. Water phosphorus forms (mineral, total and organic), nitrogen forms (ammonia, total and organic), iron and manganese were measured (Nanocolor spectrophotometer by Macherey-Nagel (GmbH&Co. KG, Düren, Germany), Spectroquant Prove 100 by Merck (KGaA, Darmstadt, Germany), and IL 550 TOC-TN analyzer by HACH Inc. (Loveland, CO, USA)). Lake water used for experiments was also analyzed (mineral P and total phosphorus (TP) by molybdenum blue method using Nanocolor spectrophotometer by Macherey-Nagel (GmbH&Co. KG, Düren, Germany); pH and conductivity by HQ 40 d multi probe by HACH Inc. (Loveland, CO, USA)).

2.4. Sorption Laboratory Experiment

Two grams of fresh sediment aliquots (in triplicates) were placed in the 50 cm^3 Falcon type centrifuge tubes and phosphate solutions (25 cm^3) were added (P concentrations of 0.00; 0.15; 0.30; 0.60; 1.20; 2.40 mg dm^{-3}). Two drops of chloroform were added to inhibit bacteria activity [30–32]. The centrifuge tubes were shaken in an orbital shaker at 250 rpm (Innova 40 incubator by New Brunswick Scientific Co. Inc., Edison, NJ, USA) at the constant temperature of 20 °C. After 24 h of equilibration, the solutions were centrifuged at 4000 rpm for 10 min (Rotina 420, Andreas Hettich GmbH&Co. KG, Tuttlingen, Germany) and the supernatants were decanted and filtered through a 0.45 um pore filter into clean and dried glass beakers and analyzed for phosphate P. The P adsorbed on sediment samples was calculated using the difference between the initial and equilibrium concentration.

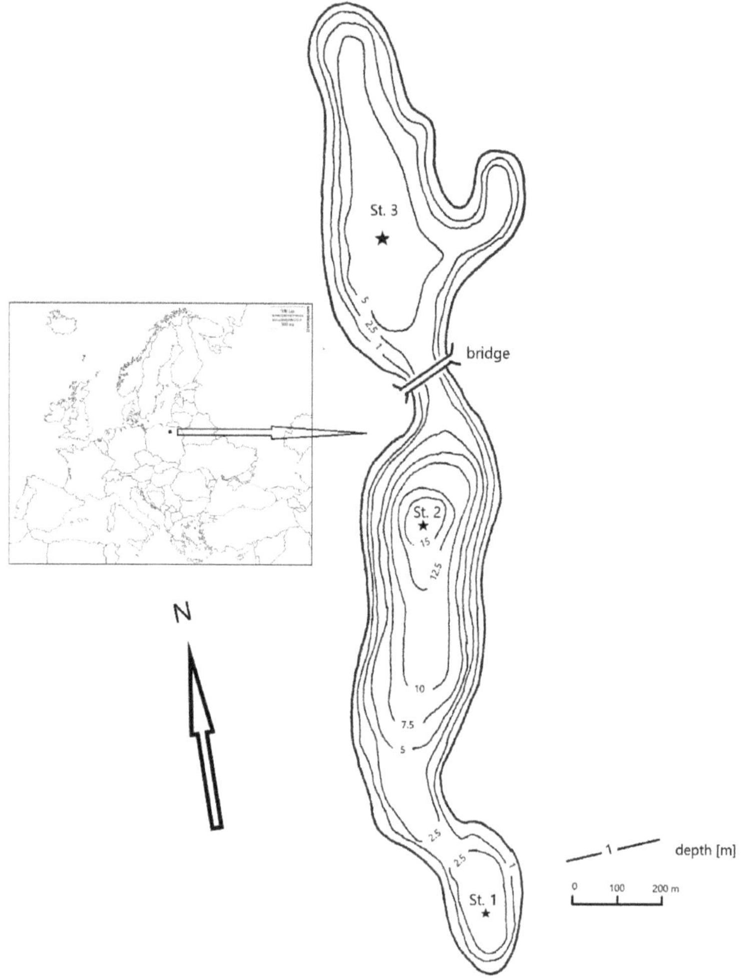

Figure 1. Location of research stations on Długie Lake (source: Inland Fisheries Institute in Olsztyn, d-maps.com).

2.5. Estimation of Sorption Parameters

Obtained results were fitted to the Langmuir sorption model,

$$S = \frac{S_{max} \times k \times C}{1 + k\,C} \tag{1}$$

and double Langmuir model,

$$S = \frac{S_1 \times k_1 \times C}{1 + k_1 \times C} + \frac{S_2 \times k_2 \times C}{1 + k_2 \times C} \tag{2}$$

where:

C—phosphorus concentration after the 24-h equilibration procedure (mg dm^{-3});

S—total phosphorus adsorbed by sediment (solid phase), (mg kg^{-1} dw);

$$S = S_0 + S' \tag{3}$$

S_0—native sorbed phosphorus
the phosphorus pool, which was desorbed at an initial concentration of 0 mg P dm^{-3} (mg kg^{-1} dw);
S'— phosphorus amount adsorbed during experiment, (mg kg^{-1} dw);
S_{max}—maximum sorption capacity of sediment in the Langmuir equation (mg kg^{-1} dw);
k—constant related to bonding energy in the Langmuir equation (dm^3 mg^{-1}).
S_1—maximum sorption capacity of sediment in the double Langmuir equation (type I active sites) (mg kg^{-1} dw);
S_2—maximum sorption capacity of sediment in the double Langmuir equation (type II active sites) (mg kg^{-1} dw);
S_{max2}—the total sorption capacity of sediment in the double Langmuir equation (the sum of S_1 and S_2 maximum sorption capacities) (mg kg^{-1} dw)

$$S_{max2} = S_1 + S_2 \tag{4}$$

k_1, k_2—constants related to the bonding energy in the double Langmuir equation (dm^3 mg^{-1}).

These equations coefficients (k, S_{max}, S_1, k_1, S_2, k_2, S_{max2}) were estimated using a non-linear estimation method [33–36] via the Statistica software package 13.0 (Tibco Software Inc., Palo Alto, CA, USA) [37].

The coefficient of determination R^2 was assumed to be the measure of the curve fitting at the determined parameters to the experimental data.

The Freundlich model coefficients also were assessed using the non-linear estimation method [33–37]:

$$S = K_f \times C^{\frac{1}{n}} \tag{5}$$

where:

S, C—as in the Equation (1);
K_f—Freundlich sorption constant (dm^3 kg^{-1});
$1/n$—a constant which characterizes the heterogeneity of the adsorption process.

EPC_0 parameter was assessed using the Freundlich equation with correction for desorbed phosphorus ($-S_0$) at the initial experimental concentration (0 mg P dm^{-3}) [38]:

$$S = \left(K_f \times C^{\frac{1}{n}}\right) - S_0 \tag{6}$$

The Gibbs free energy change was calculated using formula [39]:

$$-\Delta Gads = RT \ln K_d \tag{7}$$

where:

K_d—division coefficient (dm^3 kg^{-1});
R—gas constant (J mol^{-1} K^{-1});
T—temperature (K).

2.6. Sediment Analyses

In non-centrifuged sediment samples the water content and solid matter were measured after drying at 105 °C (Barnstead Thermolyne 62700 Furnace, Barnstead International, Dubuque, IA, USA) until a constant weight was reached.

The sediment chemical composition included organic matter as a loss of ignition at 550 °C after carbonate regeneration (using CO_2 saturated deionized water), and carbonates (as CO_2) after ignition at 1000 °C (Barnstead Thermolyne 62700 Furnace, Barnstead International, Dubuque, IA,

USA). The sediment samples were mineralized with a mixture of H_2SO_4, $HClO_4$ and HNO_3 (1 + 2 + 3). After mineralization, the sample was filtered through ash-free filter No 390. The remains on the filter were treated as silica and mineralized at 900 °C. In the filtrate the consecutive parameters were measured spectrophotometrically (Spectroquant Prove 100 by Merck KGaA, Darmstadt, Germany): iron, aluminum, manganese, calcium, magnesium. Total nitrogen (TN) was analyzed by Kjeldahl method (using a BÜCHI K-425 unit, B-24 distillation unit, BÜCHI Labortechnik AG, Flawil, Switzerland).

Sediment phosphorus fractions were analyzed according to scheme proposed by [30]. Mineral P content in the extracts was analyzed with the molybdenum blue method (Nanocolor spectrophotometer by Macherey—Nagel, GmbH&Co. KG, Düren, Germany)

3. Statistical Analysis

The results were subjected to statistical analysis using the Statistica 13.0 Software package [37]. The multiple regression analysis [40] was performed in order to identify chemical factors (independent variables—bottom sediment chemical components: Si, organic matter, Fe, Al, Mn, Ca, Mg, Mn), which are significantly connected to the estimated sorption parameters (dependent variables: S_{max}, k, S_1, k_1, S_2, k_2, S_{max2}, 1/n, K_f, EPC_0, S_0). That analysis allowed to obtain the linear models type:

$$Y = B_0 \pm B_1 \ X_1 \ \pm \ B_2 \ X_2 \ \pm \ldots \ \pm \ B_i \ X_i + \ E_{ij} \tag{8}$$

where:

Y—dependent variable (in the present research, the particular sorption characteristic in the Freundlich, Langmuir or double Langmuir model);
B_0—constant (intercept);
$B_1 \ldots B_i$—regression coefficients;
$X_1 \ldots X_i$—independent variables (bottom sediment chemical components: Si, organic matter, Fe, Al, Mn, Ca, Mg, Mn);
E_{ij}—residual component;
R—multiple correlation coefficient;
R^2—multiple determination coefficient.

All results were subjected to log transformation in order to approximate of the data to the normal distribution.

4. Results

4.1. Sorption Parameters

The relationship between the equilibrium concentration (C_e) and the amount of adsorbed phosphates is shown in Figures 2–4. The amount of adsorbed phosphorus (S) increased with increasing concentrations of equilibrium phosphorus (C_e). The surface layer of the sediment usually showed a greater sorption capacity than the deeper layers. The largest amount of phosphates was adsorbed by sediments from the layer of 5–10 cm at Station 2—356.39 mg P kg^{-1} dw, while the weakest sorption capacity with respect to phosphorus was shown by sediment taken at St. 1—202.47 mg P kg^{-1} dw (sediment layer 0–5 cm).

The obtained real adsorption results fitted well to all tested adsorption models. The values of the determination coefficient R^2 were in the range from 0.93 (Freundlich model for sediment layers, 10–15 cm and 15–20 cm; and double Langmuir model for sediment layer, 10–15 cm at St. 1) to 1.00 (double Langmuir model for sediment layer, 0–5 cm, St. 1). The double Langmuir model usually showed the best fit to experimental data, in that R^2 values usually were higher for that model than for two other models (Tables 1 and 2).

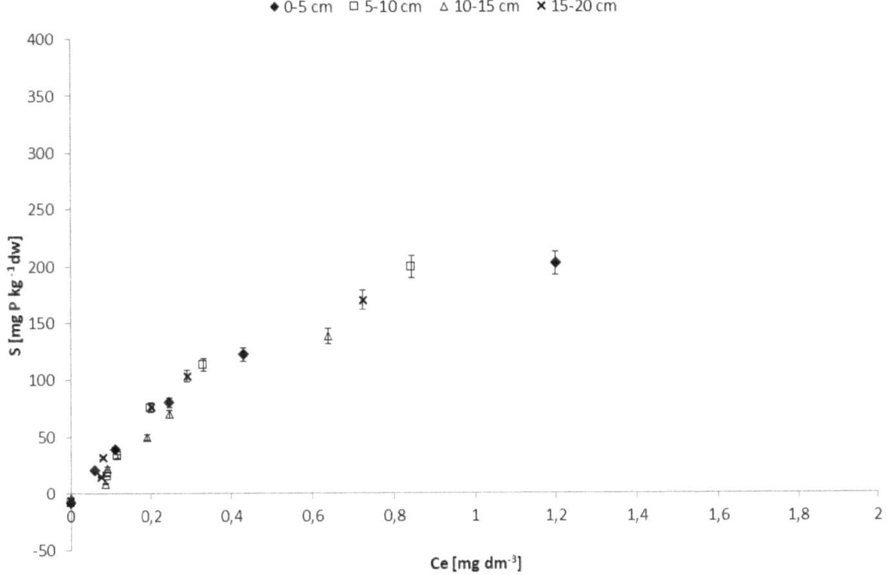

Figure 2. The P adsorption isotherms for sediment taken at St. 1.

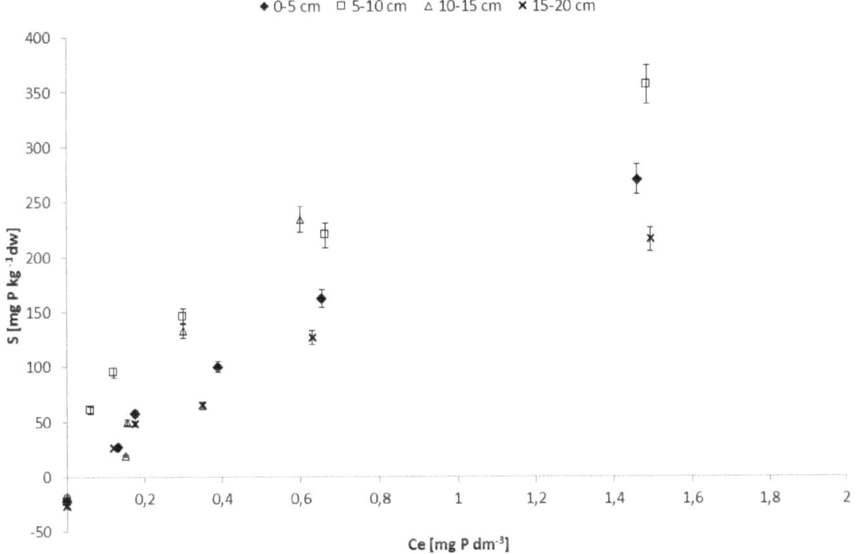

Figure 3. The P adsorption isotherms for sediment taken at St. 2.

The values of the maximum P sorption capacity (S_{max}) obtained from the Langmuir model were varied. The highest value of S_{max} (524.2 mg P kg^{-1} dw) was noted for the sediment layer 15–20 cm (St. 2) (Table 1). The sediment taken at St. 3 had the lowest S_{max} (between 241.7 and 279.0 mg P kg^{-1} dw) (Table 1). Also, k coefficient values (the constant from Langmuir equation referring to the binding energy, expressed in dm^3 mg^{-1}) were varied, but generally the highest values were observed for the surficial sediment layers (0–5 cm) at all three research stations. The highest k value was calculated for sediment taken at St. 3 (3.74 dm^3 mg^{-1} - sediment layer 0–5 cm), whilst the lowest −0.62 dm^3 mg^{-1}—at St. 1 (sediment layer 10–15 cm) (Table 2).

The values of the correction factor from the Freundlich equation (1/n), which characterize the heterogeneity of the adsorption process ranged from 0.4276 (St. 3, sediment layer, 0–5 cm) to 0.8476 (St. 1, sediment layer, 10–15 cm) (Table 2). The minimum value of the partition coefficient from the Freundlich equation (K_f) was assessed for bottom sediment taken at St. 3 (sediment layer 15–20 cm) and it amounted to 116.0 dm^3 kg^{-1}, whilst the maximum K_f value was calculated for the surficial sediment layer, (0–5 cm) taken at the deepest St. 2 (309.7 dm^3 kg^{-1}) (Table 2).

The assessed P equilibrium concentration (EPC_0) ranged from 0.001 mg P dm^{-3} (St. 3, sediment layer, 0–5 cm) to 0.059 mg P dm^{-3} (at the same station, sediment layer, 10–15 cm).

"Native sorbed phosphorus" (S_0) values were the highest for sediment layer 10–15 cm taken at St. 3—27.36 mg kg^{-1} dw—and the lowest for sediment taken at St. 1 (sediment layer 10–15 cm)—4.62 mg kg^{-1} dw.

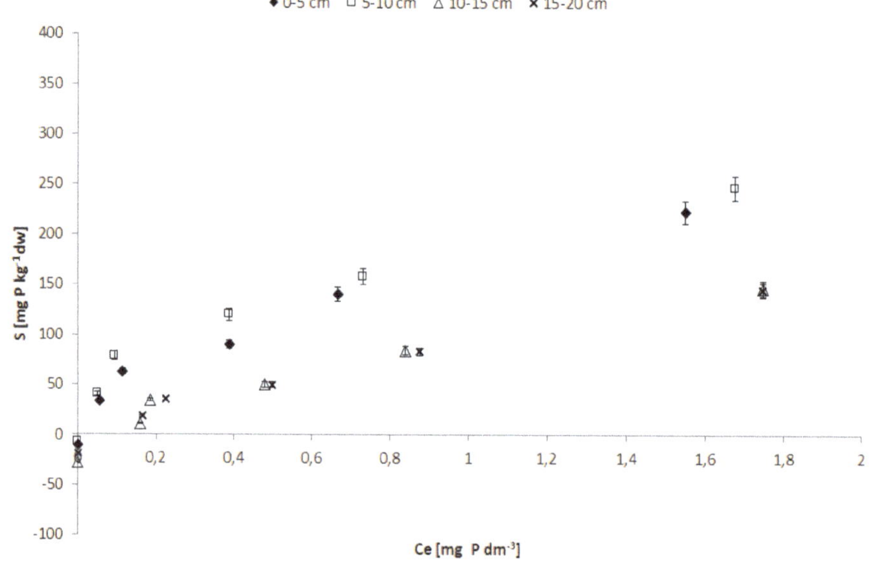

Figure 4. The P adsorption isotherms for sediment taken at St. 3.

Table 1. The assessed values of Langmuir sorption model characteristics for profundal bottom sediment of Długie Lake.

| Station | Sediment Layer | Langmuir Model | | | Double Langmuir Model | | | | | |
		S_{max} (mg kg^{-1})	k (dm^3 mg^{-1})	R^2	S_1 (mg kg^{-1})	k_1 (dm^3 mg^{-1})	S_2 (mg kg^{-1})	k_2 (dm^3 mg^{-1})	S_{max2} (mg kg^{-1})	R^2
St. 1	0–5 cm	320.0	1.60	0.99	319.91	1.53	3.11	11.32	323.0	1.00
	5–10 cm	470.5	0.94	0.96	229.35	0.93	241.11	0.94	470.5	0.96
	10–15 cm	497.7	0.62	0.93	218.30	1.02	927.69	0.10	1146.0	0.93
	15–20 cm	325.4	1.61	0.95	163.36	1.62	161.97	1.61	325.3	0.95
St. 2	0–5 cm	375.9	1.05	0.98	534.3	0.24	106.61	3.47	640.9	0.99
	5–10 cm	458.8	2.39	0.94	109.8	32.43	1178.18	0.202	1288.0	0.99
	10–15 cm	508.3	0.91	0.99	1212.89	0.08	224.1	1.92	1437.0	0.99
	15–20 cm	524.2	0.71	0.97	16.4	12.65	601.74	0.501	618.14	0.98
St. 3	0–5 cm	269.2	3.74	0.93	470.3	0.357	65.86	27.7	535.9	0.99
	5–10 cm	279.0	2.44	0.95	98.66	16.04	2066.9	0.047	2165.6	0.99
	10–15 cm	241.7	1.27	0.94	1155.6	0.067	54.7	12.206	1210.3	0.97
	15–20 cm	254.2	0.94	0.95	38.2	20.81	1303.9	0.061	1342.1	0.98

Table 2. The assessed values of the Freundlich sorption model characteristics, "native sorbed phosphorus" values and Gibbs free energy for profundal bottom sediment of Długie Lake.

Station	Sediment Layer		Freundlich Model				
		$1/n$	K_f (dm^3 kg^{-1})	R^2	EPC_0 (mg dm^{-3})	S_0 (mg kg^{-1})	ΔG_{ads}
St. 1	0–5 cm	0.5875	192.9	0.98	0.004	8.02	−13.72
	5–10 cm	0.7545	239.4	0.95	0.012	8.32	−13.48
	10–15 cm	0.8476	210.2	0.93	0.011	4.62	−12.92
	15–20 cm	0.6840	222.2	0.93	0.006	6.36	−13.78
St. 2	0–5 cm	0.4771	309.7	0.99	0.004	22.96	−15.17
	5–10 cm	0.6429	231.9	0.98	0.027	22.63	−13.32
	10–15 cm	0.6090	183.5	0.98	0.020	16.81	−12.98
	15–20 cm	0.7045	212.0	0.97	0.033	19.31	−12.93
St. 3	0–5 cm	0.4276	199.7	0.98	0.001	7.21	−14.49
	5–10 cm	0.4712	187.8	0.98	0.002	10.53	−14.06
	10–15 cm	0.5432	126.9	0.97	0.059	27.36	−11.23
	15–20 cm	0.5898	116.0	0.97	0.046	18.92	−11.40

Assessed mean Gibbs' free energy change values were negative and ranged from −11.23 kJ mol^{-1} (St. 3, sediment layer, 15–20 cm) to −15.17 kJ mol^{-1} (St. 2, sediment layer, 0–5 cm). They often increased with sediment depth (Table 2).

The conductivity, pH, and phosphate concentration of the lake water used for experiments were 260 μS cm^{-1}, 7.71, and 0.007 mg P–PO$_4$ dm^{-3}, respectively. These values were taken into consideration during the experiment's final results calculation.

The multiple regression analysis [40] was performed in order to identify sediment chemical components (especially Al content, as a factor used for lake restoration), which are significantly connected to estimated sorption parameters. The results are shown in Table 3. The sorption characteristics were dependent on OM, Fe, Mn, Si and Mg. The sorption capacities from Langmuir models (S_{max}, S_2 and S_{max2}) were significantly dependent on Al content in sediment. Also, the constants $1/n$, k_2 and EPC_0 parameter were significantly connected to Al content in sediment (Table 3).

Table 3. Multiple regression analysis of P sorption parameters depending on bottom sediment chemical composition ($n = 12$, $p \leq 0.1$, raw data was log transformed).

Fitted Model Equation	R	R^2
$S_{max} = 1.969\ Fe_{sed} + 1.644\ Al_{sed} + E_{ij}$	0.885	0.782
k—non significant		
$S_0 = 5.289 − 3.622\ Si_{sed} + 2.792\ Fe_{sed} − 4.471\ Mn_{sed} + E_{ij}$	0.967	0.935
$1/n = 0.389\ Al_{sed} + E_{ij}$	0.952	0.907
$K_f = 1.514 + 1.063\ Fe_{sed} − 1.145\ Mg_{sed} + E_{ij}$	0.670	0.450
$EPC_0 = −0.463 − 0.327\ Mn_{sed} + 0.166\ Mg_{sed} + 0.14\ OM_{sed} − 0.095\ Al_{sed} + E_{ij}$	0.924	0.854
S_1—non significant		
k_1—non significant		
$S_2 = 13.253\ Al_{sed} + E_{ij}$	0.695	0.480
$k_2 = −14.566\ Al_{sed} + E_{ij}$	0.875	0.767
$S_{max2} = −42.279 + 7.371\ Al_{sed} + 6.985\ OM_{sed} + 8.523\ Si_{sed} + E_{ij}$	0.965	0.931

(Fe$_{sed}$: iron content in sediment, Al$_{sed}$: aluminum content in sediment, Mn$_{sed}$: manganese content in sediment, OM$_{sed}$: organic matter content in sediment, Mg$_{sed}$: magnesium content in sediment, Si: silica content in sediment; all in mg g^{-1} dw).

4.2. Water

During the research, dissolved oxygen was present in the near-bottom water at two shallow stations (9.5 mg O_2 dm^{-3} at St. 1 and 9.6 mg O_2 dm^{-3} at St. 3), whilst anoxia was noted at the deepest part of Długie Lake (St. 2).

The phosphorus form concentration in the water medium of the water—sediment interface was different at particular research stations. At two shallow stations (St. 1 and St. 3), the noted concentration of TP, min P and org P was lower than at the deepest St. 2. The minimum TP value was observed at St. 3 (1.02 mg P dm^{-3}, surficial layer 0–5 cm) whilst the maximum TP amount (4.1 mg P dm^{-3}) was noted at the deepest layer of analyzed water (15–20 cm) (Figure 2). The organic P form dominated quantitatively in the near-bottom water at all three research stations and in interstitial water at two shallow stations (St. 1 and 3). The mineral P form share was generally higher in the interstitial water at St. 2. It is worth noting that the clear decrease of P concentration (both P forms—mineral and organic) occurred in the layer of 10–15 cm at the St. 1 and 2, whilst concentration rose with the sediment depth at St. 3 (Figure 5).

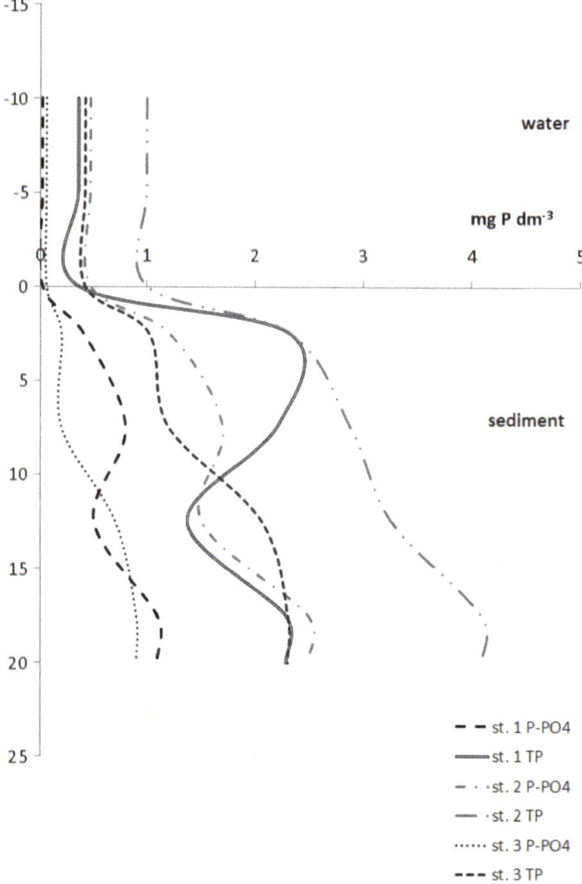

Figure 5. The total phosphorus (TP) and mineral P profiles in the water–sediment interface of Długie Lake.

Nitrogen compound concentrations were varied at the research stations. At the deepest St. 2, the nitrogen concentration in near-bottom water was the highest (20.43 mg N dm^{-3}), and it was much lower at the two shallow stations (0.68 mg N dm^{-3} and 2.50 mg N dm^{-3} at St. 1 and 3, respectively). In the near-bottom water, the organic form of N quantitatively dominated, whilst the share of mineral N (ammonia) rose with the sediment depth in the interstitial water at all research stations. The TN concentration observed in the interstitial water was the highest at the deepest Station 2 (max 30.3 mg N dm^{-3}, layer 5–10 cm), and the lowest TN amount occurred at St. 1 (min 2.06 mg N dm^{-3}, layer 0–5 cm) (Figure 6).

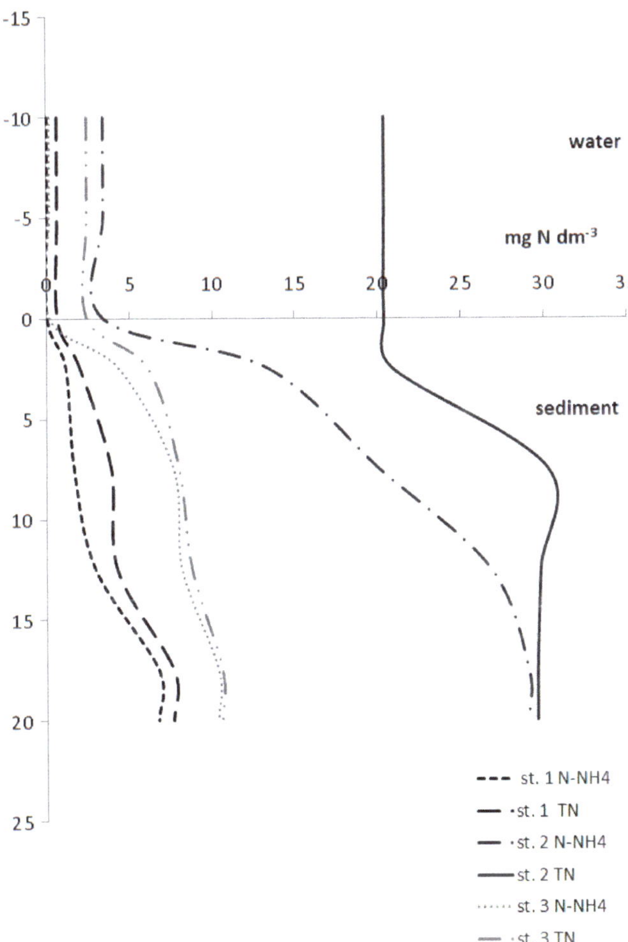

Figure 6. The total nitrogen (TN) and ammonia profiles in the water-sediment interface of Długie Lake.

Fe and Mn amounts observed in the water–sediment interface ranged from 0.02 mg Fe dm^{-3} and 0.04 mg Mn dm^{-3} (near-bottom water at St. 1) to 3.60 mg Fe dm^{-3} and 3.50 mg Mn dm^{-3} (interstitial water at St. 2, layer 15–20 cm) (Figure 7).

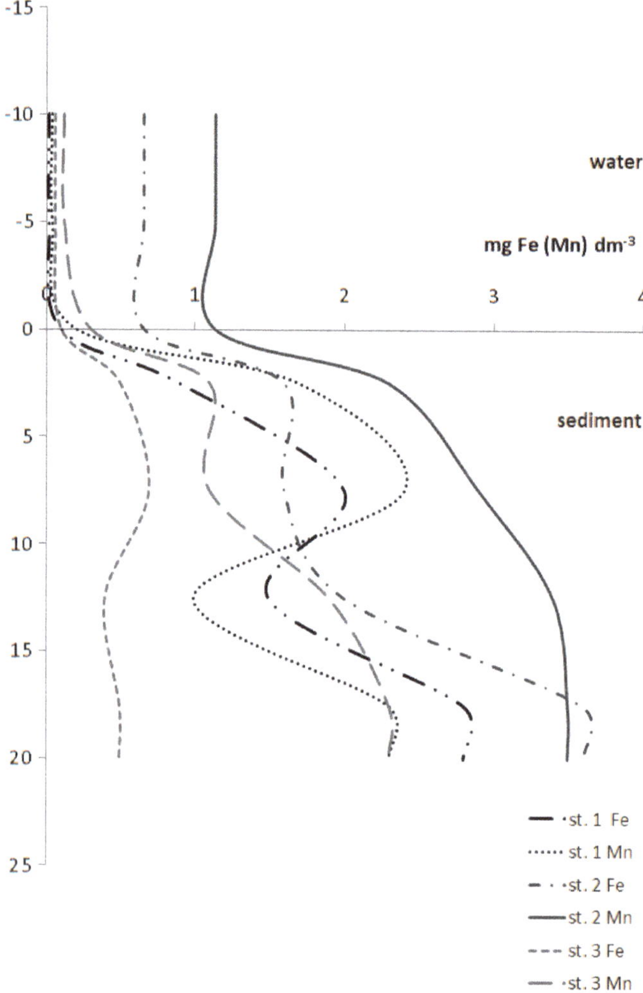

Figure 7. The Fe and Mn profiles in the water-sediment interface of Długie Lake.

4.3. Sediment Chemical Composition

The sediment of Długie Lake was highly hydrated: the percentage of dry weight was mainly below the level of 10% dw (except for the two deeper sediment layers at St. 1, at 15.59% dw and 11.24% dw) (Table 4). This can be classified as the mixed silica–organic type (St. 1) and organic–silica type (St. 2). The share of organic matter exceeded 50% dw at St. 3 only (except the deepest sediment layer 15–20 cm), and in general, the sediment can be classified as organic type. The rest of the analyzed components occurred in low amounts, not exceeding several percentage of dw (Table 4). The aluminum content, particularly considered as a factor used for phosphorus inactivation, was the highest in the deeper sediment layers (10–15 cm at St. 1 and 2 and 5–10 cm at St. 3), whilst the maximum Al content was noted in sediment taken at St. 3 (18.95 ± 1.72 mg Al g⁻¹ dw in the layer 5–10 cm) and St. 2 (18.90 ± 1.33 mg Al g⁻¹ dw in the layer 10–15 cm) (Table 4).

Table 4. The mean (±SD) values of main components of Długie Lake sediment (in mg g^{-1} d.w.).

Station	Sediment Layer	OM	Si	IC	Fe	Al	Ca	Mg	Mn	TN	% dw
St. 1	0–5 cm	326.26 ± 12.51	220.61 ± 3.94	9.25 ± 0.61	22.58 ± 2.46	12.75 ± 0.85	18.40 ± 2.09	5.94 ± 0.51	0.36 ± 0.04	16.16 ± 0.65	7.58
	5–10 cm	311.40 ± 12.33	227.22 ± 3.99	10.08 ± 0.72	26.56 ± 2.54	13.08 ± 0.81	21.70 ± 2.15	5.18 ± 0.63	0.31 ± 0.03	15.45 ± 0.46	9.86
	10–15 cm	305.43 ± 12.26	228.72 ± 4.23	10.19 ± 0.82	21.04 ± 2.02	14.57 ± 0.89	18.39 ± 2.18	6.10 ± 0.92	0.29 ± 0.03	15.09 ± 0.56	15.59
	15–20 cm	296.49 ± 12.18	229.08 ± 4.24	10.72 ± 0.88	21.82 ± 2.08	12.87 ± 0.81	22.28 ± 2.03	5.10 ± 0.99	0.26 ± 0.03	14.63 ± 0.72	11.24
St. 2	0–5 cm	498.06 ± 23.77	137.25 ± 10.27	9.65 ± 0.59	18.91 ± 1.26	15.39 ± 1.53	17.40 ± 1.21	4.05 ± 1.02	0.28 ± 0.07	27.71 ± 0.72	3.10
	5–10 cm	463.74 ± 22.58	145.60 ± 10.83	9.82 ± 0.63	19.97 ± 1.28	17.33 ± 1.25	16.00 ± 1.15	5.44 ± 0.99	0.27 ± 0.04	26.29 ± 0.81	4.33
	10–15 cm	455.18 ± 23.24	146.07 ± 11.03	10.38 ± 0.78	20.58 ± 1.33	18.90 ± 1.33	16.49 ± 1.18	6.94 ± 0.92	0.41 ± 0.04	26.97 ± 0.68	5.35
	15–20 cm	442.68 ± 22.86	161.84 ± 11.26	8.97 ± 0.62	21.96 ± 1.44	16.16 ± 1.28	14.49 ± 1.03	8.15 ± 1.20	0.39 ± 0.05	25.99 ± 0.24	6.02
St. 3	0–5 cm	511.92 ± 13.57	135.29 ± 7.83	8.81 ± 0.25	15.27 ± 1.63	15.77 ± 1.62	16.11 ± 1.28	5.43 ± 0.51	0.49 ± 0.12	29.91 ± 1.05	3.48
	5–10 cm	511.20 ± 13.78	133.61 ± 7.96	8.86 ± 0.27	14.21 ± 1.28	18.95 ± 1.72	19.09 ± 1.36	4.30 ± 0.23	0.28 ± 0.09	29.36 ± 0.99	4.61
	10–15 cm	502.98 ± 12.98	135.94 ± 8.03	9.24 ± 0.18	16.47 ± 1.33	16.79 ± 1.27	17.16 ± 1.32	5.34 ± 0.33	0.26 ± 0.06	29.08 ± 0.81	5.87
	15–20 cm	482.78 ± 12.23	150.41 ± 8.23	9.31 ± 0.21	17.94 ± 1.81	15.32 ± 1.27	16.78 ± 1.22	5.00 ± 0.56	0.23 ± 0.09	27.46 ± 0.90	6.38

4.4. Phosphorus Fractions

The total phosphorus content in the bottom sediment of Długie Lake was high and ranged from 3.786 ± 0.210 mg P g^{-1} dw (St. 1, sediment layer 0–5 cm) to 6.505 ± 0.220 mg P g^{-1} dw (St. 2, sediment layer 10–15 cm). The largest pool of sediment P was stored as a NaOH–nrP fraction (phosphorus bound to organic matter). The maximum amount of this P fraction was noted for sediment on St. 2 (2.338 ± 0.043 mg P g^{-1} dw, 35.9% TP, sediment layer 10–15 cm), whilst the lowest of NaOH–nrP amount occurred at St. 1 (0.862 ± 0.012 mg P g^{-1} dw, 22.7%TP, sediment layer 15–20 cm). The NaOH–rP fraction (P bound mainly to Al) was the second P fraction in terms of quantity. Its maximum amount was noted at St. 2 (2.298 ± 0.073 mg P g^{-1} dw, 35.3%TP, sediment layer 10–15 cm) and the lowest was in the surficial sediment layer (0–5 cm) at the St. 3 (0.612 ± 0.007 mg P g^{-1} dw). The HCl–P and res–P fractions included from 11.5%TP to 28.3%TP and from 10.6%TP to 26.3%TP, respectively. The most mobile P fractions (NH$_4$Cl–P and BD–P) occurred in the smallest amounts among the all P fractions, occupying ca. 1%TP and 2.4–3.6% TP, respectively.

5. Discussion

Artificial mixing was the first method of restoration, which was implemented on Długie Lake [5,6,29]. The direct impact of this method on the bottom zone was the improvement oxic conditions in the near-bottom water of aerated parts of Długie Lake (St. 2 and 3). Oxygen presence in that water stratum favored P binding by the sediment (the TP in water decreased and an increase of sediment TP was also observed, especially in the first period of aeration) [29]. Long-term artificial mixing caused the depletion of iron and manganese in the lake water, and this phenomenon was the limit for further P removal from Długie Lake water. In the present study, this stage of Długie Lake restoration was represented in the deepest layer of analyzed sediment cores (15–20 cm). The observed values of the sorption capacity of this sediment layer were mainly the lowest (Figures 2–4), and this was caused probably by diagenetic processes, which limit the direct P adsorption abilities [41].

The restoration of Długie Lake by the phosphorus inactivation method with the use of polyaluminum chloride PAX 18 caused an increase in the aluminum content in bottom sediment at all sites studied. Before P inactivation, Al amounts in sediment did not exceed ca. 12–14 mg Al g^{-1} dw [42]. Maximum amounts of this element were detected in the sediment layer 10–15 cm (14.57 ± 0.89 mg Al g^{-1} dw and 18.90 ± 1.33 mg Al g^{-1} dw at St. 1 and 2, respectively) and 5–10 cm (18.95 ± 1.72 mg Al g^{-1} dw at St. 3) (Table 4). This fact is in accordance with observations by other authors [17,23]. Sediments from these layers also had the highest NaOH–rP fraction contents (phosphorus bound to aluminum oxides and hydroxides: 1.028 ± 0.008, 2.298 ± 0.073 and 0.735 ± 0.003 mg P g^{-1} dw, at St. 1, 2 and 3, respectively (Table 5), which is undoubtedly the result of the restoration activities.

Oxic conditions in the lake in the bottom zone were varied: two shallow parts were well oxygenated, while the deepest lake part was anoxic for most of the year. In spite of that fact, the mineral phosphorus concentration noted in 2018 (15 years after the termination of the lake restoration procedure) amounted to 0.46 mg P dm^{-3} (Figure 5), and organic phosphorus was the dominant P form at all research stations. Before starting the restoration procedures on Długie Lake in 1987, the maximum phosphate concentration in the near-bottom water exceeded 2.80 mg P dm^{-3}, and in 1999 (the year without artificial mixing), this was 1.2 mg P dm^{-3} [29].

The research by [43] showed that the phosphorus inactivation implemented on Długie Lake did not have a direct impact on the nitrogen compound content in the lake. Total nitrogen amounts were lower compared to the period before restoration, but this was a result of primary production limitation due to the phosphorus level decreasing. The nitrogen compound amounts observed during the present research in the near-bottom and interstitial water, as well as iron and manganese contents, were regulated by the oxygen level in the near-bottom water.

Table 5. The mean (±SD) values of phosphorus fractions and total phosphorus in Długie Lake sediment (in mg g^{-1} d.w.).

Station	Sediment Layer	NH$_4$Cl–P	BD–P	NaOH–rP	NaOH–nrP	HCl–P	res–P	TP
St. 1	0–5 cm	0.035 ± 0.006	0.136 ± 0.018	0.957 ± 0.006	1.263 ± 0.018	0.767 ± 0.018	0.628 ± 0.022	3.786 ± 0.210
	5–10 cm	0.025 ± 0.005	0.108 ± 0.016	0.893 ± 0.006	1.300 ± 0.012	0.879 ± 0.012	0.789 ± 0.019	4.129 ± 0.280
	10–15 cm	0.024 ± 0.005	0.104 ± 0.016	1.028 ± 0.008	1.371 ± 0.011	1.006 ± 0.014	0.788 ± 0.013	4.186 ± 0.160
	15–20 cm	0.034 ± 0.007	0.138 ± 0.018	0.924 ± 0.007	0.862 ± 0.012	1.074 ± 0.014	0.764 ± 0.015	3.796 ± 0.120
St. 2	0–5 cm	0.040 ± 0.003	0.162 ± 0.015	1.598 ± 0.062	1.948 ± 0.022	0.626 ± 0.015	0.778 ± 0.022	5.151 ± 0.340
	5–10 cm	0.045 ± 0.005	0.206 ± 0.016	2.100 ± 0.088	2.018 ± 0.025	0.882 ± 0.013	0.787 ± 0.030	6.237 ± 0.320
	10–15 cm	0.042 ± 0.004	0.215 ± 0.018	2.298 ± 0.073	2.338 ± 0.043	0.921 ± 0.020	0.692 ± 0.018	6.307 ± 0.220
	15–20 cm	0.040 ± 0.003	0.164 ± 0.013	0.839 ± 0.043	2.274 ± 0.032	0.836 ± 0.011	1.033 ± 0.015	5.187 ± 0.280
St. 3	0–5 cm	0.048 ± 0.004	0.137 ± 0.002	0.612 ± 0.007	1.860 ± 0.012	0.602 ± 0.012	1.164 ± 0.009	4.423 ± 0.130
	5–10 cm	0.051 ± 0.003	0.132 ± 0.003	0.735 ± 0.003	2.150 ± 0.014	0.552 ± 0.009	1.028 ± 0.011	4.572 ± 0.132
	10–15 cm	0.045 ± 0.003	0.130 ± 0.003	0.659 ± 0.006	2.219 ± 0.010	0.539 ± 0.008	1.099 ± 0.009	4.768 ± 0.180
	15–20 cm	0.046 ± 0.004	0.131 ± 0.002	0.650 ± 0.003	2.113 ± 0.011	0.602 ± 0.011	0.939 ± 0.008	4.481 ± 0.160

The experimentally obtained P adsorption results showed that the sediment taken at St. 2 had the highest sorption capacity. The largest amount of phosphates was adsorbed by sediments from the surface layer (5–10 cm) at Station 2, at 356.39 mg P kg^{-1} dw, while the weakest sediment sorptive abilities with respect to phosphorus were shown by sediment taken at St. 1, at 202.47 mg P kg^{-1} dw (Figures 2–4). These values are within the range recorded for the sediments of other Olsztyn lakes [44] or around the world [45–50]. The pH level during experiments was favorable for P adsorption processes onto sediment. The research conducted by [44] on five urban lakes, located in Olsztyn showed that alkaline pH (9.0) decreased the abilities of P adsorption in sediment, in which organic matter is a main component binding P. Assessed mean Gibbs' free energy change values were negative and ranged from −11.23 kJ mol^{-1} (St. 3, sediment layer 15–20 cm) to −15.17 kJ mol^{-1} (St. 2, sediment layer 0–5 cm). They often increased with sediment depth (Table 1). Assessed ΔG_{ads} values were similar in range to values observed in [46,48]. The negative ΔG_{ads} values noted during experiments confirm that P sorption was a spontaneous reaction [46] and the order of magnitude of ΔG_{ads} was typical for physical adsorption processes.

The best match of real phosphorus adsorption data to the adsorption model was obtained for the double Langmuir model, as was indicated by the highest values of the determination coefficient R^2 (Tables 1 and 2). This may indicate that phosphorus sorption by the tested sediments most likely occurs through phosphate binding at two types of active sorption sites. A similar phenomenon was also observed by [33] with regards to phosphate adsorption on chitosan and modified chitosan. Holford et al. [51] and Limousin et al. [52] claim that the double Langmuir model was the best model in characterizing phosphate adsorption on soils. Natural bottom sediment can be treated as a multi-component system, with different active sorption centers, and that fact seems to cause a better fit of observed adsorption results to the double Langmuir equation.

The phosphorus sorption characteristics were dependent on OM, Fe, Mn, Si and Mg, which is in accordance with numerous research works [38,44,45,53–55] (Table 3). The multiple regression analysis revealed that the sorption capacity from Langmuir models (S_{max}, S_2 and S_{max2}) was significantly dependent on Al content in sediment (Table 3). Thus, it seems to be possible, that the modification of sediment sorption capacity using an aluminum coagulant created additional active sorption sites, as well as the second type sites described in the double Langmuir equation. The constants 1/n, k_2 and EPC_0 parameter were significantly connected to Al content in sediment as well (Table 3). The EPC_0 parameter informs us about the concentration at which there is an equilibrium between processes of phosphate sorption and release by bottom sediments [26,38,53,56]. Pan et al. [49,50] described the dual nature of particles (suspended solids or bottom sediment), which can be sink or source of P, depending on relationship between EPC_0 and P concentration in the water. The negative dependence between EPC_0 and aluminum content seems to confirm that the restoration technique used in Długie Lake significantly influenced the increase of P sorption properties.

Taking into consideration the fact that the theoretical sorption capacity assessed using Langmuir models is higher than real S_0 values observed during the experiment ("native sorbed phosphorus"), as well as a sum of highly mobile phosphorus fractions (NH$_4$Cl–P + BD–P), which represents sediment "native exchangeable P" [57], the sediment of Długie Lake theoretically should bind P, because the % of sediment saturation by P (%DSP) for S_{max} (Langmuir model) ranged between ca 25% DSP (St. 1, layer 10–15 cm) and 72% DSP (St. 3, layer 10–15 cm). For the double Langmuir model, the % DSP with regards to S_{max2} was much lower (between ca. 8% DSP at St. 3, sediment layer 5–10 cm, and 53% DSP at St. 1 for the deepest sediment layer).

Lewandowski et al. [23] also observed a higher sorption capacity for sediment layers modified by aluminum sulphate, which was used for the restoration of the Süsser See Lake in Germany. These authors as well as Cooke et al. [17] and Rydin [24] maintain that phosphorus bound during inactivation remains permanently buried in the sediment. Reitzel et al. [27] mention that the Al flocs created during restoration undergo an aging process, which can decrease the sorption capacity, compared to freshly formed flocs. However, present research can confirm the thesis that the "active" Al-modified sediment layer of Długie Lake theoretically should still bind phosphorus under stable pH conditions until the sorption active sites are exhausted.

6. Conclusions

The present research revealed the following:

1. The double Langmuir model matched the P adsorption experimental data of Długie Lake sediment best (the highest R^2 values). This fact may indicate that phosphorus adsorption in the tested sediments most likely occurs through phosphate binding at two types of active sorption sites.

2. Phosphate adsorption by the investigated lake sediment during experiments was significantly connected to aluminum content in sediment, as was indicated by the multiple regression equations obtained for the following adsorption parameters: S_{max}, S_{max2}, S_2, k_2, $1/n$ and EPC_0. A modification of sediment sorption capacity using aluminum coagulant probably increased the number of additional active sorption sites, as well as the second type of sites described by the double Langmuir equation.

3. The fact that the theoretical sorption capacity assessed using Langmuir models is higher than S_0, as well as a sum of highly mobile phosphorus fractions ($NH_4Cl–P + BD–P$), confirm that the both kinds of sediment of Długie Lake (the "active" layer and layers created after ending restoration procedures) still should bind P.

4. The relatively low amounts of phosphates, noted in the near-bottom water of Długie Lake, even in anoxia, confirm that the aluminum-modified sediment layer still can control internal P loading in the lake.

Author Contributions: Conceptualization, R.A.; Investigation, R.A., J.G. and M.Ł.; Methodology, K.P. and R.T.; Software, J.T.; Supervision, J.G.; Writing—review and editing, R.A.

Funding: This study was funded with the funds of the statutory subject Problem Group No 38 UPB, titled "Inland water ecosystems, and the protection and restoration of lakes"'. Subject No 0806.0802 "Improvement of the water reservoirs protection and restoration methods". Project financially co-supported by Minister of Science and Higher Education in the range of the program entitled "Regional Initiative of Excellence" for the years 2019–2022, Project No. 010/RID/2018/19, amount of funding 12,000,000 PLN." The funders had no role in the design of the study; in the collection, analyses, or interpretation of data; in the writing of the manuscript, or in the decision to publish the results.

Acknowledgments: The authors wish to thank Tomasz Jóźwiak for helpful discussions concerning the adsorption processes. They also wish to thank two anonymous Reviewers whose critical and very valuable comments have helped to improve the manuscript.

Conflicts of Interest: The authors declare no conflict of interest.

References

1. Zetland, D. *The End of Abundance. Economic Solutions to Water Scarcity*; digital edition 1.2; Aguanomics Press: Mission Viejo, CA, USA; Amsterdam, The Netherlands, 2011.

2. Zetland, D. *Living with Water Scarcity*; digital edition; Aguanomic Press: Mission Viejo, CA, USA; Amsterdam, The Netherlands; Vancouver, BC, Canada, 2014.

3. Ahtiainen, M.; Sandman, O.; Tynni, R. Sysmäjärvi—A lake polluted by mining wastewater. *Hydrobiologia* **1983**, *103*, 303–308. [CrossRef]

4. Auer, M.T.; Johnson, N.A.; Penn, M.R.; Effler, S.W. Measurement and verification of rates of sediment phosphorus release for a hypereutrophic urban lake. *Hydrobiologia* **1993**, *253*, 301–309. [CrossRef]

5. Grochowska, J.; Augustyniak, R.; Łopata, M. How durable is the improvement of environmental conditions in a lake after the termination of restoration treatments. *Ecol. Eng.* **2017**, *104*, 23–29. [CrossRef]

6. Grochowska, J.; Augustyniak, R.; Łopata, M.; Parszuto, K.; Tandyrak, R.; Płachta, A. From saprotrophic to clear water status: The restoration path of a degraded urban lake. *Waterair Soil Pollut.* **2019**, *230*, 94. [CrossRef]

7. Hamilton, D.; Collier, K.; Quinn, J.; Howard-Williams, C. (Eds.) *Lake Restoration Handbook, A New Zealand Perspective*; Springer International Publishing: Berlin, Germany, 2018; 599p.

8. Xu, H.; Brown, D.G.; Moore, M.R.; Currie, W.S. Optimizing spatial land management to balance water quality and economic returns in a Lake Erie watershed. *Ecol. Econ.* **2018**, *145*, 104–114. [CrossRef]

9. Xu, H.; Liu, Z.; Wang, L.; Wan, H.; Jing, C.; Jiang, J.; Wu, J.; Qi, J. Trade-offs and spatial dependency of rice production and environmental consequences at community level in Southeastern China. *Environ. Res. Lett.* **2018**, *13*, 024021. [CrossRef]

10. Baker, D.B.; Confesor, R.; Ewing, D.E.; Johnson, L.T.; Kramer, J.W.; Merryfield, B.J. Phosphorus loading to Lake Erie from the Maumee, Sandusky and Cuyahoga rivers: The importance of bioavailability. *J. Great Lakes Res.* **2014**, *40*, 502–517. [CrossRef]

11. Khare, Y.; Naja, G.M.; Stainback, G.A.; Martinez, C.J.; Paudel, R.; Van Lent, T. A phased assessment of restoration alternatives to achieve phosphorus water quality targets for Lake Okeechobee, Florida, USA. *Water* **2019**, *11*, 327. [CrossRef]

12. Klapper, H. Technologies for lake restoration. *J. Limnol.* **2003**, *62* (Suppl. 1), 73–90. [CrossRef]

13. Özkundakci, D.; Hamilton, D.P.; Gibbs, M.M. Hypolimnetic phosphorus and nitrogen dynamics in a small, eutrophic lake with a seasonally anoxic hypolimnion. *Hydrobiology* **2010**. [CrossRef]

14. Pan, G.; Miao, X.; Bi, L.; Zhang, H.; Wang, L.; Wang, L.; Wang, Z.; Chen, J.; Ali, J.; Pan, M.; et al. Modified Local Soil (MLS) technology for harmful algal bloom control, sediment remediation, and ecological restoration. *Water* **2019**, *11*, 1123. [CrossRef]

15. Mackay, E.B.; Maberly, S.C.; Pan, G.; Reitzel, K.; Bruere, A.; Corker, N.; Douglas, G.; Egemose, S.; Hamilton, D.; Hatton-Ellis, T.; et al. Geoengineering in lakes: Welcome attraction or fatal distraction? *Inland Waters* **2014**, *4*, 349–356. [CrossRef]

16. Søndergaard, M. Nutrient Dynamics in Lakes—With Emphasis on Phosphorus, Sediment and Lake Restoration. Ph.D. Thesis, National Environmental Research Institute, University of Aarhus, Aarhus, Denmark, 2007; 276p.

17. Cooke, G.D.; Welch, E.B.; Peterson, S.A.; Nichols, S.A. *Restoration and Management of Lakes and Reservoirs*; CRC Press, Taylor & Francis Group, LLC: Boca Raton, FL, USA, 2005.

18. Douglas, G.B.; Hamilton, D.P.; Robb, M.S.; Pan, G.; Spears, B.M.; Lurling, M. Guiding principles for the development and application of solid-phase phosphorus adsorbents for freshwater ecosystems. *Aquat. Ecol.* **2016**, *50*, 385–405. [CrossRef]

19. Pan, G.; Yang, B.; Wang, D.; Chen, H.; Tian, B.; Zhang, M.; Yuan, X.; Chen, J. In-lake algal bloom removal and submerged vegetation restoration using modified local soils. *Ecol. Eng.* **2011**, *37*, 302–308. [CrossRef]

20. Bryl, Ł.; Wiśniewski, R. Efekty inaktywacji fosforu w osadach dennych Jeziora Wolsztyńskiego (The effects of phosphorus inactivation in the bottom sediment of Wolsztyńskie Lake). In *Ochrona i Rekultywacja Jezior (Lakes Protection and Restoration)*; Ryszard, W., Ed.; Polskie Zrzeszenie Inżynierów i Techników Sanitarnych (Polish Association of Civil Engineers and Technicians): Toruń, Poland, 2015; pp. 71–83. (In Polish)

21. Kajak, Z. *Eutrofizacja Jezior (Lakes' Eutrophication)*; PWN Warszawa: Warszawa, Poland, 1979. (In Polish)

22. Kajak, Z. *Hydrobiologia-Limnologia (Hydrobiology-Limnology)*; PWN Warszawa: Warszawa, Poland, 2001. (In Polish)

23. Lewandowski, J.; Schauser, I.; Hupfer, M. Long term effect of phosphorus precipitations with alum in hypereutrophic Lake Süsser See (Germany). *Water Res.* **2003**, *37*, 3194–3204. [CrossRef]

24. Rydin, E. Potentially mobile phosphorus in Lake Erken sediment. *Water Res.* **2000**, *34*, 2037–2042. [CrossRef]

25. Wiśniewski, R. Inaktywacja fosforanów w osadach dennych jako metoda redukcji symptomów eutrofizacji (Phosphate inactivation in the bottom sediment as a method of eutrophication symptoms reduction). *Bibl. Monit. Środ. Łódź* **1995**, *5*, 189–201. (In Polish)

26. Wiśniewski, R. Zróżnicowanie stanu i stopnia degradacji jezior. Wartość diagnostyczna badań osadów (The differentiation of lakes status and degradation. The diagnostic value of sediment research). In *Ochrona i Rekultywacja Jezior (Lakes' Protection and Restoration)*; Wiśniewski, R., Piotrowiak, J., Eds.; Polskie Zrzeszenie Inżynierów i Techników Sanitarnych (Polish Association of Civil Engineers and Technicians): Toruń, Poland, 2007. (In Polish)

27. Reitzel, K.; Jensen, H.S.; Egemose, S. pH dependent dissolution of sediment aluminum in six Danish lakes treated with aluminum. *Water Res.* **2013**, *47*, 1409–1420. [CrossRef]

28. Van Hullebush, E.; Auvray, F.; Deluchat, V.; Chazal, P.; Baudu, M. Phosphorus fractionation and short-term mobility in the surface sediment of a polymictic shallow lake treated with a low dose of alum (Courtille Lake, France). *Water Air Soil Pollut.* **2003**, *146*, 75–91. [CrossRef]

29. Brzozowska, R.; Gawrońska, H. Influence of a multi-year artificial aeration of a lake using destratification method on the sediment-water phosphorus exchange. *Arch. Environ. Prot.* **2005**, *31*, 71–88.

30. Hongthanat, N.; Kovar, J.L.; Thompson, M.L. Sorption indices to estimate risk of soil phosphorus loss in the Rathbun Lake Watershed, Iowa. *Soil Sci.* **2011**, *176*, 237–244. [CrossRef]

31. Hongthanat, N.; Kovar, J.L.; Thompson, M.L.; Russell, J.R.; Isenhart, T.M. Phosphorus source—Sink relationships of stream sediments in the Rathbun Lake watershed in southern Iowa, USA. *Environ. Monit. Assess.* **2016**, *188*, 453–467. [CrossRef] [PubMed]

32. Zhou, A.M.; Tang, H.X.; Wang, D.S. Phosphorus adsorption on natural sediments: Modeling and effects of pH and sediment composition. *Water Res.* **2005**, *39*, 1245–1254. [CrossRef] [PubMed]

33. Filipkowska, U. *Wykorzystanie Właściwości Adsorpcyjnych Materiałów Odpadowych do Usuwania Barwników z Roztworów Wodnych (The Use of Adsorption Abilities of Waste Materials for Dye Removal from Aqueous Solutions)*; Wyd. Nauk. Komitetu Inżynierii Środowiska PAN: Lublin, Poland, 2011.

34. Filipkowska, U.; Jóźwiak, T.; Szymczyk, P. Application of cross-linked chitosan for phosphate removal from aqueous solutions. *Prog. Chem. Appl. Chitin Its Deriv.* **2014**, *XIX*, 5–14. [CrossRef]

35. Jóźwiak, T.; Filipkowska, U.; Szymczyk, P.; Kuczajowska-Zadrożna, M.; Mielcarek, A. The use of cross-linked chitosan beads for nutrients (nitrate and orthophosphate) removal from a mixture of P–PO$_4$, N–NO$_2$ and N–NO$_3$. *Int. J. Biol. Macromol.* **2017**, *104*, 1280–1293. [CrossRef]

36. Szymczyk, P.; Filipkowska, U.; Jóźwiak, T.; Kuczajowska-Zadrożna, M. Phosphate removal from aqueous solutions by chitin and chitosan in flakes. *Prog. Chem. Appl. Chitin Derivatives* **2016**, *XXI*. [CrossRef]

37. Tibco Software Inc. *Statistica Software Package 13.0.*; Tibco Software Inc.: Palo Alto, CA USA, 2018.

38. Łukawska–Matuszewska, K.; Voght, R.D.; Xie, R. Phosphorus pools and internal loading in a eutrophic lake with gradients in sediment geochemistry created by land use in the watershed. *Hydrobiologia* **2013**, *713*, 183–197. [CrossRef]

39. *Fizykochemiczne Metody Analizy w Chemii Środowiska. Cz. 2 (Physical and Chemical Analytical Methods in the Environmental Chemistry*; Kosobucki, P.; Buszewski, B. (Eds.) Part. 2; Wyd. Nauk. UMK: Toruń, Poland, 2016. (In Polish)

40. Zar, J.H. *Biostatistical Analysis*; Prentice Hall: Upper Saddle River, NJ, USA, 1984.

41. Böstrom, B.; Andersen, J.M.; Fleisher, S.; Jannson, M. Exchange of phosphorus across the water-sediment interface. *Hydrobiologia* **1988**, *170*, 229–244. [CrossRef]

42. Brzozowska, R. Wpływ wieloletniego sztucznego napowietrzania na przykładzie Jeziora Długiego na wymianę związków biogennych między osadami a wodą (The Influence of Multi-Year Artificial Aeration, on the Example of Długie Lake on the Nutrient Exchange between Sediment and Water. Ph.D. Thesis, UWM in Olsztyn, Olsztyn, Poland, 2003. (In Polish).

43. Grochowska, J.; Brzozowska, R. Influence of different recultivation methods on durability of nitrogen compounds changes in the waters of an urban lake. *Water Environ. J.* **2015**. [CrossRef]

44. Augustyniak, R. *Wpływ Czynników Fizyczno-Chemicznych i Mikrobiologicznych na Zasilanie Wewnętrzne Fosforem wód Wybranych Jezior Miejskich (The Influence of Physical, Chemical and Microbiological Factors on the Phosphorus Internal Loading to the Water of Selected Urban Lakes)*; Polish Academy of Sciences, Environmental Engineering Committee Publishing House: Lublin, Poland, 2018; 242p. (In Polish)

45. Cyr, H.; Mc Cabe, S.K.; Nürnberg, G.K. Phosphorus sorption experiments and the potential for internal phosphorus loading in littoral areas of a stratified lake. *Water Res.* **2009**, *43*, 1654–1666. [CrossRef]

46. Huang, L.; Fu, L.; Jin, C.; Gielen, G.; Lin, X.; Wang, H.; Zhang, Y. Effect of temperature on phosphorus sorption to sediments from shallow eutrophic lakes. *Ecol. Eng.* **2011**, *37*, 1515–1522. [CrossRef]

47. Huang, W.; Wang, K.; Du, H.; Wang, T.; Wang, S.; Yangmao, Z.; Jiang, X. Characteristic of phosphorus sorption at the sediment–water interface in Dongting lake, a Yangtzee-connected lake. *Hydrol. Res.* **2016**, *47* (Suppl. 1), 225–237. [CrossRef]

48. Huang, W.; Chen, X.; Wang, K.; Jiang, X. Seasonal characteristics of phosphorus sorption by sediment from plain lakes with different trophic statuses. *R. Soc. Open Sci.* **2018**, *5*, 172237. [CrossRef] [PubMed]

49. Pan, G.; Krom, M.D.; Herut, B. Adsorption-desorption of phosphate on airborne dust and riverborne particulates in east Mediterranean seawater. *Environ. Sci. Technol.* **2002**, *36*, 3519–3524. [CrossRef] [PubMed]

50. Pan, G.; Krom, M.; Zhang, M.; Zhang, X.; Wang, L.; Dai, L.; Sheng, Y.; Mortimer, R.J.G. Impact of suspended inorganic particles on phosphorus cycling in the Yellow River (China). *Environ. Sci. Tech.* **2013**, *47*, 9685–9692. [CrossRef]

51. Holford, I.C.R.; Wedderburn, R.W.M.; Mattingly, G.E.G. A Langmuir two-surface equation as a model for phosphate adsorption by soils. *J. Soil Sci.* **1974**, *25*, 242–255. [CrossRef]

52. Limousin, G.; Gaudet, J.-P.; Charlet, L.; Szenknect, S.; Barthes, V.; Krimissa, M. Sorption isotherms: A review on physical bases, modeling and measurement. *Appl. Geochem.* **2007**, *22*, 249–275. [CrossRef]

53. Froelich, P.N. Kinetic control of dissolved phosphate in natural rivers and estuaries: A primer on the phosphate buffer mechanism. *Limnol. Oceanogr.* **1988**, *33*, 649–668. [CrossRef]

54. Bartoszek, L.; Koszelnik, P. Assessment of phosphorus retention in the bottom sediments of the Solina-Myczkowce complex of reservoirs. *Rocz. Ochr. Środowiska* **2016**, *18*, 213–230.

55. Orihel, D.M.; Baulch, H.M.; Casson, N.J.; North, R.L.; Parsons, C.T.; Seckar, D.C.M.; Venkiteswaran, J.J. Internal phosphorus loading in Canadian fresh waters: A critical review and data analysis. *Can. J. Fish. Aquat. Sci.* **2017**, *74*, 2005–2029. [CrossRef]

56. House, W.A.; Denison, F.H. Factors influencing the measurement of equilibrium phosphate concentrations in river sediments. *Water Res.* **2002**, *34*, 1187–1200. [CrossRef]

57. James, W.F. Internal phosphorus loading contributions from deposited and resuspended sediment to the Lake of the Woods. *Lake Reserv. Manag.* **2017**, *33*, 347–359. [CrossRef]

Article

A Phased Assessment of Restoration Alternatives to Achieve Phosphorus Water Quality Targets for Lake Okeechobee, Florida, USA

Yogesh Khare [1,*], Ghinwa Melodie Naja [1], G. Andrew Stainback [1], Christopher J. Martinez [2], Rajendra Paudel [1] and Thomas Van Lent [1]

[1] Everglades Foundation, Science Department, 18001 Old Cutler Road Suite 625 Palmetto Bay, FL 33157, USA; mnaja@evergladesfoundation.org (G.M.N.); astainback@evergladesfoundation.org (G.A.S.); rpaudel@evergladesfoundation.org (R.P.); tvanlent@evergladesfoundation.org (T.V.L.)

[2] Department of Agricultural and Biological Engineering, University of Florida, Gainesville, FL 32611, USA; chrisjm@ufl.edu

* Correspondence: ykhare@evergladesfoundation.org; Tel.: +1-305-251-0001

Received: 17 January 2019; Accepted: 12 February 2019; Published: 14 February 2019

Abstract: Achieving total phosphorus (TP) total maximum daily loads (TMDL) for Lake Okeechobee (Florida, FL, USA), a large freshwater lake, is a key component of the greater Everglades ecosystem restoration and sustainability of south Florida. This study was aimed at identification of a cost-effective restoration alternative using four TP control strategies—Best Management Practices (BMPs), Dispersed Water Management (DWM), Wetland Restoration, and Stormwater Treatment Areas (STAs)—to achieve a flow-weighted mean TP concentration of 40 μg/L at lake inflow points, through a phased scenario analysis approach. The Watershed Assessment Model was used to simulate flow and phosphorus dynamics. The 10-year (1998–2007) 'Base' scenario calibration indicated 'acceptable' to 'good' performance with simulated annual average flows and TP load of 2.64×10^9 m^3 and 428.6 metric tons, respectively. Scenario results showed that TP load reduction without STAs would be around 11–40% with respect to Base compared to over 75% reduction requirement to achieve TMDL, indicating STAs as a necessary component to achieve restoration. The most cost-effective alternative to achieve TP target consisted of implementation of nutrient management BMPs, continuation of existing DWM projects, and the construction of ~200 km^2 of STAs for a total project cost of US $4.26 billion.

Keywords: Lake Okeechobee; restoration; watershed modeling; TMDL; best management practices; dispersed water management; cost-effective

1. Introduction

Nutrient impairment of freshwater lakes and rivers has been recognized as a major worldwide problem with ecological, health, and economic impacts [1]. Symptoms of impairment and eutrophication range from harmful algal blooms, depleted oxygen, fish kills, and ecosystem collapse [2,3]. These ecological impacts have also a direct economic impact as communities around the world are using those freshwaters for drinking, recreational, and scenic purposes. The cost of freshwater eutrophication in England and Wales has been estimated to be $105–160 million per year mainly from reduced value of waterfront properties, reduced recreational value, commercial fisheries, and increased drinking water treatment costs [4]. Dodds et al. [5] estimated the economic damages from eutrophication is costing the United States up to $4.3 billion a year.

Watershed approaches to manage water resource quality and quantity and restore impaired waterbodies have been used since the late 1980s. During the last decades, several initiatives were implemented in the United States at a watershed level to curtail nutrient problems. These strategies range from implementing Best Management Practices (BMPs) at the source level to major downstream cleanup efforts of impaired waterbodies. Scavia et al. [6] investigated several phosphorus load reduction strategies for the Maumee River watershed and reported that there are several restoration options to meet the 40% reduction in phosphorus loadings to Lake Erie. To determine the effectiveness of these strategies, monitoring nutrients at the source level and modeling at the watershed level to determine pollution hot-spots and potential changes under future climate are necessary to determine the impacts of these initiatives [7–10]. These types of studies should also be accompanied by identification of optimal combination of nutrient control strategies considering overall implementation cost to present an accurate and reliable assessment of how much it would take to reach the target for informed decision making [11–13].

Lake Okeechobee (LO) is classified as impaired by the state of Florida. This lake covers an area of 1900 km^2 and is a major shallow subtropical fresh waterbody providing flood control, water supply, ecological, and recreational benefits to south Florida [14]. During the last decades, the lake received increasing total phosphorus (TP) loads from agricultural and urban activities in its watershed, resulting in an increase in-lake annual average TP concentrations from ~50 µg/L in mid 1970s to over 100 µg/L by 2000 [15]. Since the 1980s, LO has witnessed severe algal bloom events with a three-fold increase in the algae bloom frequency [16,17]. The state of Florida has undertaken several initiatives for the LO watershed, listing projects necessary to decrease the TP loadings. In 1987 and after several severe large blooms covering over 300 km^2 of the lake's surface, the Florida Legislature created the Surface Water Improvement and Management (SWIM) Act to protect, restore, and maintain Florida's impaired surface waterbodies [18]. The interim SWIM plan for LO was completed in 1989 and an overall 40% reduction in TP loading was set as a target to be achieved by July 1992 [19]. This plan was developed after 15 years of data collection, field studies, and modeling efforts [20] and was updated several times during the 1990s and early 2000s. In 2001, a phosphorus Total Maximum Daily Load (TMDL) of 140 metric tons (mtons) per year (including 35 mtons through atmospheric deposition) was established for LO to achieve the in-lake TP concentration of 40 µg/L [21]. In 2000, the Florida Legislature passed the Lake Okeechobee Protection Act (LOPA) directing the state agencies to implement a comprehensive long-term program to restore and protect LO and its downstream receiving waters. The LOPA mandated that the lake TMDL be met by January 1, 2015. Lake Okeechobee Watershed Protection Plans and construction projects were developed consequently. In 2007, the Florida legislature passed the Northern Everglades and Estuaries Protection Program (NEEPP), promoting a comprehensive watershed approach to protect LO and the Caloosahatchee and St. Lucie rivers and estuaries. NEEPP required the development of a detailed technical plan for Phase II of the Lake Okeechobee Watershed Construction Project, which was completed in February 2008 [22]. This plan identified the projects and other actions that would be required to meet TMDL. In December 2014, FDEP developed a Basin Management Action Plan (BMAP) adopting a phased approach for the implementation of strategies to achieve 43% TP reduction during the first 10 years. Despite this long regulatory history and with more than 30 years since the SWIM Act was passed by the legislature, no reduction in TP loading to the lake has occurred since 1990. Albeit some projects and regulatory programs have been implemented in the watershed, the recent five-year average TP load (TPL) was 531 mtons (1 May 2013 to 30 April 2017), which is 3.8 times the TMDL. It implies that to achieve TMDL for LO, TPL from the contributing watershed needs to be curtailed by more than 75%. During the last decade, basins on the northern side of the LO consistently contributed over 90% of the TPL from non-atmospheric sources [15] and hence need prioritized attention in TMDL planning.

All the state agencies' restoration plans for LO are mainly based on the adoption of regulatory programs, implementation of urban and agricultural BMPs, maintaining research and monitoring programs, construction of reservoir assisted Stormwater Treatment Areas (STAs), dispersed water management (DWM), and wetland restoration (WR) to be implemented at different levels [23]. BMPs are conservation practices implemented in the watershed to target phosphorus at the farm level. STAs are constructed wetlands operated to filter and remove phosphorus downstream of the farms. DWM refers to practices to retain water on private or public lands. WR involves improving the functionality of degraded wetlands or of land parcels classified as pre-drainage era wetlands. The purpose of this paper is to perform a feasibility assessment of restoration scenarios by using, combining, and scaling-up the above listed features to reach the LO TMDL. The optimal combination of conservation practices and restoration strategies to reach the LO target has to be determined with a clear understanding of the associated implementation costs. In this study, the northern LO (NLO) basins (Figure 1) were simulated along with restoration scenarios representing different levels of BMPs, DWM, WR, and STAs that could be implemented to decrease the TP loadings to LO. This was done while also estimating the costs of implementing each scenario to lead to the selection of the most cost-effective restoration alternative. In this regional scale analysis, restoration alternatives were formed in phases, carry-forwarding only the cost-effective scenario to the next phase, to make this assessment manageable.

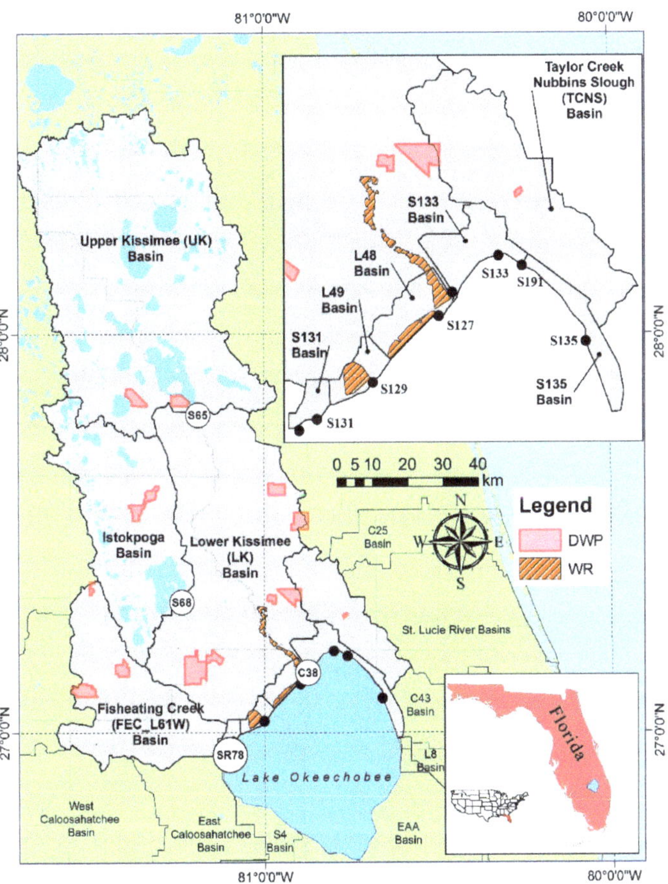

Figure 1. Location and details of Northern Lake Okeechobee basins. The zoomed-in inset map shows the details of the smaller basins with the outlet structures. SR78, C38, S68, S65, S131, S129, S127, S133, S191, and S135 indicate model calibration locations.

2. Materials and Methods

2.1. Study Area

The region of interest for the present study includes the northern basins contributing to LO inflows (Figure 1) with a total drainage area of ~10,600 km^2. The individual basins are (1) the Kissimmee Chain of Lakes/Upper Kissimmee (UK) basin, (2) the Lower Kissimmee (LK) basin, (3) the Lake Istokpoga basin, (4) the Fisheating Creek basin including the L61W drainage area (FEC_L61W), (5) Taylor Creek–Nubbin Slough (TCNS), (6) S135, (7) S133, (8) S131, (9) L48, and (10) L49. The Kissimmee River forms the main drainage channel in this area. There are several smaller creeks, sloughs, agricultural ditches, flood control canals, and structures that route runoff from this region to LO. The NLO region is well monitored for rainfall, flow, and water quality (details can be found on the South Florida Water Management District website https://apps.sfwmd.gov/WAB/EnvironmentalMonitoring/index.html). The average annual rainfall in UK is approximately 1255 mm, while the rest of the basins receive 1120 mm annually [24,25] with 70% of the rainfall occurring in the wet season (May to October). Correspondingly, most of the TP transport also takes place during the same period posing a challenge to non-point source TP management [26,27]. About 41.1% of the study area is under agricultural land use (Table 1) with improved and unimproved cattle pastures covering 29.1% of the land cover. Other agricultural activities include citrus groves, sugarcane, row and field crops, sod farms, and tree plantations. Urban and developed areas cover 13.4% of this region and are predominantly located in the northern portion of UK basin around the Orlando metropolitan area. Wetlands cover 19.4% of the NLO basins along with 7.3% of forests and 7.9% of water bodies that include the Kissimmee Chain of lakes and Lake Istokpoga. Immokalee and Smyrna are the dominant soil types covering approximately 65% NLO basins' extent. Both soils are Spodosols characterized by sandy texture, poor drainage, and low phosphorus retention capacity [28].

Table 1. Land use (LU) summary in the northern Lake Okeechobee watershed.

LU Category	Percent
Agriculture	*41.1*
Dairy and Animal Feeding Operations	0.7
Pastures	29.1
Improved Pasture	22.5
Unimproved Pasture	5.7
Woodland Pasture	0.5
Intensive Pasture	0.2
Others	0.1
Other Agriculture	11.3
Urban and Developed	*13.4*
Residential	10.8
Commercial, Industrial, and other Developed	2.6
Natural	*45.5*
Wetlands	19.4
Forests	7.3
Scrub, Brush, and Barren	10.8
Lakes, Streams, and Water Bodies	7.9

2.2. Watershed Modeling

The individual NLO basins were modeled for hydrologic and phosphorus water quality dynamics using the Watershed Assessment Model (WAM) [29]. The Base conditions (1998–2007) and three of the four restoration strategies considered in this study (BMPs, DWM, and WR) were modeled with WAM. The stormwater treatment area (STA) requirements to achieve a TP concentration target of 40 µg/L for each restoration scenario modeled in WAM was determined using the Dynamic Model for Stormwater

Treatment Area (DMSTA) [30]. Both models have been widely used in Florida as decision support tools while formulating and evaluating alternatives for restoration plans [15,23,31–34].

2.2.1. Watershed Assessment Model: Model Setup and Re-calibration

WAM is a spatially distributed model designed to handle the hydrologic conditions typical to Florida (abundance of wetlands, high groundwater table conditions, hydric soils, canal networks that experience periodic flow reversals, and complex water control structure operations). These hydrologic conditions in south Florida are unique and, unlike WAM, most of the widely used watershed scale models (e.g., Soil and Water Analysis Tool, SWAT; Hydrologic Simulation Program FORTRAN, HSPF) fail to successfully simulate them [35]. WAM simulates the earthbound part of the hydrological cycle to calculate basin level flow and water quality dynamics [29]. Unique combinations of soil, land use, rainfall, irrigation, and waste utility zones are identified, grouped together, and simulated using one of the three source level models. The Everglades Agricultural Area Model (EAAMOD) [36] is used for agronomic crops on hydric soils while open water, mining, aquaculture, and wetlands are simulated using a simple water balance approach model called 'Special Case' developed for WAM [37]. EAAMOD was identified as the most appropriate field scale water quality model for agriculture on hydric soils in central and south Florida [38]. All other unique cell combinations are simulated using the Groundwater Loading Effects of Agricultural Management Systems model (GLEAMS) [39]. Daily flows and water quality constituents from these field scale models calculated at each grid cell (user defined, but typically and in this study 0.01 km^2) are then adjusted for retention/detention (R/D) systems, if present, before routing simulated daily outputs to the stream network and ultimately to the watershed outlet. WAM uses six elevation-based flowpaths/distance grids to attenuate daily source level surface and sub-surface flows and loads accounting for features en route to the watershed outlet. Overland and in-stream water quality attenuation follows generalized exponential decay type equations (Equations (1) and (2)), respectively.

$$C_{out} = C_b + (C_{in} - C_b)e^{-(aq^{-b})d} \tag{1}$$

$$C_{out} = C_b + (C_{in} - C_b)e^{-a\frac{\tau}{R}} \tag{2}$$

where C_b = background nutrient concentration (mg/L), C_{out} = outflow from flowpath (overland) or stream reach (instream) nutrient concentration (mg/L), C_{in} = inflow from source cell (overland) or upstream reach (instream) nutrient concentration (mg/L), a and b = empirical water quality attenuation exponent and multiplier, respectively (for in-stream attenuation $b = 0$), q = flow (m^3 s^{-1}), d = overland flowpath distance (m), R = hydraulic radius of stream, τ = time interval.

A 10-year simulation period (1998–2007) was selected for this study to cover a range of conditions (wet and dry periods, see Figure S1 for details) prior to the implementation of any major water quality improvement project (e.g., BMPs) in the region, while maintaining longevity of the study. WAM has been already used to model the NLO basins in phosphorus budget analysis studies [26,40,41]. These existing model setups were obtained from the SFWMD (personal communication) for all basins shown in Figure 1, except the UK setup, which was obtained from [42]. The above listed studies have detailed information on WAM input datasets used in the modeling (land use, soil, rainfall, hydrography, elevation, water control structures and their operations, etc.). Re-calibration was performed for all the individual basins through minor adjustments to the location of the calibration gauges, evapotranspiration, and phosphorus attenuation exponent and multiplier parameters for streams and dominant land uses [35,43] to account for marginally different simulation periods in earlier studies (1999–2007 in [40] and 2002–2007 in [42]). Two goodness of fit (GOF) measures, Nash–Sutcliffe Efficiency (NSE) Equation (3) and Percentage Bias (PBIAS) Equation (4), were calculated for monthly flows and TPL at 10 locations (one gauge per basin) [44] (Figure 1). While NSE is a good indicator of ability of the model to capture long-term hydrologic and water quality dynamics, PBIAS is a good measure of the model deviation and of the average model behavior. Both these GOFs are recommended evaluation measures by the American Society of Agricultural and Biological Engineers standards on hydrologic and water quality modeling guidelines [45] with clearly defined criteria for model performance classification. Since the focus of this study was to assess the relative change in average flows and TPL along with an economic valuation of restoration alternatives, re-calibration was performed considering the entire simulation period.

$$NSE = 1 - \frac{\sum_{i=1}^{N}(O_i - P_i)^2}{\sum_{i=1}^{N}(O_i - \overline{O})^2} \tag{3}$$

$$PBIAS = \frac{\sum_{i=1}^{N}(O_i - P_i)}{\sum_{i=1}^{N} O_i} \times 100 \tag{4}$$

where N = number of time steps, O_i = observation at the ith time step, P_i = simulated value at the ith time step, $\bar{O}$ = mean of observations.

2.2.2. Dynamic Model for Stormwater Treatment Area (DMSTA)

The DMSTA2 is a transient wetland flow and phosphorus cycling model used to design treatment wetlands [30,46]. Daily timeseries of rainfall, evapotranspiration, inflow, and phosphorus concentration constitute DMSTA2 inputs while wetland size, hydraulics, and vegetation type constitute primary design variables. Treatment areas are typically divided into cells connected in series or parallel pathways forming a treatment network. Phosphorus dynamics is simulated considering soil, biota, and water column components. Inter-compartmental fluxes, uptake, and release of phosphorus by vegetative biomass are accounted through simplified processes. The vegetation-specific calibration parameters for DMSTA2 have been developed for different vegetation types using data from full-scale STAs, lakes, reservoirs, and natural wetlands in south Florida [46,47]. DMSTA2 is a reliable STA design and restoration alternatives evaluation tool used by state and federal agencies [32,33,48]. In this study, STA inflow and phosphorus concentration timeseries were created by combining the daily WAM outputs (flow and phosphorus concentration timeseries) at the basin outlets. Rainfall and evapotranspiration timeseries were defined based on measurements in this region.

2.3. Restoration Scenarios

Twenty restoration alternatives/scenarios (including Base) were formulated by combining four phosphorus control strategies (BMP, DWM, WR, and STA). The salient features and corresponding WAM parameterization of these restoration strategies are described below. BMP types and DWM parameterizations were applied uniformly across respective land uses in all NLO basins. It is noteworthy that identification of pollution hot-spots and performing spatial optimization of BMPs has been demonstrated in many recent modeling studies (e.g., [49–53]). However, due to the large size of the investigated region (10,600 km^2) and focus on overall restoration cost to achieve TMDL, no explicit spatial optimization of these restoration strategies was performed. TP loading hot-spots were identified for this region (Figures S2 and S3, Table S1) and were found to be consistent with previous studies [26,40,42]. The analysis presented in this work can be used as a starting point for further optimization of restoration projects.

2.3.1. Best Management Practices

Best Management Practices (BMPs) implemented in NLO basins have been classified into three types [54,55]. Type I BMPs are implemented by the producer/land owner, typically at a low cost, and primarily involve nutrient management actions (e.g., fertilizer selection, application rate and timing, record keeping) [56]. Type II BMPs require structural modifications (fencing, wetland/stormwater R/D, improved irrigation system) and are typically funded through cost-share programs (after the producer/landowner had already implemented type I BMPs). Type III BMPs are more expensive and involve edge of the field stormwater R/D with chemical technologies, e.g., aluminum sulphate and iron chloride, etc. (after the producer/landowner had already implemented types I and II BMPs) [57]. Based on these BMP types, three levels of restoration—(1) BMP1 = Type I, (2) BMP12 = Type I + Type II, and (3) BMP123 = Type I + Type II + Type III—were considered. BMP parameterization for land uses in the NLO basins are summarized in Table 2. While BMPs are formulated and implemented for phosphorus as well as other nutrients, only the phosphorus-related practices were the focus of this study. Fertilization rates, irrigation control, culvert operations, and wetland/stormwater R/D BMPs were simulated mechanistically. Effects of all other BMPs (at the field scale), which included buffer strips, critical area fencing, edge of the field chemical treatment, etc., were incorporated in the model by adopting BMPs effectiveness factors, i.e., phosphorus reduction factors [54]. Not all individual BMPs were applied to all land uses and their categorization into type I, II, and III also differed in some cases. Additional details on BMPs can be found in the literature [54], which was used as the basis for BMP scenario formulation.

Table 2. Summary of WAM parameterization for restoration strategies (a) Best Management Practices (BMP), and (b) Disperse Water Management (DWM).

Land Use	Percent P Fertilizer Reduction	Water Management	Increase in Retention/Detention (m^3/ha)	Percent P Reduction under Other Type II BMPs	Percent Reduction under EOF Chemical Treatment and Other Type III BMPs	Increase in Storage (m^3/ha) LRD	HRD
	Type I	Type II	Type II	Type II	Type III	DWM	
Citrus	12	N	260	0	45	0	0
Commercial and Services	10	N	40	0	0	0	0
Coniferous Plantation	33	N	50	0	50	0	0
Dairies	100	Y	300	0	50	0	0
Field Crops	20	Y	200	0	40	0	0
Fruit Orchards	30	N	100	0	0	0	0
High Density Residential	30	N	100	0	65	0	0
Improved Pasture	100	N	100	5	50	100	600
Industrial	5	N	100	0	0	0	0
Intensive Pasture	100	N	100	5	50	100	600
Low Density Residential	25	N	250	0	65	0	0
Managed Landscape	10	N	100	0	0	0	0
Medium Density Residential	30	N	100	0	65	0	0
Multiple Dwelling Units	30	N	100	0	65	0	0
Ornamental Nurseries	30	N	1000	5	50	0	0
Row Crops	30	Y	300	0	50	0	0
Sod Farms	20	Y	250	0	50	0	0
Sugarcane	30	Y	400	0	50	0	0
Transportation Corridors	100	N	40	0	0	0	0
Tree Nurseries	30	Y	150	0	0	0	0
Unimproved Pasture	100	N	30	2	45	100	600
Woodland Pasture	100	N	10	2	40	100	600

2.3.2. Dispersed Water Management

Dispersed Water Management (DWM) is a practice to retain water on private or public lands using existing or minimal new infrastructure to facilitate restoration efforts in the Northern Everglades [58]. Apart from providing dispersed storage, some DWM projects can help improving water quality (nutrient retention in wetlands) while also providing improved ecological habitat [59,60]. In this work, three types of DWM alternatives were considered. The first alternative was based on existing projects and the remaining two alternatives were based on hypothetical R/D levels. DWM was modeled using depth-based wet stormwater R/D ponds. WAM calculates daily runoff and percolation from the pond based on a water balance approach (rainfall, evapotranspiration, percolation). Nutrient outflow concentrations are calculated using wetland attenuation parameters with stream routing algorithm Equation (2).

As of 2016, there were 21 dispersed water projects (DWPs) operating or under construction in the study area (Table S2), estimated to provide a total of 34×10^6 m^3 of average annual storage using the Potential Water Retention Model (PWRM) developed by the [60,61]. Corresponding R/D pond parameters (storage ratio and pond depth) were inversely quantified to match the PWRM estimates for annual flow reductions [61]. In this study, DWPs were modeled as an alternative since most of the DWPs within the study area, located on privately owned pasturelands [62], were built after

2007. This concept of using pastures for water storage was extended to create two hypothetical DWM levels, (a) Low Retention/Detention (LRD) and (b) High Retention/Detention (HRD). LRD and HRD alternatives were applied to improved, unimproved, woodland, and intensive pastures land uses (covering 29% of the NLO watershed area). Storage ratio and pond depths were defined based on prior studies (e.g., [40]) and expert opinion (D. Bottcher, SWET Inc., personal communication) to determine the practically viable setting for agricultural stormwater R/D. This parameterization translated to approximately 1% and 6% of pastureland being converted to R/D ponds of 0.5 m depth for DWM purposes under LRD and HRD, respectively.

2.3.3. Wetland Restoration

Wetland Restoration (WR) involves rehydration of drained wetlands to provide storage, nutrient assimilation, habitat improvement for wetland wildlife, and creating recreation opportunities. Historically, wetlands covered a large area of the NLO watershed, much of which was converted into agricultural land in the 20th century. However, under present land use, it is neither feasible nor possible to restore all pre-development wetlands. State and federal agencies shortlisted four areas for WR projects through a multi-criteria selection process, namely (a) Kissimmee River Wetlands (LK basin), (b) Paradise Run Wetlands (LK and L48 basins), (c) Lake Okeechobee West Wetlands (L48 basin), and (d) IP-10 Wetlands (L49 basin), cumulatively covering an area of 52 km^2 [63]. These four projects constituted one WR alternative in this study. WAM uses 'Special Case' sub-module to simulate wetlands by maintaining the complete water balance based on rainfall, evapotranspiration, crop coefficient, percolation, and maximum inundation depth [29]. Nutrient concentrations in runoff are calculated based on user specified event mean concentrations (EMC) for each wetland type. WR projects were parameterized in WAM by first changing all land uses within WR project areas to wetland and setting a maximum inundation depth to 2 m to mimic the common wetland parametrization in WAM [37], thus increasing storage capacity and hydroperiod.

2.3.4. Stormwater Treatment Area

Stormwater Treatment Areas (STAs) were sized to meet a long-term annual TP flow weighted mean concentration (FWMC) of 40 μg/L at the LO inflow, a TP level similar to the LO pelagic zone TP concentration goal [21]. Walker and Havens [17] showed how the relationship between near-shore TP concentration of 40 μg/L and algae bloom frequency was used to derive this goal. Two assumptions were made when designing the STAs: (1) STAs were to be located close to LO to avoid prolonged 'no flow' periods, a major STA maintenance aspect [47] and (2) outflows could be diverted from NLO basins to the STA locations, thus enabling the grouping of the basin WAM outflows and TPLs and restrict the number of STAs to a lower number (five for this analysis). No further analysis about land availability or suitability was performed. Previous STA design studies have indicated that phosphorus removal efficiency is size dependent with very small treatment areas not working optimally [64]. Each STA was designed as a network of 3 to 5 parallel pathways, with each pathway consisting of an emergent aquatic vegetation cell followed by a submerged aquatic vegetation cell. Such a design has been found effective in removing phosphorus [48]. Currently, there are three working STA projects located in the TCNS (0.6 and 3.2 km^2) and S135 (3.7 km^2) basins with a total effective treatment area of 7.5 km^2 designed to remove 16.1 mtons of TPL. Though two of these STAs in the TCNS became operational around 2007 (designed to remove 7.1 mtons of TPL), they have been facing operational challenges, while only the first phase of the third STA, known as Lakeside Ranch STA, (designed to remove 9.1 mtons of TPL) was completed in 2013. The second phase of Lakeside Ranch STA (~3.3 km^2), designed to remove 10 mtons of TPL annually, is anticipated to become fully operational in 2020 [15]. Hence, the three STAs were not incorporated in the Base scenario, while impact of only the Lakeside Ranch STA was demonstrated for the optimal restoration scenario selected in this study. The Lakeside Ranch STA was considered in this study because of its high TPL removal performance and of its location close to LO to avoid prolonged 'no flow' periods, a main assumption of the present study.

2.4. Cost Model

The cost of implementing different combinations of BMPs, DWM, WR, and STAs to achieve the LO inflow TP FWMC of 40 µg/L were determined to find the most cost-effective restoration scenario. The following cost minimization equations were used (Equations (5) and (6)):

$$Min(Z) = \sum_{n=1}^{3} \sum_{i1=1}^{22} C_{i1,n}^{B} L_{i1,n} A_{i1,n}$$
$$+ \sum_{m=1}^{2} \sum_{i2=1}^{4} C_{i2,m}^{D1} L_{i2,m} A_{i2,m} + \sum_{j=1}^{21} C^{D2} V_j + \sum_{k=1}^{5} C^{S} L_k + \sum_{l=1}^{4} C_{l}^{W} \tag{5}$$

$$\text{FWMC TP} \leq 40\ \mu g/L \tag{6}$$

Subjected to the constraint:

Where $i1$ = land use type, n = BMP type (I, II, III), $C_{i1,n}^{B}$ = unit cost of BMP ($/kg/acre/year) type n and for land use $i1$, $L_{i1,n}$ = TPL reduction (kg) from land use $i1$ for n types of BMPs, $A_{i1,n}$ = area (acres) for land use $i1$ and n types of BMPs, $i2$ = land use type for DWM, m = DWM type (LRD, HRD), $C_{i2,m}^{D1}$ = unit cost of DWM ($/kg/acre/year) for DWM type m and land use $i2$, $L_{i2,m}$ = TPL reduction (kg) from land use $i2$ for DWM type m, $A_{i2,m}$ = area (acres) for land use $i2$ for DWM type m, j = DWP project index (1 to 21), C^{D2} = unit cost of DWP ($/acre-ft/year), V_j = runoff storage volume (acre-ft) for DWP project j, k = STA project index (1 to 5), C^{S} = unit cost of STA ($/kg of TP removed/year), L_k = TP Load reduction (kg) by STA project k, l = WR project index (1 to 4), and C_{l}^{W} = cost of WR project l ($/year). Since project durations assumed in unit cost estimations differed from one another, all costs were converted to 2014 dollars using the Consumer Price Index (CPI) from the Bureau of Labor Statistics and were discounted using a 4% real discount rate.

The effectiveness of BMP implementation and hence the cost is spatial in nature (due to land use and soil variability) and was incorporated into the cost estimation model by quantifying the TPL reduction from individual land parcels in WAM. The unit costs for BMPs (Table 3) were based on [54]. The average storage-based payment rate for projects on private ranchlands were used to determine the costs of DWPs [58]. Considering the uncertainties in potential storage estimates [65,66] and the fact that the LRD and HRD projects modeled in this work were aimed at TPL reduction assessment instead of runoff storage, stormwater R/D BMP unit costs were used for LRD and HRD cost evaluation. WR project costs were adopted from [67] preliminary cost estimates. The determination of the STA costs was based on the costs of two STA projects in the TCNS basin completed in the late 2000s and reported in [68]. Land acquisition cost was updated to $15,000/acre based on more recent estimates of land cost [69]. Table 4 reports unit costs and other related assumptions for DWM, STA, and WR.

To reduce the number of combinations and permutations of the different phosphorus control strategies, a phased assessment approach was adopted in this study (Figure 2). First, WAM re-calibration was performed to establish the Base scenario. In Phase I, the three BMP scenarios—BMP1, BMP12, and BMP123—were assessed. In Phase II, the most cost-efficient scenario of Phase I was considered to build five new scenarios based on DWP and DWM projects. Very large cost differences between Phase I scenarios (details presented in Results) made this approach justifiable. In Phase III, one new scenario was formulated by combining the most cost-efficient Phase II scenario with WR projects. Each of these scenarios were evaluated with and without STAs. From the potential 100+ different combinations, this phased approach reduced the total investigated scenarios to 20, thus also reducing the model simulation resource requirements.

Table 3. Summary of unit costs associated with Best Management Practices [1].

Land Use	Project Type	Amortized Capital Cost ($/kg of TP removed/acre/year)	Amortized O&M Cost ($/kg of TP removed/acre/year)	Total Annual Cost [2] ($/kg of TP removed/acre/year)
Citrus, Fruit Orchards	Type I	0.0	0.0	0.0
	Type II	98.5	24.6	153.9
	Type III	147.7	36.9	230.8
Commercial, High Den. Res., Med. Den. Res., Low Den. Res., Multi-Dwelling, Industrial, Transportation Corridors	Type I	0.0	0.0	0.0
	Type II	5904.8	1476.2	9226.2
	Type III	2699.1	674.8	4217.3
Conifer	Type I	0.0	0.0	0.0
	Type II	1439.6	359.9	2249.4
	Type III	460.0	115.0	718.7
Dairies	Type I	9.1	2.3	14.2
	Type II	409.5	102.4	639.8
	Type III	141.2	35.3	220.7
Field Crops	Type I	16.8	4.2	26.3
	Type II	35.0	8.7	54.7
	Type III	70.0	17.5	109.3
Improved Pasture, Intensive Pasture	Type I	57.0	14.3	89.1
	Type II	116.6	29.2	182.2
	Type III	129.6	32.4	202.5
Ornamental Nurseries, Tree Nurseries	Type I	3.9	1.0	6.1
	Type II	60.9	15.2	95.2
	Type III	89.4	22.4	139.7
Row Crops	Type I	2.6	0.6	4.0
	Type II	45.4	11.3	70.9
	Type III	58.3	14.6	91.1
Sod Farms	Type I	5.2	1.3	8.1
	Type II	66.1	16.5	103.3
	Type III	108.8	27.2	170.1
Sugarcane	Type I	0.0	0.0	0.0
	Type II	308.4	77.1	481.9
	Type III	348.6	87.1	544.6
Unimproved Pasture	Type I	27.2	6.8	42.5
	Type II	72.6	18.1	113.4
	Type III	106.3	26.6	166.0
Woodland Pasture	Type I	84.2	21.1	131.6
	Type II	281.2	70.3	439.3
	Type III	241.0	60.3	376.6
Managed Landscape	Type I	5.2	1.3	8.1
	Type II	66.1	16.5	103.3
	Type III	108.8	27.2	170.1

[1]—BMP unit costs are based on [54] Letter Report. Total project cost (2014 present value) is calculated assuming 4% discount rate and 20-year project period. [2]—Total annual cost includes contingency cost (25% of Initial and O&M costs).

Table 4. Summary of costs associated with Disperse Water Management (DWM), Stormwater Treatment Areas (STA), and Wetland Restoration (WR).

Restoration Strategy	Land Use	Project Type	Amortized Capital Cost	Amortized O&M Cost	Total Annual Cost [1]
DWM	Improved and Intensive Pasture	LRD, HRD [2]	168.4	42.1	263.2
	Unimproved Pasture	LRD, HRD	268.2	67.1	419.1
	Woodland Pasture	LRD, HRD	338.2	84.5	528.4
	N.A.	DWP [3]			297.9
STA [4]			448.4	112.1	700.5
WR [5]					3265.3

[1]—Total Annual cost includes contingency. [2]—LRD and HRD unit costs ($/kg of TP removed/acre/year) are P reduction based costs, estimated based on R/D BMP in [54] Letter Report. Assumed project duration = 20 years, contingency = 25%, and discount rate = 4%. [3]—DWP unit costs ($/acre-ft of runoff reduced/year) are based on payment rates for existing Disperse Water Management Projects on public ranchlands [58]. Assumed project duration = 10 years, contingency = 25%, and discount rate = 4%. [4]—STA unit costs ($/kg of TP removed/year) are based on actual Taylor Creek and Nubbin Slough STA project costs (updated for land acquisition). Assumed project duration = 50 years, contingency = 25%, and discount rate = 4%. [5]—WR unit costs ($/acre/year) are based on estimates presented in Lake Okeechobee Watershed Restoration Project [67]. Assumed project duration = 20 years, discount rate = 4%. Since [67] estimates included contingency costs, no further contingency was added.

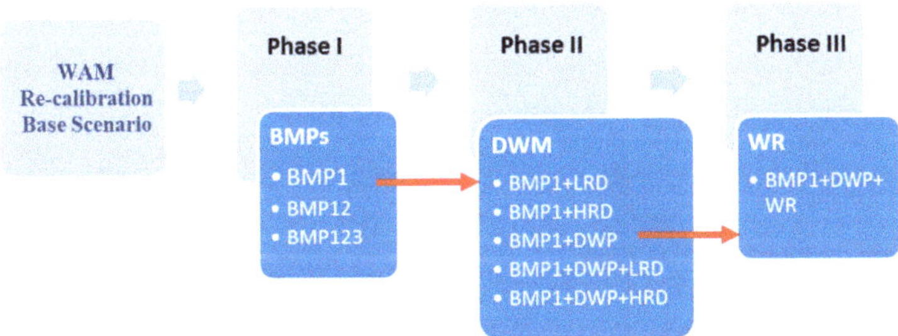

Figure 2. Schematic showing restoration scenario formulation. Note that each of the scenarios presented in this schematic was also evaluated for STA size requirement.

3. Results

3.1. Watershed Modeling and TP Reduction Strategies

3.1.1. Re-calibration

A summary of the NLO basin re-calibration GOF results is presented in Table 5. NSE values for monthly flows varied between 0.33 (SR78) to 0.94 (S65) with a mean NSE of 0.61 (arithmetic average of individual NSEs). The NSE values for monthly TPLs ranged between 0.26 (S131) and 0.81 (S133) with a mean NSE of 0.54. PBIAS for monthly flows varied between −8% (S129) and +11% (S191) (mean PBIAS = +0.5%), while monthly TPLs PBIAS was in the range −7% (S65) to +19% (S191) with a mean PBIAS of +6%. If we were to classify WAM simulation performance based on the recommendations of [44,45] for acceptable modeling results (NSE Flow > 0.5, NSE TPL > 0.35; abs(PBIAS Flow) < 15%, abs(PBIAS TPL) < 30%), three basins (S131, FEC_L61W, and UK) would have simulated flows with unsatisfactory NSEs, and two basins (S131 and L49) would have simulated TPLs with unsatisfactory NSEs. In all cases, the corresponding PBIAS values were well within the acceptable limit indicating that WAM was not able to satisfactorily capture flow or TPL dynamics at those locations but was able to simulate the average system behavior over the period of study. Since the purpose of this study was to assess the average TPL reductions under different scenarios, usage of the re-calibrated WAM setups were justified for this exploratory feasibility study.

Table 5. Re-calibration summary of northern Lake Okeechobee watershed basins.

Basin	Area (km²)	Re-Calibration Location	Monthly Flow NSE	Monthly Flow PBIAS	Monthly TPL NSE	Monthly TPL PBIAS
S131	28.9	S131	0.49	7%	0.26	10%
L48	84.0	S127	0.58	−3%	0.45	12%
L49	48.9	S129	0.56	−8%	0.32	8%
S133	103.8	S133	0.84	−1%	0.81	9%
S135	73.2	S135	0.67	−7%	0.58	8%
FEC_L61W	1205.0	SR78	0.33	7%	0.49	4%
Istokpoga	1578.6	S68	0.54	5%	not enough data	
Taylor Creek-Nubbin Slough	487.0	S191	0.79	11%	0.78	19%
Upper Kissimmee	4162.8	S65	0.36	−4%	0.48	−7%
Lower Kissimmee	2810.9	C38	0.94	−2%	0.68	−6%

3.1.2. WAM Results for the Base and Restoration Scenarios

The re-calibrated Base scenario simulation results indicated that the NLO basins contributed an annual average of 2644×10^6 m^3 of flows and 428.6 mtons of TPL to LO during 1998–2007 (Figure 3a,b). The two largest NLO basins (UK and LK) contributed 40% and 26% of the total water inflows to LO, respectively, while their TPL contributions were 18% and 37%, respectively. The TCNS basin contributed 4% of the total flows to LO and was responsible for 19% of the TPL. These results are consistent with the monitoring data and with previous modeling efforts showing that TCNS, S154, S65E, S65D, and Indian Prairie basins (the latter four are part of LK in this study) are the highest TP contributors to LO in terms of load per unit area [15,70]. Figure 4 summarizes flows, TPLs, and FWMC for the Base and restoration scenarios, pre-STA values. The Base scenario inter-annual variability at LO inflows was large with flows ranging from 749×10^6 m^3 to 4436×10^6 m^3 (Figure 4a), TPLs ranging from 95 mtons to 675 mtons (Figure 4b), and FWMC ranging from 124 µg/L to 286 µg/L (Figure 4c) with an average of 173 µg/L. The reduction in flows and TPLs (%) when compared to the Base conditions are also presented on the secondary vertical axes of Figure 4a,b.

Under the restoration scenarios, flows ranged between 2644×10^6 m^3 (BMP1) and 2515×10^6 m^3 (BMP1+DWP+HDR) indicating that flows reduced by less than 5% relative to the Base scenario (Figure 4a). The BMP123 scenario resulted in the highest TPL reduction (39.3%) with a TPL contribution of 260 mtons to LO (Figure 4b). The second and third best scenarios in terms of TPL reduction were BMP1+DWP+HRD and BMP1+HRD with an average annual TPL of 307 mtons (28.3% reduction) and 312 mtons (27.3% reduction), respectively. TPL reduction for the other six scenarios ranged between 11% and 22.3%. Under the Base scenario, the annual average inflow TP FWMC to LO was 173 µg/L, 4.3 times the target value of 40 µg/L (Figure 4c). Under the restoration scenarios, changes in FWMCs followed the reduction in TPLs. BMP123 contributed inflows to LO at 105 µg/L (2.6 times the target), the lowest among all scenarios. Under this scenario, the TP target concentration of 40 µg/L could not be achieved even for a single year. For the other scenarios, annual average FWMC ranged between 130 and 154 µg/L.

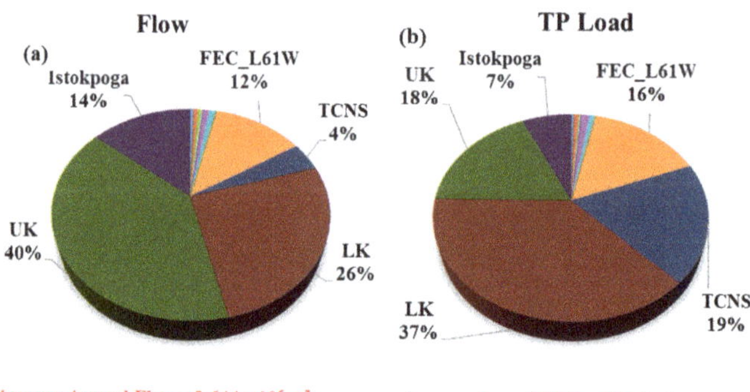

Figure 3. Simulated Average Annual (**a**) Flows and (**b**) Total Phosphorus Loads and their distributions from the North of Lake Okeechobee basins (1998–2007).

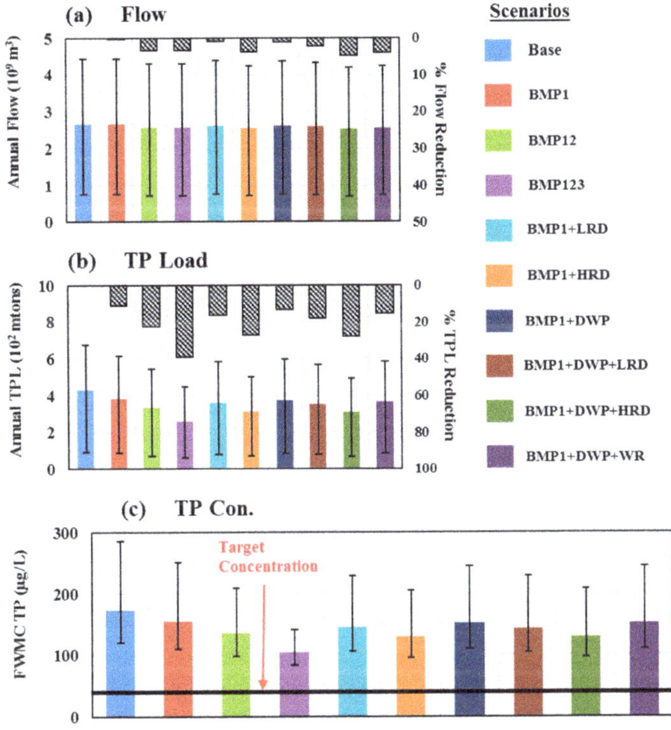

Figure 4. Summary of pre-STA (**a**) Average annual flows and flow reduction (%) when compared to the Base, (**b**) Average annual Total Phosphorus Loads (TPL) and TPL reduction (%) when compared to the Base, and (**c**) Average annual flow weighted mean TP concentration of water from northern Lake Okeechobee (NLO) basins. Error bars indicate the modeled minimum to maximum annual values during 1998–2007. Shaded bars represent % reduction.

3.1.3. STA Sizing

WAM results indicated that restoration scenarios using BMPs, DWM, and WR were not sufficient and STAs are necessary to achieve the target inflow TP concentration for LO. The total STA size requirement ranged from 142 km^2 for the BMP123 scenario and 229 km^2 for the Base scenario with the corresponding average annual TPL reduction ranging from 155 and 323 mtons, respectively (Figure 5). The other restoration scenarios required STAs within a relatively narrower area range of 172 to 206 km^2. The STA size requirement in the NLO region is comparable to the currently operating STAs south of LO covering an area of approximately 227 km^2. Figure 5 also indicated a good linear relationship ($R^2 = 0.99$) between the average annual TPL removal and STA sizes. The same figure also shows that under the BMP123 scenario, STAs contributed 48% to the TPL removal requirement while the relative STA contribution was more than 60% for all other scenarios, indicating the importance of the STA technology in LO water quality restoration.

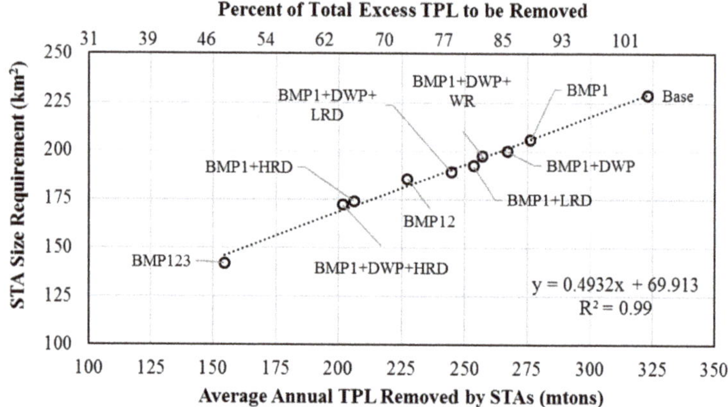

Figure 5. Summary of total STA size requirements for NLO basins to achieve long term annual average TP FWMC of 40 µg/L at Lake Okeechobee inflow. 'x' in the regression equation is average annual TPL removed by STAs in mtons, while 'y' is the total STA size in km². The total excess TPL to be removed is 323 mtons (average annual).

3.2. Assessment of Cost-Effectiveness

The schematic of this restoration scenario-based project cost-effectiveness assessment is shown in Figure 6a. The only constraint used for minimizing the costs was the attainment of the LO inflow TP target of 40 µg/L, making the area to the left of the vertical line the only "feasible region" (shaded area). In other words, any scenario laying to the right side of the constraint line is an unacceptable combination of restoration strategies. As previously noted, scenarios based on combining BMPs, DWMs, and WR alone could not reduce TP concentrations to or below 40 µg/L. Accordingly, all these scenarios were placed in the non-feasible region (Figure 6b), while the scenarios with STAs fell on the constraint line to form a set of feasible solutions. However, the total costs varied substantially from one scenario to another. The highest total costs (2014 present value) were estimated around US $7.6 and $10.3 billion for BMP12+STA and BMP123+STA, respectively. The remaining eight scenarios with STAs had total costs varying in a relatively narrow range (US $4.26 to 4.75 billion), forming a set of cost-effective feasible solutions. The high pre-STA costs of BMP12 and BMP123 implementation (~US $4.1 and $7.9 billion, respectively) justified the phased approach adopted in this work.

The BMP1+DWP+STA was determined as the most cost-efficient scenario among the ones investigated in this study with a total cost of US $4.26 billion. The cost breakdown (Figure 6c and Table 6) indicated that type I BMPs, DWP (existing dispersed water management projects), and STAs (200 km²) were responsible for 4.8% (US $0.2 billion), 0.4% (US $17 million), and 94.8% (US $4.04 billion) of the total project cost, respectively. The corresponding TPL removal contributions of those restoration strategies were 14.5% (46.7 mtons), 2.9% (9.3 mtons), and 82.6% (267 mtons), respectively. For all other BMP-based feasible scenarios, type I BMPs contributed around 14.5% to the required TPL removal while the associated cost was relatively lower (4.0 to 4.8%). The relative TPL removal and cost of DWM varied substantially among scenarios (2.9 to 23.1% and 0.4 to 30.2%, respectively) depending on the type of DWM. The relatively high cost (65.6 to 100%) and TPL removal contribution (62.5 to 100%) of STAs to the overall project indicate the importance of accurately estimating the STA size and the associated unit cost of TP removal.

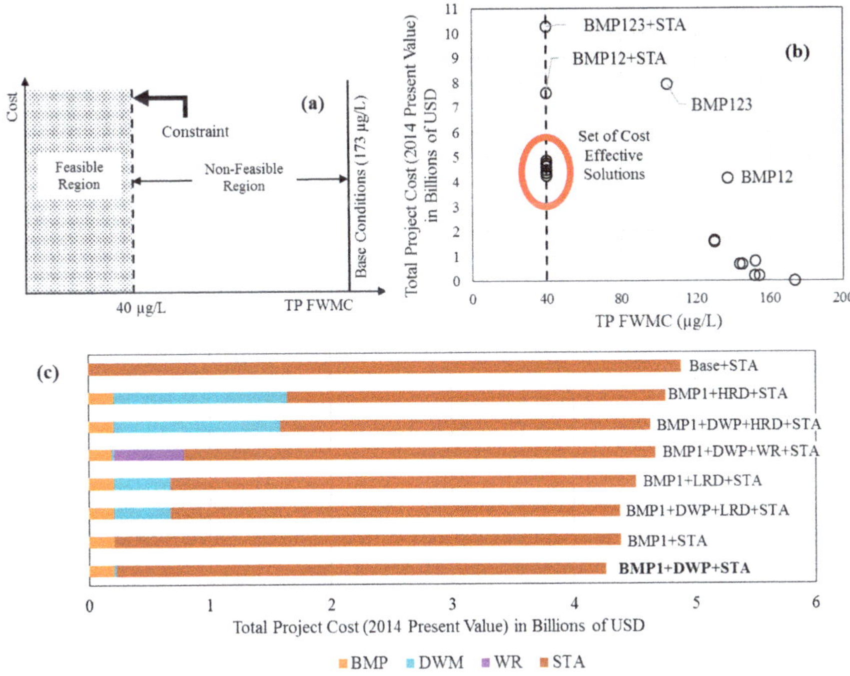

Figure 6. (**a**) Schematic of cost minimization assessment, (**b**) Summary of NLO restoration scenarios' cost feasibility analysis, and (**c**) Cost break-up for the set of cost-effective feasible restoration scenarios.

Table 6. Relative contributions of restoration strategies to TPL removal and Total Project Cost.

	% Total TPL Reduced				% Total Project Cost			
	BMPs	DWM	WR	STAs	BMPs	DWM	WR	STAs
BMP1+DWP+STA	14.5	2.9	0.0	82.7	4.8	0.4	0.0	94.8
BMP1+STA	14.5	0.0	0.0	85.5	4.6	0.0	0.0	95.4
BMP1+DWP+LRD+STA	14.5	9.8	0.0	75.8	4.6	10.8	0.0	84.6
BMP1+LRD+STA	14.5	7.1	0.0	78.5	4.5	10.4	0.0	85.1
BMP1+DWP+WR+STA	14.5	2.9	3.0	79.6	4.0	0.4	12.4	83.2
BMP1+DWP+HRD+STA	14.5	23.1	0.0	62.5	4.4	29.7	0.0	66.0
BMP1+HRD+STA	14.5	21.7	0.0	63.8	4.3	30.2	0.0	65.6
Base+STA	0.0	0.0	0.0	100.0	0.0	0.0	0.0	100.0

It is worth noting that there is a cost estimation uncertainty on all the projects listed above depending on the cost methodology adopted. For the DWM cost calculation, there is an additional uncertainty associated with the runoff reduction volume calculations using PWRM for DWP, for example, [71] raised concerns about the accuracy of PWRM. Based on modeling studies, albeit restricted to only a few wetland-pasture study sites in south Florida, [72] and [66] showed that storage estimates are highly sensitive to wetland evapotranspiration and topographic data. These parameters are not fully or adequately included in the inputs of PWRM model. Consequently, the actual payment rates for water storage do not use accurate storage estimates. Goswami and Shukla [65] concluded that the actual storage potential in NLO basins through dispersed storage (i.e., wetland restoration on ranchlands) is only about 14% of the required storage estimated by the state agencies. Runoff reduction achieved under BMP1+LRD scenarios in the present study was similar to the storage projections in [65]. In this work, DWM (i.e., LRD and HRD) was explored as phosphorus control strategy rather than using it for creating water storage. Hence, the corresponding cost calculations were based on TPL reduction potential i.e., stormwater R/D BMP rates. For a large-scale implementation of DWM projects in the NLO basin, a detailed feasibility study of DWM storage efficiency, nutrient removal, and costs are required.

The most cost-effective scenario identified in this work consisted of (a) implementation of nutrient/fertilizer management BMPs throughout the NLO region, (b) continuation of existing NLO DWM projects, and (c) construction of 200 km^2 STAs for a total project cost of ~US \$4.26 billion (2014 present value). Taking into account the Lakeside Ranch STA with an effective treatment area of 7 km^2, there is a need for 193 km^2 of STAs with a total project cost of \$4.15 billion for future STAs needs + BMP type I + existing DWM projects.

4. Discussion

4.1. Current State Action Plan and the Present Study Implications

The 2014 BMAP is a restoration roadmap for water quality improvements in the NLO basin [73]. It details projects being (since 2009) or to be implemented to improve LO water quality and develops a monitoring strategy to track improvements. The plan is to follow a phased implementation of the strategies to achieve the LO TMDL. During the first phase of the BMAP, the projects are spread over a ten-year timeframe with a projected TPL reduction of 148 mtons (around 43% of the total necessary TP reduction to achieve the TMDL). Around 59 mtons of the expected TPL reductions will come from implementation of type I (31 mtons TPL reduction) and type II (28 mtons TPL reduction) BMPs. TPL reduction attributed to BMP implementation before 2009 was not accounted for in the BMAP. The other BMAP projects listed were DWM, WR, and STAs providing 19 mtons, 24 mtons, and 30 mtons of additional TPL reductions, respectively. The remaining 16 mtons of TPL removal would be through other treatment technologies. In this first BMAP phase, 40% of the expected TPL reduction will be originating from implementing BMPs with no long-term vision of how the Lake Okeechobee TMDL would be achieved.

Xu et al. [74] developed a procedure to determine the economic efficiency of conservation practices to reduce TPLs from the Sandusky watershed to Lake Erie and concluded that relying on traditional conservation practices alone is not sufficient in meeting the phosphorus targets. Gaddis et al. [75] investigated the TPLs from the Stevens Brook watershed and demonstrated that the maximum achievable annual phosphorus load reduction for the watershed is 46%, a significant step (but not enough) toward achieving the TMDL. These results are consistent with the findings of this paper demonstrating that STAs (above and beyond standard BMPs) would need to be implemented to reach the target. There is a need to establish a NLO procedure to assess the effectiveness and cost efficiency of nutrient control strategies using a holistic temporal and spatial approach.

Based on the current technologies (BMPs, DWM, WR and STAs), this paper assessed the most cost-effective plan to reach the LO target. The total cost of the most cost-effective plan would be around US \$4.26 billion. This total estimate is substantially greater than the state of Florida agencies funding requests and expenditures in this watershed. Since 2000 to date, the State spent a total of around US \$200 to \$300 million [15,76]. In the BMAP document and excluding the Kissimmee River restoration, the total cost for reducing the TPL to LO by 43% was around \$151 million with many features having unknown costs [73]. Based on the present study, the total cost for full implementation of BMP1 and BMP12 are around US \$203 million and US \$4 billion, respectively. A realistic estimate of the long-term total costs for reaching the LO TMDL needs to be developed and peer reviewed. This should be done along with an evaluation of the return on investment by quantifying the ecosystem services that a restored watershed would bring to the economy [77]. The high cost for restoring LO should not halt or delay restoration progress. To the contrary, state agencies should incentivize innovative practices to decrease/offset the cost of local and regional efforts to remove TP from surface waters, e.g., bioenergy production and novel landscape management [78–80], without delaying the immediate implementation of TP removal strategies delineated above. Finding knowledge gaps and incentivizing research and development in this area is critical to surpass the current level of treatment and lower capital and operational costs.

4.2. Challenges

An important challenge that was not accounted for in this study is the problem of legacy phosphorus in the NLO basins [81] directly impacting the time lag for observing any decreased water quality trend in the watershed. Hence the timeframe for improved inflow TP concentrations through the implementation of restoration strategies (with the exception of STAs) should not be interpreted as 10 years. Reddy et al. [82] estimated that phosphorus built-up in LO basin soils is enough to sustain ~500 mtons/year of TPL to LO for at least 22 to 55 years even after complete curtailment of all other phosphorus input sources. Phosphorus enriched sediments within LO are also likely to sustain high TP concentrations in the lake for 1 to 3 decades beyond the basin legacy phosphorus removal timeline. Similarly, other freshwater bodies in the NLO basin that are currently still assimilating phosphorus may become sources of phosphorus once their sediments reach a phosphorus sorption capacity through sustained high inflow TPLs or would buffer outflow TP concentrations from declining until an equilibrium is reached with the lowered inflow loads [55]. For example, [83] showed that the Kissimmee Chain of Lakes and Lake Istokpoga would become sources of TP in 7–25 years under 0 to 25% inflow TPL reductions. This means that restoration success is mainly dependent on STAs during the first years of any water quality project implementation in the NLO basin and monitoring for changes in legacy phosphorus in watershed soils is essential to assess the success of BMPs, DWMs, and WR implementation [84]. At the same time, cost-effective, sustainable, and region-specific in-lake treatment options also need to be explored in order to reduce bio-availability of phosphorus to control algae blooms and transport to downstream locations [3,85].

Sustained and long-term efforts on the scale of decades are generally required for restoring impaired waterbodies [86]. Another important challenge not addressed in the present study is the effect of climate change on nutrient loadings. Even if the currently implemented watershed control strategies can reach the TMDLs under the current climate, it is predicted that non-point source pollution losses will increase due to increased precipitation and high intensity rainfall events. This will generate excess amounts of nutrients that will require additional conservation measures or new technologies to reach the phosphorus targets [87–90]. Xu et al. [91] evaluated impacts of climatic change on the economic efficiency of watershed control strategies for phosphorus reduction and reported that the performance of strategies optimized for the current climate was degraded significantly under projected future climate conditions. This is an important factor to consider when developing long term plans in watersheds with known TP legacy problems. The future climate projections remain uncertain in Florida due to an inaccurate representation of local phenomenon in global climate models [92]. There is a consensus that air temperature will increase by 1.5 °C by 2060 [93]. However, the less accurate precipitation projections are indicating that annual rainfall in central Florida will remain unchanged with a substantial change in seasonality [94]. This could trigger changes in cropping patterns, agricultural practices, and future land uses. For the NLO basin, a robust plan accounting for future climate projections (and allied changes) exacerbating the nutrient pollution problems can be tackled if and when accurate climate projections become available.

5. Conclusions

In this study, a heuristic phased assessment of restoration strategies scaled to a basin-wide implementation was conducted to determine the cost-effective restoration alternative to achieve the LO TP TMDL target of 40 μg/L for inflows from the NLO basins. It was found that restoration alternatives based only on BMPs, DWM, and WR, irrespective of the degree of implementation, will not reduce TPLs to the LO target unless combined with STAs. The most cost-effective scenario identified in this work consisted of (a) implementation of nutrient/fertilizer management BMPs throughout the NLO region, (b) continuation of existing NLO DWM projects, and (c) construction of 200 km^2 STAs for a total project cost of ~US \$4.26 billion. Provided that the Lakeside Ranch STA performs as designed, the need for STAs would reduce to 193 km^2 and the project cost would reduce to \$4.15 billion. Seven other scenarios (with STAs) had higher cost estimates but within 15% of the cost-effective scenario. The findings of this study provide a preliminary holistic assessment of the level of restoration projects and associated costs required to achieve the LO target. In order to decrease cost and increase efficiency of regional efforts to remove TP from surface waters, state agencies should incentivize research and development to find and develop innovative practices and a technological breakthrough to control large-scale phosphorus contamination.

Supplementary Materials: The following are available online at http://www.mdpi.com/2073-4441/11/2/327/s1, Figure S1: Historical Palmer Drought Severity Index for central-south Florida that includes Lake Okeechobee, the Everglades Agricultural Area, and the Everglades, Figure S2: Maps of land use and source level phosphorus loading, Figure S3: Phosphorus hot-spots in northern Lake Okeechobee basins under Base (1998–2007) conditions, Table S1: Land use wise source level phosphorus loading summary for the northern Lake Okeechobee basins, Table S2: Summary of existing dispersed water management projects modeled under the DWP scenario.

Author Contributions: Y.K. developed the methodology, reated WAM setups of restoration scenarios, performed actual model simulations, analysis of results, and manuscript writing (original draft and manuscript revisions). G.M.N. conceptualized this study and also contributing directly to writing (original draft, review and editing). G.A.S. worked on cost estimation analysis and writing (reviewing and editing). C.J.M. and R.P. contributed verification of results as well as to writing (reviewing and editing). T.V.L. work on project conceptualizing and writing (review and editing).

Funding: This research received no external funding.

Acknowledgments: Any opinions, findings, conclusions, or recommendations expressed in the material are those of the author(s) and do not necessarily reflect the views of the Everglades Foundation.

Conflicts of Interest: The authors declare no conflict of interest.

References

1. Sondergaard, M.; Jeppesen, E. Anthropogenic impacts on lake and stream ecosystems, and approaches to restoration. *J. Appl. Ecol.* **2007**, *44*, 1089–1094. [CrossRef]
2. Pan, G.; Yang, B.; Wang, D.; Chen, H.; Tian, B.; Zhang, M.; Yuan, X.; Chen, J. In-lake algal bloom removal and submerged vegetation restoration using modified local soils. *Ecol. Eng.* **2011**, *37*, 302–308. [CrossRef]
3. Douglas, G.B.; Hamilton, D.P.; Robb, M.S.; Pan, G.; Spears, B.M.; Lurling, M. Guiding principles for the development and application of solid-phase phosphorus adsorbents for freshwater ecosystems. *Aquat. Ecol.* **2016**, *50*, 385–405. [CrossRef]
4. Pretty, J.N.; Mason, C.F.; Nedwell, D.B.; Hine, R.E.; Leaf, S.; Dils, R. Environmental costs of freshwater eutrophication in England and Wales. *Environ. Sci. Technol.* **2003**, *37*, 201–208. [CrossRef] [PubMed]
5. Dodds, W.K.; Bouska, W.W.; Eitzmann, J.L.; Pilger, T.J.; Pitts, K.L.; Riley, A.J.; Schloesser, J.T.; Thornbrugh, D.J. Eutrophication of US Freshwaters: Analysis of Potential Economic Damages. *Environ. Sci. Technol.* **2009**, *43*, 12–19. [CrossRef] [PubMed]
6. Scavia, D.; Kalcic, M.; Muenich, R.L.; Read, J.; Aloysius, N.; Bertani, I.; Boles, C.; Confesor, R.; DePinto, J.; Gildow, M.; et al. Multiple models guide strategies for agricultural nutrient reductions. *Front. Ecol. Environ.* **2017**, *15*, 126–132. [CrossRef]
7. Merriman, K.R.; Daggupati, P.; Srinivasan, R.; Toussant, C.; Russell, A.M.; Hayhurst, B. Assessing the Impact of site-specific BMPs using a spatially explicit, field-scale SWAT model with edge-of-field and tile hydrology and water-quality data in the Eagle Creek Watershed, Ohio. *Water-Sui* **2018**, *10*, 1299. [CrossRef]
8. Motsinger, J.; Kalita, P.; Bhattarai, R. Analysis of best management practices implementation on water quality using the soil and water assessment tool. *Water-Sui* **2016**, *8*, 145. [CrossRef]
9. Shrestha, N.K.; Allataifeh, N.; Rudra, R.; Daggupati, P.; Goel, P.K.; Dickinson, W.T. Identifying threshold storm events and quantifying potential impacts of climate change on sediment yield in a small upland agricultural watershed of ontario. *Hydrol. Process.* **2019**. [CrossRef]
10. Collick, A.S.; Fuka, D.R.; Kleinman, P.J.A.; Buda, A.R.; Weld, J.L.; White, M.J.; Veith, T.L.; Bryant, R.B.; Bolster, C.H.; Easton, Z.M. Predicting phosphorus dynamics in complex terrains using a variable source area hydrology model. *Hydrol. Process.* **2015**, *29*, 588–601. [CrossRef]

11. Geng, R.Z.; Yin, P.H.; Gong, Q.R.; Wang, X.Y.; Sharpley, A.N. BMP optimization to improve the economic viability of farms in the upper watershed of Miyun Reservoir, Beijing, China. *Water-Sui* **2017**, *9*, 633. [CrossRef]

12. Gren, I.M.; Soderqvist, T.; Wulff, F. Nutrient reductions to the Baltic Sea: Ecology, costs and benefits. *J. Environ. Manag.* **1997**, *51*, 123–143. [CrossRef]

13. Valcu-Lisman, A.M.; Gassman, P.W.; Arritt, R.; Campbell, T.; Herzmann, D.E. Cost-effectiveness of reverse auctions for watershed nutrient reductions in the presence of climate variability: An empirical approach for the Boone River watershed. *J. Soil Water Conserv.* **2017**, *72*, 280–295. [CrossRef]

14. Havens, K.E.; James, R.T. The phosphorus mass balance of Lake Okeechobee, Florida: Implications for eutrophication management. *Lake Reserv. Manag.* **2005**, *21*, 139–148. [CrossRef]

15. SFWMD. *South Florida Environmental Report*; South Florida Water Management District: West Palm Beach, FL, USA, 2018.

16. Havens, K.E. Relationships of Annual Chlorophyll a Means, Maxima, and Algal Bloom Frequencies in a Shallow Eutrophic Lake (Lake Okeechobee, Florida, USA) AU - Havens, Karl E. *Lake Reserv. Manag.* **1994**, *10*, 133–136. [CrossRef]

17. Havens, K.E.; Walker, W.W. Development of a total phosphorus concentration goal in the TMDL process for Lake Okeechobee, Florida (USA). *Lake Reserv. Manag.* **2002**, *18*, 227–238. [CrossRef]

18. Florida Administrative Code. *Surface Water Improvement and Management Act*; Florida Administrative Code: Tallahassee, FL, USA, 1990.

19. Reckhow, K.H.; Korfmacher, K.S.; Aumen, N.G. Decision analysis to guide lake okeechobee research planning. *Lake Reserv. Manag.* **1997**, *13*, 49–56. [CrossRef]

20. Goldstein, A.L.; Ritter, G.J. A Performance-based regulatory program for phosphorus control to prevent the accelerated eutrophication of Lake-Okeechobee, Florida. *Water Sci. Technol.* **1993**, *28*, 13–26. [CrossRef]

21. Florida Department of Environmental Protection. *Total Maximum Daily Load for Total Phosphorus Lake Okeechobee, Florida*; FDEP: Tallahassee, FL, USA, 2001.

22. SFWMD; FDEP; FDACS. Lake Okeechobee Watershed Construction Project: Phase II Technical Plan. Available online: https://www.sfwmd.gov/sites/default/files/documents/lakeo_watershed_construction%20proj_phase_ii_tech_plan.pdf (accessed on 10 September 2016).

23. Florida Department of Environmental Protection. *2016 Progress Report for the Lake Okeechobee Basin Management Action Plan-Final*; Florida Department of Environmental Protection: Tallahassee, FL, USA, 2017.

24. South Florida Water Management District. *South Florida Environmental Report*; South Florida Water Management District: West Palm Beach, FL, USA, 2005.

25. South Florida Water Management District. *South Florida Environmental Report*; South Florida Water Management District: West Palm Beach, FL, USA, 2008.

26. Chebud, Y.; Naja, G.M.; Rivero, R. Phosphorus run-off assessment in a watershed. *J. Environ. Monit.* **2011**, *13*, 66–73. [CrossRef]

27. Jawitz, J.W.; Mitchell, J. Temporal inequality in catchment discharge and solute export. *Water Resour. Res.* **2011**, *47*. [CrossRef]

28. Flaig, E.G.; Reddy, K.R. Fate of phosphorus in the Lake Okeechobee watershed, Florida, USA: Overview and recommendations. *Ecol. Eng.* **1995**, *5*, 127–142. [CrossRef]

29. Bottcher, A.B.; Whiteley, B.J.; James, A.I.; Hiscock, J.G. Watershed Assessment Model (Wam): Model Use, Calibration, and Validation. *T. Asabe* **2012**, *55*, 1367–1383. [CrossRef]

30. Walker, W.W.; Kadlec, R.H. Dynamic model for stormwater treatment areas- model version 2 documentation update. Available online: http://www.wwwalker.net/dmsta/index.htm (accessed on 10 May 2016).

31. Corrales, J.; Naja, G.M.; Bhat, M.G.; Miralles-Wilhelm, F. Modeling a phosphorus credit trading program in an agricultural watershed. *J. Environ. Manag.* **2014**, *143*, 162–172. [CrossRef] [PubMed]

32. Naja, M.; Childers, D.L.; Gaiser, E.E. Water quality implications of hydrologic restoration alternatives in the Florida Everglades, United States. *Restor. Ecol.* **2017**, *25*, S48–S58. [CrossRef]

33. South Florida Water Management District. *Restoration Strategies Regional Water Quality Plan*; South Florida Water Management District: West Palm Beach, FL, USA, 2012.

34. Khare, Y.P.; Naja, G.M.; Paudel, R.; Martinez, C.J. A watershed scale assessment of phosphorus remediation strategies for achieving water quality restoration targets in the western everglades. *Ecol. Eng.* **2019**, submitted for publication.

35. Khare, Y.P. *Hydrologic and Water Quality Model Evaluation with Global Sensitivity Analysis: Improvements and Applications*; University of Florida: Gainesville, FL, USA, 2014.

36. SWET, I. *EAAMOD Technical and User Manuals*; University of Florida: Gainesville, FL, USA, 2008.

37. SWET, I. Watershed Assessment Model (WAM) Documentation. Available online: http://www.swet.com/documentation/ (accessed on 5 May 2017).

38. Zhang, J.; Gornak, S.I. Evaluation of field-scale water quality models for the Lake Okeechobee regulatory program. *Appl. Eng. Agric.* **1999**, *15*, 441–447. [CrossRef]

39. Knisel, W.G. *GLEAMS: Groundwater Loading Effects of Agricultural Management Systems*; University of Georgia: Tifton, GA, USA, 1993.

40. Henningson, D.; Richardson, I.; Soil and Water Engineering Technology Inc. *WAM Enhancements and Application in the LAKE Okeechobee Watershed: Final Report (Contract# 3600001244)*; South Florida Water Management District: West Palm Beach, FL, USA, 2009.

41. He, Z.L.; Hiscock, J.G.; Merlin, A.; Hornung, L.; Liu, Y.L.; Zhang, J. Phosphorus budget and land use relationships for the Lake Okeechobee Watershed, Florida. *Ecol. Eng.* **2014**, *64*, 325–336. [CrossRef]

42. Corrales, J.; Naja, G.M.; Bhat, M.G.; Miralles-Wilhelm, F. Water quality trading opportunities in two sub-watersheds in the northern Lake Okeechobee watershed. *J. Environ. Manag.* **2017**, *196*, 544–559. [CrossRef]

43. Khare, Y.; Martinez, C.J.; Munoz-Carpena, R.; Bottcher, A.; James, A. Effective Global Sensitivity Analysis for High-Dimensional Hydrologic and Water Quality Models. *J. Hydrol. Eng.* **2019**, *24*. [CrossRef]

44. Moriasi, D.N.; Gitau, M.W.; Pai, N.; Daggupati, P. Hydrologic and Water Quality Models: Performance Measures and Evaluation Criteria. *Trans. ASABE* **2015**, *58*, 1763–1785.

45. American Society of Agricultural & Biological Engineers (ASABE). *Guidelines for Calibrating, Validating, and Evaluating Hydrologic and Water Quality (H/WQ) Models*; American Society of Agricultural & Biological Engineers: St. Joseph, MI, USA, 2017.

46. Walker, W.W.; Kadlec, R.H. Modeling phosphorus dynamics in everglades wetlands and stormwater treatment areas. *Crit. Rev. Env. Sci. Tec.* **2011**, *41*, 430–446. [CrossRef]

47. Chen, H.J.; Ivanoff, D.; Pietro, K. Long-term phosphorus removal in the Everglades stormwater treatment areas of South Florida in the United States. *Ecol. Eng.* **2015**, *79*, 158–168. [CrossRef]

48. Kadlec, R.H. Large constructed wetlands for phosphorus control: A review. *Water-Sui* **2016**, *8*, 243. [CrossRef]

49. Kaini, P.; Artita, K.; Nicklow, J.W. Optimizing structural best management practices using SWAT and genetic algorithm to improve water quality goals. *Water. Resour. Manag.* **2012**, *26*, 1827–1845. [CrossRef]

50. Kalcic, M.M.; Frankenberger, J.; Chaubey, I. Spatial optimization of six conservation practices using swat intile-drained agricultural watersheds. *J. Am. Water Resour. Assoc.* **2015**, *51*, 956–972. [CrossRef]

51. Rodriguez, H.G.; Popp, J.; Maringanti, C.; Chaubey, I. Selection and placement of best management practices used to reduce water quality degradation in Lincoln Lake watershed. *Water Resour. Res.* **2011**, *47*. [CrossRef]

52. Geng, R.Z.; Wang, X.Y.; Sharpley, A.N.; Meng, F.D. Spatially-distributed cost-effectiveness analysis framework to control phosphorus from agricultural diffuse pollution. *PloS ONE* **2015**, *10*. [CrossRef]

53. Arabi, M.; Govindaraju, R.S.; Hantush, M.M. Cost-effective allocation of watershed management practices using a genetic algorithm. *Water Resour. Res.* **2006**, *42*. [CrossRef]

54. South Florida Water Management District. *Phosphorous Reduction Performance and Implementation Costs under BMPs and Technologies in the Lake Okeechobee Protection Plan Area-Letter Report*; South Florida Water Management District: West Plam Beach, FL, USA, 2006.

55. South Florida Water Management District. *Nutrient Loading Rates, Reduction Factors and Implementation Costs Associated with BMPs and Technologies*; South Florida Water Management District: West Palm Beach, FL, USA, 2008.

56. Florida Department of Agriculture and Consumer Services. Agricultural Best Management Practices. Available online: https://www.freshfromflorida.com/Business-Services/Water/Agricultural-Best-Management-Practices (accessed on 21 June 2017).

57. Johns, G.M.; O'Dell, K. Benefit-cost analysis to develop the lake okeechobee protection plan. *Fla. Water Resour. J.* **2004**, *34*, 34–38.

58. Beirnes, J.T.; Bhagudas, J. *Audit of Dispersed Water Management Program*; South Florida Water Management District: West Palm Beach, FL, USA, 2014.

59. Dunne, E.J.; Smith, J.; Perkins, D.B.; Clark, M.W.; Jawitz, J.W.; Reddy, K.R. Phosphorus storages in historically isolated wetland ecosystems and surrounding pasture uplands. *Ecol. Eng.* **2007**, *31*, 16–28. [CrossRef]

60. South Florida Water Management District. *South Florida Environmental Report, Chapter 8: Lake Okeechobee Watershed Protection Program Annual Update*; South Florida Water Management District: West Palm Beach, FL, USA, 2016.
61. South Florida Water Management District. *User Manual for the Potential Water Retention Model (PWRM)*; South Florida Water Management District: West Palm Beach, FL, USA, 2012.
62. South Florida Water Management District. *South Florida Environmental Report, Chapter 8A: Northern Everglades and Estuaries Protection Program-Annual Progress Report*; South Florida Water Management District: West Palm Beach, FL, USA, 2017.
63. United States Army Corps of Engineers; South Florida Water Management District. Lake Okeechobee Restoration Project–Project Delivery Team Meeting. 23 June 2017. Available online: http://www.saj.usace.army.mil/Missions/Environmental/Ecosystem-Restoration/Lake-Okeechobee-Watershed-Project/ (accessed on 15 November 2017).
64. Wetland Solutions, I. *Development of Design Criteria for Stormwater Treatment Areas (STAs) in the Northern Lake Okeechobee Watershed*; South Florida Water Management District: West Palm Beach, FL, USA, 2009.
65. Goswami, D.; Shukla, S. Effects of passive hydration on surface water and groundwater storages in drained ranchland wetlands in the Everglades Basin in Florida. *J. Irrig. Drain. Eng.* **2015**, *141*. [CrossRef]
66. Guzha, A.C.; Shukla, S. Effect of topographic data accuracy on water storage environmental service and associated hydrological attributes in South Florida. *J. Irrig. Drain. Eng.* **2012**, *138*, 651–661. [CrossRef]
67. United States Army Corps of Engineers; South Florida Water Management District. Lake Okeechobee Restoration Project–Project Delivery Team Meeting. 15 August 2017. Available online: http://www.saj.usace.army.mil/Missions/Environmental/Ecosystem-Restoration/Lake-Okeechobee-Watershed-Project/ (accessed on 15 November 2017).
68. Hazen, S. Compilation of Benefit and Costs of STA and Reservoir Projects in the South Florida Water Management District. Available online: http://www.fresp.org/pdfs/Compilation%20of%20STA%20and%20REZ%20Benefits%20Costs%20HandS%2011_2011.pdf (accessed on 24 January 2018).
69. Torres, R. *Lake Okeechobee Watershed Project (Budgetary Scoping) 2nd Round*; United States Army Corps of Engineers: Jacksonville, FL, USA, 2017.
70. Boggess, C.F.; Flaig, E.G.; Fluck, R.C. Phosphorus budget-basin relationships for Lake Okeechobee tributary basins. *Ecol. Eng.* **1995**, *5*, 143–162. [CrossRef]
71. Horne, C. *Mixed-use at the Landscape Scale: Integrating Agricultural and Water Management as a Case Study for Interdisciplinary Planning*; Massachusetts Institute of Technology: Cambridge, MA, USA, 2010.
72. Wu, C.L.; Shukla, S.; Shrestha, N.K. Evapotranspiration from drained wetlands with different hydrologic regimes: Drivers, modeling, and storage functions. *J. Hydrol.* **2016**, *538*, 416–428. [CrossRef]
73. Florida Department of Environmental Protection. *Lake Okeechobee Basin Management Action Plan -Final Report*; Florida Department of Environmental Protection: Tallahassee, FL, USA, 2014.
74. Xu, H.; Brown, D.G.; Moore, M.R.; Currie, W.S. Optimizing spatial land management to balance water quality and economic returns in a Lake Erie Watershed. *Ecol. Econ.* **2018**, *145*, 104–114. [CrossRef]
75. Gaddis, E.J.B.; Voinov, A.; Seppelt, R.; Rizzo, D.M. Spatial optimization of best management practices to attain water quality targets. *Water Resour. Manag.* **2014**, *28*, 1485–1499. [CrossRef]
76. South Florida Water Management District. *South Florida Environmental Report, Chapter 8: Lake Okeechobee Watershed Protection Program Annual and Three-Year Update*; South Florida Water Management District: West Palm Beach, FL, USA, 2014.
77. McCormick, D. Measuring the Economic Benefits of America's Everglades Restoration: An Economic Evaluation of Ecosystem Services Affiliated with the World's Largest Ecosystem. Available online: https://www.evergladesfoundation.org/wp-content/uploads/sites/2/2017/12/Report-Measuring-Economic-Benefits-Exec-Summary.pdf (accessed on 3 December 2017).
78. Xu, H.; Wu, M.; Ha, M.E. Recognizing economic value in multifunctional buffers in the lower Mississippi river basin. *Biofuels Bioprod. Biorefin.* **2019**, *13*, 55–73. [CrossRef]
79. Dale, V.H.; Kline, K.L.; Buford, M.A.; Volk, T.A.; Smith, C.T.; Stupak, I. Incorporating bioenergy into sustainable landscape designs. *Renew. Sustain. Energy Rev.* **2016**, *56*, 1158–1171. [CrossRef]
80. Pan, G.; Lyu, T.; Mortimer, R. Comment: Closing phosphorus cycle from natural waters: re-capturing phosphorus through an integrated water-energy-food strategy. *J. Environ. Sci.-China* **2018**, *65*, 375–376. [CrossRef]

81. Graham, W.; Angelo, M.; Frazer, T.K.; Frederick, P.C.; Havens, K.E.; Reddy, K.R. *Options to Reduce Options to Reduce Options to Reduce Options to Reduce Options to Reduce High Volume Freshwater Flows to the St. Lucie and Caloosahatchee Estuaries and Move Water from Lake Okeechobee to the Southern Everglades-An independent technical review*; University of Florida Water Institute: Gainesville, FL, USA, 2015.

82. Reddy, K.R.; Newman, S.; Osborne, T.Z.; White, J.R.; Fitz, H.C. Phosphorous cycling in the greater Everglades Ecosystem: Legacy phosphorous implications for management and restoration. *Crit. Rev. Environ. Sci. Technol.* **2011**, *41*, 149–186. [CrossRef]

83. Belmont, M.A.; White, J.R.; Reddy, K.R. Phosphorus sorption and potential phosphorus storage in sediments of Lake Istokpoga and the upper chain of Lakes, Florida, USA. *J. Environ. Qual.* **2009**, *38*, 987–996. [CrossRef] [PubMed]

84. Nair, V.D.; Clark, M.W.; Reddy, K.R. Evaluation of Legacy phosphorus storage and release from wetland soils. *J. Environ. Qual.* **2015**, *44*, 1956–1964. [CrossRef] [PubMed]

85. Xu, R.; Zhang, M.Y.; Mortimer, R.J.G.; Pan, G. Enhanced phosphorus locking by novel Lanthanum/Aluminum-Hydroxide Composite: Implications for eutrophication control. *Environ. Sci. Technol.* **2017**, *51*, 3418–3425. [CrossRef] [PubMed]

86. Osmond, D.; Meals, D.; Hoag, D.; Arabi, M.; Luloff, A.; Jennings, G.; McFarland, M.; Spooner, J.; Sharpley, A.; Line, D. Improving conservation practices programming to protect water quality in agricultural watersheds: Lessons learned from the National Institute of Food and Agriculture-Conservation Effects Assessment Project. *J. Soil Water Conserv.* **2012**, *67*, 122a–127a. [CrossRef]

87. Bosch, N.S.; Evans, M.A.; Scavia, D.; Allan, J.D. Interacting effects of climate change and agricultural BMPs on nutrient runoff entering Lake Erie. *J. Great Lakes Res.* **2014**, *40*, 581–589. [CrossRef]

88. Jayakody, P.; Parajuli, P.B.; Cathcart, T.P. Impacts of climate variability on water quality with best management practices in sub-tropical climate of USA. *Hydrol. Process.* **2014**, *28*, 5776–5790. [CrossRef]

89. Renkenberger, J.; Montas, H.; Leisnham, P.T.; Chanse, V.; Shirmohammadi, A.; Sadeghi, A.; Brubaker, K.; Rockler, A.; Hutson, T.; Lansing, D. Effectiveness of best management practices with changing climate in a Maryland Watershed. *Trans. ASABE* **2017**, *60*, 769–782. [CrossRef]

90. Xie, H.; Chen, L.; Shen, Z.Y. Assessment of agricultural best management practices using models: Current issues and future perspectives. *Water-Sui* **2015**, *7*, 1088–1108. [CrossRef]

91. Xu, H.; Brown, D.G.; Steiner, A.L. Sensitivity to climate change of land use and management patterns optimized for efficient mitigation of nutrient pollution. *Clim. Chang.* **2018**, *147*, 647–662. [CrossRef]

92. Misra, V.; Mishra, A.; Li, H.Q. The sensitivity of the regional coupled ocean-atmosphere simulations over the Intra-Americas seas to the prescribed bathymetry. *Dyn. Atmos. Oceans* **2016**, *76*, 29–51. [CrossRef]

93. Obeysekera, J.; Barnes, J.; Nungesser, M. Climate sensitivity runs and regional hydrologic modeling for predicting the response of the greater Florida Everglades Ecosystem to climate change. *Environ. Manag.* **2015**, *55*, 749–762. [CrossRef] [PubMed]

94. Kirtman, B.; Misra, V.; Anandhi, A.; Palko, D.; Infanti, J. Florida's climate changes, variations, & impacts. In *Future Climate Change Scenarios for Florida*; Chassignet, J., Misra, O., Eds.; Florida Climate Institute: Gainesville, FL, USA, 2018.

MDPI

St. Alban-Anlage 66

4052 Basel

Switzerland

Tel. +41 61 683 77 34

Fax +41 61 302 89 18

www.mdpi.com

Water Editorial Office

E-mail: water@mdpi.com

www.mdpi.com/journal/water

9 783039 360420